Introductory Mathematics

Charles P. McKeague

SECOND PRINTING — March 2011
- Corrections made to answers

THIRD PRINTING — June 2012
- Misc. corrections made

*xyz*textbooks

Introductory Mathematics

Charles P. McKeague

Publisher: XYZ Textbooks

Project Manager: Matthew Hoy

Editorial Assistants: Elizabeth Andrews, Stefanie Cohen, Graham Culbertson, Rachael Hillman, Gordon Kirby, Aaron Salisbury, Katrina Smith, CJ Teuben

Composition: XYZ Textbooks

Sales: Amy Jacobs, Richard Jones

ISBN-13: 978-1-936368-04-4   /   ISBN-10: 1-936368-04-8

For product information and technology assistance, contact us at
XYZ Textbooks, 1-877-745-3499

For permission to use material from this text or product,
e-mail: **info@mathtv.com**

XYZ Textbooks
1339 Marsh Street
San Luis Obispo, CA 93401
USA

Printed in the United States of America

For your course and learning solutions, visit **www.xyztextbooks.com**

Brief Contents

Contents

7 Measurement 337

8 Geometry 379

9 Introduction to Algebra 419

10 Solving Equations   473

Preface to the Instructor

We have designed this book to help solve problems that you may encounter in the classroom.

Solutions to Your Problems

Problem: Some students may ask, "What are we going to use this for?"
Solution: Chapter and Section openings feature real-world examples, which show students how the material they are learning appears in the world around them.

Problem: Many students do not read the book.
Solution: At the end of each section, under the heading *Getting Ready for Class*, are four questions for students to answer from the reading. Even a minimal attempt to answer these questions enhances the students' in-class experience.

Problem: Some students may not see how the topics are connected.
Solution: At the conclusion of the problem set for each section are a series of problems under the heading *Getting Ready for the Next Section*. These problems are designed to bridge the gap between topics covered previously, and topics introduced in the next section. Students intuitively see how topics lead into, and out of, each other.

Problem: Some students lack good study skills, but may not know how to improve them.
Solution: Study skills and success skills appear throughout the book, as well as online at MathTV.com. Students learn the skills they need to become successful in this class, and in their other courses as well.

Problem: Students do well on homework, then forget everything a few days later.
Solution: We have designed this textbook so that no topic is covered and then discarded. Throughout the book, features such as *Getting Ready for the Next Section*, *Maintaining Your Skills*, the *Chapter Summary*, and the *Chapter Test* continually reinforce the skills students need to master. If students need still more practice, there are a variety of worksheets online at MathTV.com.

Problem: Some students just watch the videos at MathTV.com, but are not actively involved in learning.
Solution: The Matched Problems worksheets (available online at MathTV.com) contain problems similar to the video examples. Assigning the Matched Problems worksheets ensures that students will be actively involved with the videos.

Other Helpful Solutions

Blueprint for Problem Solving: Students can use these step-by-step methods for solving common application problems.

Facts from Geometry: Students see how topics from geometry are related to the math they are using.

Using Technology: Scattered throughout the book are optional exercises that demonstrate how students can use graphing calculators to enhance their understanding of the topics being covered.

Supplements for the Instructor

Please contact your sales representative.

MathTV.com With more than 6,000 videos, MathTV.com provides the instructor with a useful resource to help students learn the material. MathTV.com features videos of every example in the book, explained by the author and a variety of peer instructors. If a problem can be completed more than one way, the peer instructors often solve it by different methods. Instructors can also use the site's *Build a Playlist* feature to create a custom list of videos for posting on their class blog or website.

Online Homework XYZHomework.com provides powerful online instructional tools for faculty and students. Randomized questions provide unlimited practice and instant feedback with all the benefits of automatic grading. Tools for instructors include the following:

- Quick setup of your online class
- More than 1,500 randomized questions, similar to those in the textbook, for use in a variety of assessments, including online homework, quizzes and tests
- Text and videos designed to supplement your instruction
- Automated grading of online assignments
- Flexible gradebook
- Message boards and other communication tools, enhanced with calculator-style input for proper mathematics notation

Supplements for the Student

MathTV.com MathTV.com gives students access to math instruction 24 hours a day, seven days a week. Assistance with any problem or subject is never more than a few clicks away.

Online book This text is available online for both instructors and students. Tightly integrated with MathTV.com, students can read the book and watch videos of the author and peer instructors explaining each example. Access to the online book is available free with the purchase of a new book.

Additional worksheets A variety of worksheets are available to students online at MathTV.com's premium site. Worksheets include *Matched Problems, Multiple Choice, Find the Mistake*, and *Additional Problems*.

Online Homework XYZHomework.com provides powerful online instruction and homework practice for students. Benefits for the student include the following:

- Unlimited practice with problems similar to those in the text
- Online quizzes and tests for instant feedback on performance
- Online video examples
- Convenient tracking of class progress

Preface to the Student

I often find my students asking themselves the question "Why can't I understand this stuff the first time?" The answer is "You're not expected to." Learning a topic in mathematics isn't always accomplished the first time around. There are many instances when you will find yourself reading over new material a number of times before you can begin to work problems. That's just the way things are in mathematics. If you don't understand a topic the first time you see it, that doesn't mean there is something wrong with you. Understanding mathematics takes time. The process of understanding requires reading the book, studying the examples, working problems, and getting your questions answered.

How to Be Successful in Mathematics

1. **If you are in a lecture class, be sure to attend all class sessions on time.** You cannot know exactly what goes on in class unless you are there. Missing class and then expecting to find out what went on from someone else is not the same as being there yourself.

2. **Read the book.** It is best to read the section that will be covered in class beforehand. Reading in advance, even if you do not understand everything you read, is still better than going to class with no idea of what will be discussed.

3. **Work problems every day and check your answers.** The key to success in mathematics is working problems. The more problems you work, the better you will become at working them. The answers to the odd-numbered problems are given in the back of the book. When you have finished an assignment, be sure to compare your answers with those in the book. If you have made a mistake, find out what it is, and correct it.

4. **Do it on your own.** Don't be misled into thinking someone else's work is your own. Having someone else show you how to work a problem is not the same as working the same problem yourself. It is okay to get help when you are stuck. As a matter of fact, it is a good idea. Just be sure you do the work yourself.

5. **Review every day.** After you have finished the problems your instructor has assigned, take another 15 minutes and review a section you have already completed. The more you review, the longer you will retain the material you have learned.

6. **Don't expect to understand every new topic the first time you see it.** Sometimes you will understand everything you are doing, and sometimes you won't. That's just the way things are in mathematics. Expecting to understand each new topic the first time you see it can lead to disappointment and frustration. The process of understanding takes time. It requires that you read the book, work problems, and get your questions answered.

7. **Spend as much time as it takes for you to master the material.** No set formula exists for the exact amount of time you need to spend on mathematics to master it. You will find out as you go along what is or isn't enough time for you. If you end up spending 2 or more hours on each section in order to master the material there, then that's how much time it takes; trying to get by with less will not work.

8. **Relax.** It's probably not as difficult as you think.

1

Whole Numbers

iStockphoto.com © Nathan Watkins Photography

The table below shows the average amount of caffeine in a number of different beverages. The chart next to the table is a visual presentation of the same information. The relationship between the table and the chart is one of the things we will study in this chapter. The table gives information in numerical form, while the chart gives the same information in a geometrical way. In mathematics it is important to be able to move back and forth between the two forms. Later, in Chapter 8, we will introduce a third form, the algebraic form, in which we summarize relationships with equations.

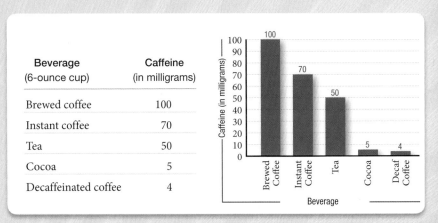

Beverage (6-ounce cup)	Caffeine (in milligrams)
Brewed coffee	100
Instant coffee	70
Tea	50
Cocoa	5
Decaffeinated coffee	4

To begin our study of basic mathematics, we will develop the rules and properties for adding, subtracting, multiplying, and dividing whole numbers.

Study Skills

Some of the students enrolled in my college math classes develop difficulties early in the course. Their difficulties are not associated with their ability to learn mathematics; they all have the potential to pass the course. Students who get off to a poor start do so because they have not developed the study skills necessary to be successful in math. Here is a list of things you can do to begin to develop effective study skills.

1. **Put Yourself on a Schedule** The general rule is that you spend two hours on homework for every hour you are in class. Make a schedule for yourself in which you set aside two hours each day to work on math. Once you make the schedule, stick to it. Don't just complete your assignments and stop. Use all the time you have set aside. If you complete an assignment and have time left over, read the next section in the book, and then work more problems.

2. **Find Your Mistakes and Correct Them** There is more to studying math than just working problems. You must always check your answers with the answers in the back of the book. When you have made a mistake, find out what it is and correct it. Making mistakes is part of the process of learning mathematics. In the prologue to *The Book of Squares*, Italian mathematician Leonardo Fibonacci (ca. 1170–ca. 1250) had this to say about the content of his book:

   > I have come to request indulgence if in any place it contains something more or less than right or necessary; for to remember everything and be mistaken in nothing is divine rather than human . . .

   Fibonacci knew, as you know, that human beings make mistakes. You cannot learn math without making mistakes.

3. **Gather Information on Available Resources** You need to anticipate that you will need extra help sometime during the course. One resource is your instructor; you need to know your instructor's office hours and where the office is located. Another resource is the math lab or study center, if it is available at your school. It also helps to have the phone numbers of other students in the class, in case you miss class. You want to anticipate a need for these resources, so now is the time to gather them together.

Place Value and Names for Numbers

The two diagrams below are known as Pascal's triangle, after the French mathematician and philosopher Blaise Pascal (1623–1662). Both diagrams contain the same information. The one on the left contains numbers in our number system; the one on the right uses numbers from Japan in 1781.

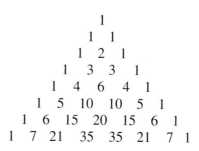

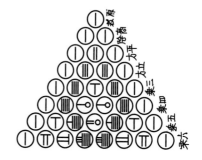

PASCAL'S TRIANGLE IN JAPAN
From Mural Chūzen's *Sampō Dōshi-mon* (1781)

Our number system is based on the number 10 and is therefore called a "base 10" number system. We write all numbers in our number system using the *digits* 0, 1, 2, 3, 4, 5, 6, 7, 8, and 9. The positions of the digits in a number determine the values of the digits. For example, the 5 in the number 251 has a different value from the 5 in the number 542.

The *place values* in our number system are as follows: The first digit on the right is in the *ones column*. The next digit to the left of the ones column is in the *tens column*. The next digit to the left is in the *hundreds column*. For a number like 542, the digit 5 is in the hundreds column, the 4 is in the tens column, and the 2 is in the ones column.

If we keep moving to the left, the columns increase in value. The following diagram shows the name and value of each of the first seven columns in our number system:

Millions Column	Hundred Thousands Column	Ten Thousands Column	Thousands Column	Hundreds Column	Tens Column	Ones Column
1,000,000	100,000	10,000	1,000	100	10	1

EXAMPLE 1   Give the place value of each digit in the number 305,964.

SOLUTION   Starting with the digit at the right, we have:

4 in the ones column, 6 in the tens column, 9 in the hundreds column, 5 in the thousands column, 0 in the ten thousands column, and 3 in the hundred thousands column.

Large Numbers

The photograph shown here was taken by the Hubble telescope in April 2002. The object in the photograph is called the *Cone Nebula*. In astronomy, distances to objects like the Cone Nebula are given in light-years, the distance light travels in a year. If we assume light travels 186,000 miles in one second, then a light-year is 5,865,696,000,000 miles; that is

NASA

5 trillion, 865 billion, 696 million miles

To find the place value of digits in large numbers, we can use Table 1. Note how the Ones, Thousands, Millions, Billions, and Trillions categories are each broken into Ones, Tens, and Hundreds. Note also that we have written the digits for our light-year in the last row of the table.

TABLE 1

Trillions			Billions			Millions			Thousands			Ones		
Hundreds	Tens	Ones	Hundreds	Tens	Ones	Hundreds	Tens	Ones	Hundreds	Tens	Ones	Hundreds	Tens	Ones
		5	8	6	5	6	9	6	0	0	0	0	0	0

EXAMPLE 2   Give the place value of each digit in the number 73,890,672,540.

SOLUTION   The following diagram shows the place value of each digit.

$$
\begin{array}{ccccccccccc}
\text{Ten Billions} & \text{Billions} & \text{Hundred Millions} & \text{Ten Millions} & \text{Millions} & \text{Hundred Thousands} & \text{Ten Thousands} & \text{Thousands} & \text{Hundreds} & \text{Tens} & \text{Ones} \\
7 & 3, & 8 & 9 & 0, & 6 & 7 & 2, & 5 & 4 & 0
\end{array}
$$

Expanded Form

We can use the idea of place value to write numbers in *expanded form*. For example, the number 542 can be written in expanded form as

$$542 = 500 + 40 + 2$$

because the 5 is in the hundreds column, the 4 is in the tens column, and the 2 is in the ones column.

Here are more examples of numbers written in expanded form.

EXAMPLE 3   Write 5,478 in expanded form.

SOLUTION $5,478 = 5,000 + 400 + 70 + 8$

We can use money to make the results from Example 3 more intuitive. Suppose you have $5,478 in cash as follows:

$5,000 $400 $70 $8

Using this diagram as a guide, we can write

$$\$5,478 = \$5,000 + \$400 + \$70 + \$8$$

which shows us that our work writing numbers in expanded form is consistent with our intuitive understanding of the different denominations of money.

EXAMPLE 4   Write 354,798 in expanded form.

SOLUTION $354,798 = 300,000 + 50,000 + 4,000 + 700 + 90 + 8$

EXAMPLE 5   Write 56,094 in expanded form.

SOLUTION   Notice that there is a 0 in the hundreds column. This means we have 0 hundreds. In expanded form, we have

$$56,094 = 50,000 + 6,000 + 90 + 4$$

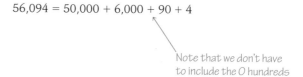

Note that we don't have to include the 0 hundreds

EXAMPLE 6   Write 5,070,603 in expanded form.

SOLUTION   The columns with 0 in them will not appear in the expanded form.

$$5,070,603 = 5,000,000 + 70,000 + 600 + 3$$

Writing Numbers in Words

The idea of place value and expanded form can be used to help write the names for numbers. Naming numbers and writing them in words takes some practice. Let's begin by looking at the names of some two-digit numbers. Table 2 lists a few. Notice that the two-digit numbers that do not end in 0 have two parts. These parts are separated by a hyphen.

TABLE 2

NUMBER	IN ENGLISH	NUMBER	IN ENGLISH
25	*Twenty-five*	30	*Thirty*
47	*Forty-seven*	62	*Sixty-two*
93	*Ninety-three*	77	*Seventy-seven*
88	*Eighty-eight*	50	*Fifty*

The following examples give the names for some larger numbers. In each case the names are written according to the place values given in Table 1.

EXAMPLE 7   Write each number in words.
 a. 452   **b.** 397   **c.** 608

SOLUTION   **a.** Four hundred fifty-two
            b. Three hundred ninety-seven
            c. Six hundred eight

EXAMPLE 8   Write each number in words.
 a. 3,561   **b.** 53,662   **c.** 547,801

SOLUTION   **a.** Three thousand, five hundred sixty-one

Notice how the comma separates
the thousands from the hundreds

            b. Fifty-three thousand, six hundred sixty-two
            c. Five hundred forty-seven thousand, eight hundred one

EXAMPLE 9   Write each number in words
            a. 507,034,005
            b. 739,600,075
            c. 5,003,007,006

SOLUTION   **a.** Five hundred seven million, thirty-four thousand, five
            b. Seven hundred thirty-nine million, six hundred thousand, seventy-five
            c. Five billion, three million, seven thousand, six

The next examples show how we write a number given in words as a number written with digits.

EXAMPLE 10   Write five thousand, six hundred forty-two, using digits instead of words.

SOLUTION   *Five thousand, six hundred forty-two*

5,   6   42 $\longrightarrow$ 5,642

EXAMPLE 11   Write each number with digits instead of words.
  a. Three million, fifty-one thousand, seven hundred
  b. Two billion, five
  c. Seven million, seven hundred seven

SOLUTION   **a.** 3,051,700

  b. 2,000,000,005

  c. 7,000,707

Sets and the Number Line

In mathematics a collection of numbers is called a *set*. In this chapter we will be working with the set of *counting numbers* and the set of *whole numbers*, which are defined as follows:

> **Note**
>
> Counting numbers are also called natural numbers.

Counting numbers = {1, 2, 3, . . .}
Whole numbers = {0, 1, 2, 3, . . .}

The dots mean "and so on," and the braces { } are used to group the numbers in the set together.

Another way to visualize the whole numbers is with a *number line*. To draw a number line, we simply draw a straight line and mark off equally spaced points along the line, as shown in Figure 1. We label the point at the left with 0 and the rest of the points, in order, with the numbers 1, 2, 3, 4, 5, and so on.

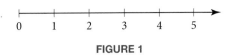

FIGURE 1

The arrow on the right indicates that the number line can continue in that direction forever. When we refer to numbers in this chapter, we will always be referring to the whole numbers.

GETTING READY FOR CLASS

After reading through the preceding section, respond in your own words and in complete sentences.

1. Give the place value of the 9 in the number 305,964.
2. Write the number 742 in expanded form.
3. Place a comma and a hyphen in the appropriate place so that the number 2,345 is written correctly in words below:

   two thousand three hundred forty five

4. Is there a largest whole number?

 SPOTLIGHT ON SUCCESS *Student Instructor Cynthia*

Each time we face our fear, we gain strength, courage, and confidence in the doing.
—Unknown

I must admit, when it comes to math, it takes me longer to learn the material compared to other students. Because of that, I was afraid to ask questions, especially when it seemed like everyone else understood what was going on. Because I wasn't getting my questions answered, my quiz and exam scores were only getting worse. I realized that I was already paying a lot to go to college and that I couldn't afford to keep doing poorly on my exams. I learned how to overcome my fear of asking questions by studying the material before class, and working on extra problem sets until I was confident enough that at least I understood the main concepts. By preparing myself beforehand, I would often end up answering the question myself. Even when that wasn't the case, the professor knew that I tried to answer the question on my own. If you want to be successful, but you are afraid to ask a question, try putting in a little extra time working on problems before you ask your instructor for help. I think you will find, like I did, that it's not as bad as you imagined it, and you will have overcome an obstacle that was in the way of your success.

Problem Set 1.1

Give the place value of each digit in the following numbers.

1. 78   **2.** 93   **3.** 45   **4.** 79

5. 348   **6.** 789   **7.** 608   **8.** 450

9. 2,378   **10.** 6,481   **11.** 273,569   **12.** 768,253

Give the place value of the 5 in each of the following numbers.

13. 458,992   **14.** 75,003,782

15. 507,994,787   **16.** 320,906,050

17. 267,894,335   **18.** 234,345,678,789

19. 4,569,000   **20.** 50,000

Write each of the following numbers in expanded form.

21. 658   **22.** 479   **23.** 68   **24.** 71

25. 4,587   **26.** 3,762   **27.** 32,674   **28.** 54,883

29. 3,462,577   **30.** 5,673,524   **31.** 407   **32.** 508

33. 30,068   **34.** 50,905   **35.** 3,004,008   **36.** 20,088,060

Write each of the following numbers in words.

37. 29   **38.** 75   **39.** 40   **40.** 90

41. 573   **42.** 895   **43.** 707   **44.** 405

45. 770   **46.** 450   **47.** 23,540   **48.** 56,708

49. 3,004   **50.** 5,008   **51.** 3,040   **52.** 5,080

53. 104,065,780   **54.** 637,008,500   **55.** 5,003,040,008   **56.** 7,050,800,001

57. 2,546,731   **58.** 6,998,454

Write each of the following numbers with digits instead of words.

59. Three hundred twenty-five

60. Forty-eight

61. Five thousand, four hundred thirty-two

62. One hundred twenty-three thousand, sixty-one

63. Eighty-six thousand, seven hundred sixty-two

64. One hundred million, two hundred thousand, three hundred

65. Two million, two hundred

66. Two million, two

67. Two million, two thousand, two hundred

68. Two billion, two hundred thousand, two hundred two

iStockPhoto/©Elemental Imaging

Applying the Concepts

69. Hot Air Balloon The first successful crossing of the Atlantic in a hot air balloon was made in August 1978 by Maxie Anderson, Ben Abruzo, and Larry Newman of the United States. The 3,100 mile trip took approximately 140 hours. What is the place value of the 3 in the distance covered by the balloon?

70. Seating Arrangements The number of different ways in which 10 people can be seated at a table with 10 places is 3,628,800. What is the place value of the 3 in this number?

71. Seating Capacity The Rose Bowl has a seating capacity of 106,721. Write this number in expanded form.

72. The illustration shows the average income of workers 18 and older by education.

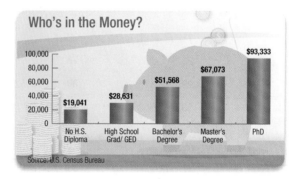

Who's in the Money?

| | | | | $93,333 |
100,000
80,000
60,000 | | | $67,073 | |
40,000 | | $51,568 | | |
20,000 | $19,041 | $28,631 | | |
0
No H.S. Diploma | High School Grad/ GED | Bachelor's Degree | Master's Degree | PhD

Source: U.S. Census Bureau

Write the following numbers in words:

a. the average income of someone with only a high school education

b. the average income of someone with a Ph.D.

Populations of Countries The table below gives estimates of the populations of some countries for the year 2000. The first column under *Population* gives the population in digits. The second column gives the population in words. Fill in the blanks.

COUNTRY	POPULATION	
	DIGITS	WORDS
73. United States	_____	Two hundred seventy-five million
74. People's Republic of China		One billion, two hundred fifty-six million
75. Japan	127,000,000	_____
76. United Kingdom	59,000,000	_____

(From U.S. Census Bureau, International Data Base)

The chart shows the number of babies born in 2006, grouped together according to the age of mothers.

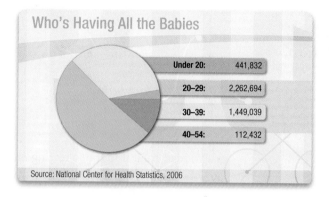

Who's Having All the Babies

Under 20:	441,832
20–29:	2,262,694
30–39:	1,449,039
40–54:	112,432

Source: National Center for Health Statistics, 2006

There is much more information available from the table than just the numbers shown. For instance, the chart tells us how many babies were born to mothers less than 30 years of age. But to find that number, we need to be able to do addition with whole numbers. Let's begin by visualizing addition on the number line.

Facts of Addition

Using lengths to visualize addition can be very helpful. In mathematics we generally do so by using the number line. For example, we add 3 and 5 on the number line like this: Start at 0 and move to 3, as shown in Figure 1. From 3, move 5 more units to the right. This brings us to 8. Therefore, $3 + 5 = 8$.

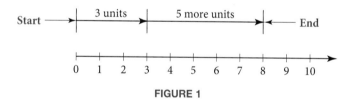

FIGURE 1

If we do this kind of addition on the number line with all combinations of the numbers 0 through 9, we get the results summarized in Table 1 on the next page.

We call the information in Table 1 our basic addition facts. Your success with the examples and problems in this section depends on knowing the basic addition facts.

Note

Table 1 is a summary of the addition facts that you *must* know in order to make a successful start in your study of basic mathematics. You *must* know how to add any pair of numbers that come from the list. You *must* be fast and accurate. You don't want to have to think about the answer to 7 + 9. You should know it's 16. Memorize these facts now. Don't put it off until later.

TABLE 1

ADDITION TABLE

	0	1	2	3	4	5	6	7	8	9
0	0	1	2	3	4	5	6	7	8	9
1	1	2	3	4	5	6	7	8	9	10
2	2	3	4	5	6	7	8	9	10	11
3	3	4	5	6	7	8	9	10	11	12
4	4	5	6	7	8	9	10	11	12	13
5	5	6	7	8	9	10	11	12	13	14
6	6	7	8	9	10	11	12	13	14	15
7	7	8	9	10	11	12	13	14	15	16
8	8	9	10	11	12	13	14	15	16	17
9	9	10	11	12	13	14	15	16	17	18

We read Table 1 in the following manner: Suppose we want to use the table to find the answer to 3 + 5. We locate the 3 in the column on the left and the 5 in the row at the top. We read *across* from the 3 and *down* from the 5. The entry in the table that is across from 3 and below 5 is 8.

Adding Whole Numbers

To add whole numbers, we add digits within the same place value. First we add the digits in the ones place, then the tens place, then the hundreds place, and so on.

EXAMPLE 1   Add 43 + 52.

SOLUTION   This type of addition is best done vertically. First we add the digits in the ones place.

$$\begin{array}{r} 43 \\ + 52 \\ \hline 5 \end{array}$$

Note

To show *why* we add digits with the same place value, we can write each number showing the place value of the digits:

$$\begin{array}{r} 43 = 4 \text{ tens} + 3 \text{ ones} \\ + 52 = 5 \text{ tens} + 2 \text{ ones} \\ \hline 9 \text{ tens} + 5 \text{ ones} \end{array}$$

Then we add the digits in the tens place.

$$\begin{array}{r} 43 \\ + 52 \\ \hline 95 \end{array}$$

EXAMPLE 2   Add 165 + 801.

SOLUTION   Writing the sum vertically, we have

$$\begin{array}{r} 165 \\ + 801 \\ \hline 966 \end{array}$$

966 ←———— Add ones place

———— Add tens place

———— Add hundreds place

Addition with Carrying

In Examples 1 and 2, the sums of the digits with the same place value were always 9 or less. There are many times when the sum of the digits with the same place value will be a number larger than 9. In these cases we have to do what is called *carrying* in addition. The following examples illustrate this process.

EXAMPLE 3   Add 197 + 213 + 324.

SOLUTION   We write the sum vertically and add digits with the same place value.

$$
\begin{array}{r}
\overset{1}{1}97 \\
213 \\
+\,324 \\
\hline
4
\end{array}
$$

When we add the ones, we get $7 + 3 + 4 = 14$
We write the 4 and carry the 1 to the tens column

$$
\begin{array}{r}
\overset{1\,1}{1}97 \\
213 \\
+\,324 \\
\hline
34
\end{array}
$$

We add the tens, including the 1 that was carried
over from the last step. We get 13, so we write
the 3 and carry the 1 to the hundreds column

$$
\begin{array}{r}
\overset{1\,1}{1}97 \\
213 \\
+\,324 \\
\hline
734
\end{array}
$$

We add the hundreds, including the 1 that was
carried over from the last step

EXAMPLE 4   Add 46,789 + 2,490 + 864.

SOLUTION   We write the sum vertically—with the digits with the same place value aligned—and then use the shorthand form of addition.

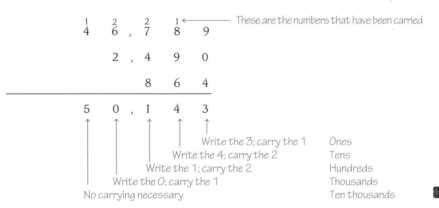

					These are the numbers that have been carried

Write the 3; carry the 1 — Ones
Write the 4; carry the 2 — Tens
Write the 1; carry the 2 — Hundreds
Write the 0; carry the 1 — Thousands
No carrying necessary — Ten thousands

Adding numbers as we are doing here takes some practice. Most people don't make mistakes in carrying. Most mistakes in addition are made in adding the numbers in the columns. That is why it is so important that you are accurate with the basic addition facts given in this chapter.

Vocabulary

The word we use to indicate addition is the word *sum*. If we say "the sum of 3 and 5 is 8," what we mean is $3 + 5 = 8$. The word *sum* always indicates addition. We can state this fact in symbols by using the letters a and b to represent numbers.

> **DEFINITION**
>
> If a and b are any two numbers, then the **sum** of a and b is $a + b$. To find the sum of two numbers, we add them.

Table 2 gives some phrases and sentences in English and their mathematical equivalents written in symbols.

Note

When mathematics is used to solve everyday problems, the problems are almost always stated in words. The translation of English to symbols is a very important part of mathematics.

TABLE 2	
IN ENGLISH	**IN SYMBOLS**
The sum of 4 and 1	$4 + 1$
4 added to 1	$1 + 4$
8 more than m	$m + 8$
x increased by 5	$x + 5$
The sum of x and y	$x + y$
The sum of 2 and 4 is 6	$2 + 4 = 6$

Properties of Addition

Once we become familiar with addition, we may notice some facts about addition that are true regardless of the numbers involved. The first of these facts involves the number 0 (zero).

Whenever we add 0 to a number, the result is the original number. For example,

$$7 + 0 = 7 \qquad \text{and} \qquad 0 + 3 = 3$$

Because this fact is true no matter what number we add to 0, we call it a property of 0.

> **Addition Property of 0**
>
> If we let a represent any number, then it is always true that
>
> $$a + 0 = a \quad \text{and} \quad 0 + a = a$$
>
> *In words:* Adding 0 to any number leaves that number unchanged.

Note

When we use letters to represent numbers, as we do when we say "If a and b are any two numbers," then a and b are called variables, because the values they take on vary. We use the variables a and b in the definitions and properties on this page because we want you to know that the definitions and properties are true for all numbers that you will encounter in this book.

A second property we notice by becoming familiar with addition is that the order of two numbers in a sum can be changed without changing the result.

$$3 + 5 = 8 \qquad \text{and} \qquad 5 + 3 = 8$$
$$4 + 9 = 13 \qquad \text{and} \qquad 9 + 4 = 13$$

This fact about addition is true for *all* numbers. The order in which you add two numbers doesn't affect the result. We call this fact the *commutative property of addition,* and we write it in symbols as follows.

Commutative Property of Addition

If a and b are any two numbers, then it is always true that

$$a + b = b + a$$

In words: Changing the order of two numbers in a sum doesn't change the result.

EXAMPLE 5   Use the commutative property of addition to rewrite each sum.

a. $4 + 6$ **b.** $5 + 9$ **c.** $3 + 0$ **d.** $7 + n$

SOLUTION   The commutative property of addition indicates that we can change the order of the numbers in a sum without changing the result. Applying this property we have

a. $4 + 6 = 6 + 4$
b. $5 + 9 = 9 + 5$
c. $3 + 0 = 0 + 3$
d. $7 + n = n + 7$

Notice that we did not actually add any of the numbers. The instructions were to use the commutative property, and the commutative property involves only the order of the numbers in a sum.

The last property of addition we will consider here has to do with sums of more than two numbers. Suppose we want to find the sum of 2, 3, and 4. We could add 2 and 3 first, and then add 4 to what we get

$$(2 + 3) + 4 = 5 + 4 = 9$$

Or, we could add the 3 and 4 together first and then add the 2

$$2 + (3 + 4) = 2 + 7 = 9$$

The result in both cases is the same. If we try this with any other numbers, the same thing happens. We call this fact about addition the *associative property of addition,* and we write it in symbols as follows.

Associative Property of Addition

If a, b, and c represent any three numbers, then

$$(a + b) + c = a + (b + c)$$

In words: Changing the grouping of three numbers in a sum doesn't change the result.

Note

This discussion is here to show why we write the next property the way we do. Sometimes it is helpful to look ahead to the property itself (in this case, the associative property of addition) to see what it is that is being justified.

EXAMPLE 6   Use the associative property of addition to rewrite each sum.

a. $(5 + 6) + 7$ **b.** $(3 + 9) + 1$ **c.** $6 + (8 + 2)$ **d.** $4 + (9 + n)$

SOLUTION   The associative property of addition indicates that we are free to regroup the numbers in a sum without changing the result.

a. $(5 + 6) + 7 = 5 + (6 + 7)$
b. $(3 + 9) + 1 = 3 + (9 + 1)$
c. $6 + (8 + 2) = (6 + 8) + 2$
d. $4 + (9 + n) = (4 + 9) + n$

The commutative and associative properties of addition tell us that when adding whole numbers, we can use any order and grouping. When adding several numbers, it is sometimes easier to look for pairs of numbers whose sums are 10, 20, and so on.

EXAMPLE 7   Add $9 + 3 + 2 + 7 + 1$.

SOLUTION $9 + 3 + 2 + 7 + 1$

$10 + 10 + 2$

22

Solving Equations

We can use the addition table to help solve some simple equations. If n is used to represent a number, then the equation

$$n + 3 = 5$$

will be true if n is 2. The number 2 is therefore called a *solution* to the equation, because, when we replace n with 2, the equation becomes a true statement:

$$2 + 3 = 5$$

Equations like this are really just puzzles, or questions. When we say, "Solve the equation $n + 3 = 5$," we are asking the question, "What number do we add to 3 to get 5?"

When we solve equations by reading the equation to ourselves and then stating the solution, as we did with the equation above, we are solving the equation by inspection.

Note

The letter n as we are using it here is a variable, because it represents a number. In this case it is the number that is a solution to an equation.

EXAMPLE 8   Find the solution to each equation by inspection.

a. $n + 5 = 9$ **b.** $n + 6 = 12$ **c.** $4 + n = 5$ **d.** $13 = n + 8$

SOLUTION   We find the solution to each equation by using the addition facts given in Table 1.

a. The solution to $n + 5 = 9$ is 4, because $4 + 5 = 9$.
b. The solution to $n + 6 = 12$ is 6, because $6 + 6 = 12$.
c. The solution to $4 + n = 5$ is 1, because $4 + 1 = 5$.
d. The solution to $13 = n + 8$ is 5, because $13 = 5 + 8$.

FACTS FROM GEOMETRY   *Perimeter*

We end this section with an introduction to perimeter. Let's start with the definition of a *polygon:*

DEFINITION

A *polygon* is a closed geometric figure, with at least three sides, in which each side is a straight line segment.

The most common polygons are squares, rectangles, and triangles. Examples of these are shown in Figure 2.

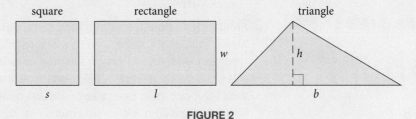

FIGURE 2

In the square, *s* is the length of the side, and each side has the same length. In the rectangle, *l* stands for the length, and *w* stands for the width. The width is usually the lesser of the two.

DEFINITION

The *perimeter* of any polygon is the sum of the lengths of the sides, and it is denoted with the letter *P*.

To find the perimeter of a polygon we add all the lengths of the sides together.

EXAMPLE 9   Find the perimeter of each geometric figure.

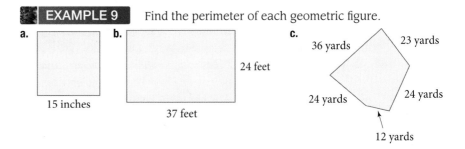

a.

15 inches

b.

24 feet

37 feet

c.

36 yards

23 yards

24 yards

24 yards

12 yards

SOLUTION   In each case we find the perimeter by adding the lengths of all the sides.

a. The figure is a square. Because the length of each side in the square is the same, the perimeter is

$P = 15 + 15 + 15 + 15 = 60$ inches

b. In the rectangle, two of the sides are 24 feet long, and the other two are 37 feet long. The perimeter is the sum of the lengths of the sides.

$P = 24 + 24 + 37 + 37 = 122$ feet

c. For this polygon, we add the lengths of the sides together. The result is the perimeter.

$P = 36 + 23 + 24 + 12 + 24 = 119$ yards

USING TECHNOLOGY

Calculators

From time to time we will include some notes like this one, which show how a calculator can be used to assist us with some of the calculations in the book. Most calculators on the market today fall into one of two categories: those with algebraic logic and those with function logic. Calculators with algebraic logic have a key with an equals sign on it. Calculators with function logic do not have an equals key. Instead they have a key labeled ENTER or EXE (for execute). Scientific calculators use algebraic logic, and graphing calculators, such as the TI-83, use function logic.

Here are the sequences of keystrokes to use to work the problem shown in Part c of Example 9.

Scientific Calculator: 36 $\boxed{+}$ 23 $\boxed{+}$ 24 $\boxed{+}$ 12 $\boxed{+}$ 24 $\boxed{=}$

Graphing Calculator: 36 $\boxed{+}$ 23 $\boxed{+}$ 24 $\boxed{+}$ 12 $\boxed{+}$ 24 $\boxed{\text{ENT}}$

GETTING READY FOR CLASS

After reading through the preceding section, respond in your own words and in complete sentences.

1. What number is the sum of 6 and 8?

2. Make up an addition problem using the number 456 that does not involve carrying.

3. Make up an addition problem using the number 456 that involves carrying from the ones column to the tens column only.

4. What is the perimeter of a polygon?

Problem Set 1.2

Find each of the following sums. (Add.)

1. $3 + 5 + 7$ **2.** $2 + 8 + 6$

3. $1 + 4 + 9$ **4.** $2 + 8 + 3$

5. $5 + 9 + 4 + 6$ **6.** $8 + 1 + 6 + 2$

7. $1 + 2 + 3 + 4 + 5$ **8.** $5 + 6 + 7 + 8 + 9$

9. $9 + 1 + 8 + 2$ **10.** $7 + 3 + 6 + 4$

Add each of the following. (There is no carrying involved in these problems.)

11. 43	**12.** 56	**13.** 81	**14.** 37	**15.** 4,281	**16.** 2,749
+ 25	+ 23	+ 17	+ 22	+ 3,016	+ 1,250

17. 3,482	**18.** 2,496	**19.** 32	**20.** 521	**21.** 6,245	**22.** 27
+ 3,005	+ 7,503	21	340	203	4,510
		+ 43	+ 135	+ 1,001	+ 342

Add each of the following. (All problems involve carrying in at least one column.)

23. 49	**24.** 85	**25.** 74	**26.** 36	**27.** 682	**28.** 439
+ 16	+ 29	+ 28	+ 46	+ 193	+ 270

29. 638	**30.** 444	**31.** 4,963	**32.** 8,291	**33.** 6,205	**34.** 8,888
+ 191	+ 595	+ 5,428	+ 7,489	+ 9,999	+ 9,999

35. 56,789	**36.** 45,678	**37.** 52,468	**38.** 13,579	**39.** 4,296	**40.** 5,637
+ 98,765	+ 87,654	+ 58,642	+ 97,531	8,720	481
				+ 4,375	+ 7,899

41. 4,994	**42.** 6,824	**43.** 12	**44.** 21	**45.** 999	**46.** 646
449	371	34	43	444	464
+ 9,449	+ 4,857	56	65	555	525
		+ 78	+ 87	+ 222	+ 252

47. 9,245	**48.** 45
672	9,876
8,341	54
+ 27	+ 6,789

Complete the following tables.

49.

First Number a	Second Number b	Their Sum $a + b$
61	38	
63	36	
65	34	
67	32	

50.

First Number a	Second Number b	Their Sum $a + b$
10	45	
20	35	
30	25	
40	15	

51.

First Number a	Second Number b	Their Sum $a + b$
9	16	
36	64	
81	144	
144	256	

52.

First Number a	Second Number b	Their Sum $a + b$
25	75	
24	76	
23	77	
22	78	

Rewrite each of the following using the commutative property of addition.

53. $5 + 9$ **54.** $2 + 1$ **55.** $3 + 8$ **56.** $9 + 2$ **57.** $6 + 4$ **58.** $1 + 7$

Rewrite each of the following using the associative property of addition.

59. $(1 + 2) + 3$ **60.** $(4 + 5) + 9$ **61.** $(2 + 1) + 6$ **62.** $(2 + 3) + 8$

63. $1 + (9 + 1)$ **64.** $2 + (8 + 2)$ **65.** $(4 + n) + 1$ **66.** $(n + 8) + 1$

Find a solution for each equation.

67. $n + 6 = 10$ **68.** $n + 4 = 7$ **69.** $n + 8 = 13$ **70.** $n + 6 = 15$

71. $4 + n = 12$ **72.** $5 + n = 7$ **73.** $17 = n + 9$ **74.** $13 = n + 5$

Write each of the following expressions in words. Use the word *sum* in each case.

75. $4 + 9$ **76.** $9 + 4$ **77.** $8 + 1$

78. $9 + 9$ **79.** $2 + 3 = 5$ **80.** $8 + 2 = 10$

Write each of the following in symbols.

81. a. The sum of 5 and 2
   b. 3 added to 8

82. a. The sum of a and 4
   b. 6 more than x

83. a. m increased by 1
   b. The sum of m and n

84. a. The sum of 4 and 8 is 12.
   b. The sum of a and b is 6.

Find the perimeter of each figure. (Note that we have abbreviated the units on each figure to save space. The abbreviation for feet is ft, inches is in., and yards is yd.) The first four figures are squares.

85.

3 in.

86.

9 in.

87.

4 ft

88.

2 ft

89.

3 yd

10 yd

90.

1 yd

5 yd

91.

5 in.   6 in.

7 in.

92.

4 in.   10 in.

12 in.

Applying the Concepts

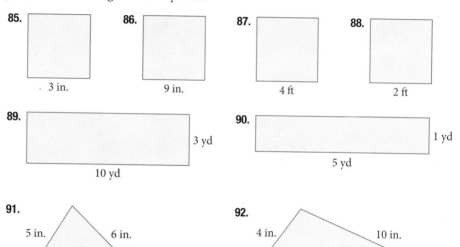
iStockPhoto/©diamirstudio

The application problems that follow are related to addition of whole numbers. Read each problem carefully to determine exactly what you are being asked to find. Don't assume that just because a number appears in a problem you have to use it to solve the problem. Sometimes you do, and sometimes you don't.

93. Gallons of Gasoline Tim bought gas for his economy car twice last month. The first time he bought 18 gallons and the second time he bought 16 gallons. What was the total amount of gasoline Tim bought last month?

94. Tallest Mountain The world's tallest mountain is Mount Everest. On May 5, 1999, it was found to be 7 feet taller than it was previously thought to be. Before this date, Everest was thought to be 29,028 feet high. That height was determined by B. L. Gulatee in 1954. What is the current height of Mount Everest?

95. Checkbook Balance On Monday Bob had a balance of $241 in his checkbook. On Tuesday he made a deposit of $108, and on Thursday he wrote a check for $24. What was the balance in his checkbook on Wednesday?

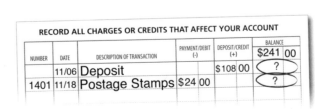

		RECORD ALL CHARGES OR CREDITS THAT AFFECT YOUR ACCOUNT			
			PAYMENT/DEBIT (-)	DEPOSIT/CREDIT (+)	BALANCE $241 00
NUMBER	DATE	DESCRIPTION OF TRANSACTION			
	11/06	Deposit		$108 00	?
1401	11/18	Postage Stamps	$24 00		?

96. **Number of Passengers** A plane flying from Los Angeles to New York left Los Angeles with 67 passengers on board. The plane stopped in Bakersfield and picked up 28 passengers, and then it stopped again in Dallas where 57 more passengers came on board. How many passengers were on the plane when it landed in New York?

Rounding Numbers, Estimating Answers, and Displaying Information

1.3

Many times when we talk about numbers, it is helpful to use numbers that have been *rounded off*, rather than exact numbers. For example, the city where I live has a population of 42,963. But when I tell people how large the city is, I usually say, "The population is about 43,000." The number 43,000 is the original number rounded to the nearest thousand. The number 42,963 is closer to 43,000 than it is to 42,000, so it is rounded to 43,000. We can visualize this situation on the number line.

Rounding

The steps used in rounding numbers are given below.

> **Steps for Rounding Whole Numbers**
>
> **1.** Locate the digit just to the right of the place you are to round to.
>
> **2.** If that digit is less than 5, replace it and all digits to its right with zeros.
>
> **3.** If that digit is 5 or more, replace it and all digits to its right with zeros, and add 1 to the digit to its left.

You can see from these steps that in order to round a number you must be told what column (or place value) to round to.

EXAMPLE 1   Round 5,382 to the nearest hundred.

SOLUTION   The 3 is in the hundreds column. We look at the digit just to its right, which is 8. Because 8 is greater than 5, we add 1 to the 3, and we replace the 8 and 2 with zeros.

EXAMPLE 2   Round 94 to the nearest ten.

SOLUTION   The 9 is in the tens column. To its right is 4. Because 4 is less than 5, we simply replace it with 0

$$94 \quad \text{is} \quad 9\mathbf{0} \quad \text{to the nearest ten}$$

Less than 5   Replaced with zero

 **EXAMPLE 3**   Round 973 to the nearest hundred.

SOLUTION   We have a 9 in the hundreds column. To its right is 7, which is greater than 5. We add 1 to 9 to get 10, and then replace the 7 and 3 with zeros:

973   is   1,0**00**   to the nearest hundred

Greater   Add 1 to   Put zeros
than 5   get 10   here

 **EXAMPLE 4**   Round 47,256,344 to the nearest million.

SOLUTION   We have 7 in the millions column. To its right is 2, which is less than 5. We simply replace all the digits to the right of 7 with zeros to get:

47,256,344   is   47,**000,000**   to the nearest million

Less than 5   Leave as is   Replaced with zeros

Table 1 gives more examples of rounding.

TABLE 1

| | Rounded to the Nearest | | |
Original Number	Ten	Hundred	Thousand
6,914	6,910	6,900	7,000
8,485	8,490	8,500	8,000
5,555	5,560	5,600	6,000
1,234	1,230	1,200	1,000

Rule: Calculating and Rounding

If we are doing calculations and are asked to round our answer, we do all our arithmetic first and then round the result. That is, the last step is to round the answer; we don't round the numbers first and then do the arithmetic.

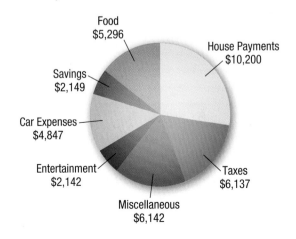 **EXAMPLE 5**   The pie chart below shows how a family earning $36,913 a year spends their money.

Food
$5,296

House Payments
$10,200

Savings
$2,149

Car Expenses
$4,847

Entertainment
$2,142

Taxes
$6,137

Miscellaneous
$6,142

a. To the nearest hundred dollars, what is the total amount spent on food and entertainment?

b. To the nearest thousand dollars, how much of their income is spent on items other than taxes and savings?

SOLUTION   In each case we add the numbers in question and then round the sum to the indicated place.

a. We add the amounts spent on food and entertainment and then round that result to the nearest hundred dollars.

Food	$5,296
Entertainment	+2,142
Total	$7,438 = $7,400 to the nearest hundred dollars

b. We add the numbers for all items except taxes and savings.

House payments	$10,200
Food	5,296
Car expenses	4,847
Entertainment	2,142
Miscellaneous	+ 6,142
Total	$28,627 = $29,000 to the nearest thousand dollars

Estimating

When we *estimate* the answer to a problem, we simplify the problem so that an approximate answer can be found quickly. There are a number of ways of doing this. One common method is to use rounded numbers to simplify the arithmetic necessary to arrive at an approximate answer, as our next example shows.

EXAMPLE 6   Estimate the answer to the following problem by rounding each number to the nearest thousand.

        4,872
        1,691
          777
     + 6,124

SOLUTION   We round each of the four numbers in the sum to the nearest thousand. Then we add the rounded numbers.

4,872	rounds to	5,000
1,691	rounds to	2,000
777	rounds to	1,000
+ 6,124	rounds to	+ 6,000
		14,000

We estimate the answer to this problem to be approximately 14,000. The actual answer, found by adding the original unrounded numbers, is 13,464.

Note The method used in Example 6 does not conflict with the rule we stated before Example 5. In Example 6 we are asked to *estimate* an answer, so it is okay to round the numbers in the problem before adding them. In Example 5 we are asked for a rounded answer, meaning that we are to find the exact answer to the problem and then round to the indicated place. In this case we must not round the numbers in the problem before adding. Look over the instructions, solutions, and answers to Examples 5 and 6 until you understand the difference between the problems shown there.

DESCRIPTIVE STATISTICS   *Bar Charts*

In the introduction to this chapter, we gave two representations for the amount of caffeine in five different drinks, one numeric and the other visual. Those two representations are shown below in Table 2 and Figure 1.

Beverage (6-Ounce Cup)	Caffeine (In Milligrams)
Brewed Coffee	100
Instant Coffee	70
Tea	50
Cocoa	5
Decaffeinated Coffee	4

TABLE 2

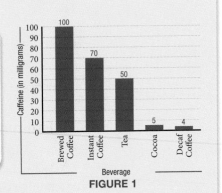

FIGURE 1

The diagram in Figure 1 is called a *bar chart*. The horizontal line below which the drinks are listed is called the *horizontal axis*, while the vertical line that is labeled from 0 to 100 is called the *vertical axis*.

USING TECHNOLOGY	*Spreadsheet Programs*

When I put together the manuscript for this book, I used a spreadsheet program to draw the bar charts, pie charts, and some of the other diagrams you will see as you progress through the book.

Figure 2 shows how the screen on my computer looked when I was preparing the bar chart for Figure 1 in this section. Notice that I also used the computer to create a pie chart from the same data.

If you have a computer with a spreadsheet program, you may want to use it to create some of the charts you will be asked to create in the problem sets throughout the book.

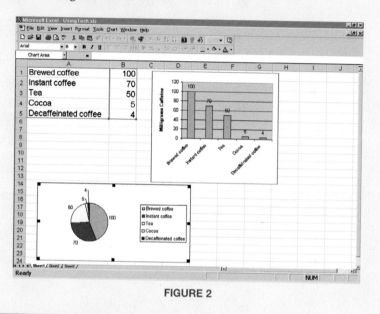

FIGURE 2

GETTING READY FOR CLASS

After reading through the preceding section, respond in your own words and in complete sentences.

1. Describe the process you would use to round the number 5,382 to the nearest thousand.

2. Describe the process you would use to round the number 47,256,344 to the nearest ten thousand.

3. Find a number not containing the digit 7 that will round to 700 when rounded to the nearest hundred.

4. When I ask a class of students to round the number 7,499 to the nearest thousand, a few students will give the answer as 8,000. In what way are these students using the rule for rounding numbers incorrectly?

Problem Set 1.3

Round each of the numbers to the nearest ten.

1. 42   **2.** 44   **3.** 46   **4.** 48   **5.** 45   **6.** 73

7. 77   **8.** 75   **9.** 458   **10.** 455   **11.** 471   **12.** 680

13. 56,782   **14.** 32,807   **15.** 4,504   **16.** 3,897

Round each of the numbers to the nearest hundred.

17. 549   **18.** 954   **19.** 833   **20.** 604   **21.** 899   **22.** 988

23. 1090   **24.** 6,778   **25.** 5,044   **26.** 56,990   **27.** 39,603   **28.** 31,999

Round each of the numbers to the nearest thousand.

29. 4,670   **30.** 9,054   **31.** 9,760   **32.** 4,444

33. 978   **34.** 567   **35.** 657,892   **36.** 688,909

37. 509,905   **38.** 608,433   **39.** 3,789,345   **40.** 5,744,500

Complete the following table by rounding the numbers on the left as indicated by the headings in the table.

Original Number	Rounded to the Nearest		
	Ten	Hundred	Thousand
41. 7,821			
42. 5,945			
43. 5,999			
44. 4,353			
45. 10,985			
46. 11,108			
47. 99,999			
48. 95,505			

Applying the Concepts

49. Average Salary Based on salary studies by *The Associated Press,* major league baseball's average player salary for the 2004 season was $2,486,609, representing a decrease of 2.7% over the previous season's average. Round the 2004 average player salary to the nearest hundred thousand.

50. Tallest Mountain The world's tallest mountain is Mount Everest. On May 5, 1999, it was found to be 7 feet taller than it was previously thought to be. Before this date, Everest was thought to be 29,028 feet high. That height was determined by B. L. Gulatee in 1954. The first measurement of Everest was in 1847. At that time the height was given as 29,002 feet. Round the current height, the 1954 height, and the 1847 height of Mount Everest to the nearest thousand.

Age of Mothers About 4 million babies were born in 2006. The chart shows the breakdown by mothers' age and number of babies. Use the chart to answer problems 51-54.

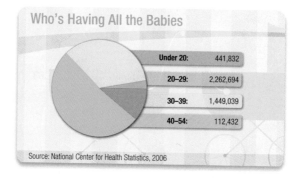

Who's Having All the Babies

Under 20:	441,832
20–29:	2,262,694
30–39:	1,449,039
40–54:	112,432

Source: National Center for Health Statistics, 2006

51. What is the exact number of babies born in 2006?

52. Using your answer from Problem 51, is the statement "About 4 million babies were born in 2006" correct?

53. To the nearest hundred thousand, how many babies were born to mothers aged 20 to 29 in 2006?

54. To the nearest thousand, how many babies were born to mothers 40 years old or older?

Business Expenses The pie chart shows one year's worth of expenses for a small business. Use the chart to answer Problems 55–58.

55. To the nearest hundred dollars, how much was spent on postage and supplies?

56. Find the total amount spent, to the nearest hundred dollars, on rent, utilities and car expenses.

57. To the nearest thousand dollars, how much was spent on items other than salaries, rent and utilities?

58. To the nearest thousand dollars, how much was spent on items other than postage, supplies, and car expenses?

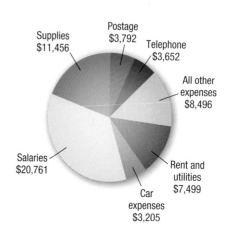

Supplies $11,456
Postage $3,792
Telephone $3,652
All other expenses $8,496
Salaries $20,761
Rent and utilities $7,499
Car expenses $3,205

Estimating Estimate the answer to each of the following problems by rounding each number to the indicated place value and then adding.

59. hundred

750
275
+ 120

60. thousand

1,891
765
+ 3,223

61. hundred

472
422
536
+511

62. hundred

399
601
744
+ 298

63. thousand

25,399
7,601
18,744
+ 6,298

64. thousand

9,999
8,888
7,777
+ 6,666

65. Caffeine Content The following table lists the amount of caffeine in five differ- ent soft drinks. Construct a bar chart from the information in the table.

CAFFEINE CONTENT IN SOFT DRINKS	
Drink	Caffeine (in milligrams)
Jolt	100
Mountain Dew	55
Coca-Cola	45
Diet Pepsi	36
7 Up	0

66. Caffeine Content The following table lists the amount of caffeine in five differ- ent nonprescription drugs. Construct a bar chart from the information in the table.

CAFFEINE CONTENT IN NONPRESCRIPTION DRUGS	
Nonprescription drug	Caffeine (in milligrams)
Dexatrim	200
No Doz	100
Excedrin	65
Triaminicin tablets	30
Dristan tablets	16

67. **Exercise** The following table lists the number of calories burned in 1 hour of exercise by a person who weighs 150 pounds. Construct a bar chart from the information in the table.

CALORIES BURNED BY A 150-POUND PERSON IN ONE HOUR	
Activity	Calories
Bicycling	374
Bowling	265
Handball	680
Jazzercise	340
Jogging	680
Skiing	544

68. **Fast Food** The following table lists the number of calories consumed by eating some popular fast foods. Construct a bar chart from the information in the table.

CALORIES IN FAST FOOD	
Food	Calories
McDonald's hamburger	270
Burger King hamburger	260
Jack in the Box hamburger	280
McDonald's Big Mac	510
Burger King Whopper	630
Jack in the Box Colossus burger	940

Subtraction with Whole Numbers

In business, subtraction is used to calculate profit. Profit is found by subtracting costs from revenue. The following double bar chart shows the costs and revenue of the Baby Steps Shoe Company during one 4-week period.

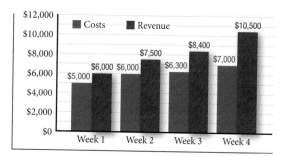

To find the profit for Week 1, we subtract the costs from the revenue, as follows:

$$\text{Profit} = \$6,000 - \$5,000$$
$$\text{Profit} = \$1,000$$

Subtraction is the opposite operation of addition. If you understand addition and can work simple addition problems quickly and accurately, then subtraction shouldn't be difficult for you.

Vocabulary

The word *difference* always indicates subtraction. We can state this in symbols by letting the letters a and b represent numbers.

DEFINITION

The **difference** of two numbers a and b is
$$a - b$$

Table 1 gives some word statements involving subtraction and their mathematical equivalents written in symbols.

TABLE 1

In English	In Symbols
The difference of 9 and 1	$9 - 1$
The difference of 1 and 9	$1 - 9$
The difference of m and 4	$m - 4$
The difference of x and y	$x - y$
3 subtracted from 8	$8 - 3$
2 subtracted from t	$t - 2$
The difference of 7 and 4 is 3	$7 - 4 = 3$
The difference of 9 and 3 is 6	$9 - 3 = 6$

The Meaning of Subtraction

When we want to subtract 3 from 8, we write

$$8 - 3, \quad 8 \text{ subtract } 3, \quad \text{or} \quad 8 \text{ minus } 3$$

The number we are looking for here is the difference between 8 and 3, or the number we add to 3 to get 8. That is:

$$8 - 3 = ? \quad \text{is the same as} \quad ? + 3 = 8$$

In both cases we are looking for the number we add to 3 to get 8. The number we are looking for is 5. We have two ways to write the same statement.

Subtraction   Addition
$$8 - 3 = 5 \quad \text{or} \quad 5 + 3 = 8$$

For every subtraction problem, there is an equivalent addition problem. Table 2 lists some examples.

TABLE 2

Subtraction		Addition
$7 - 3 = 4$	because	$4 + 3 = 7$
$9 - 7 = 2$	because	$2 + 7 = 9$
$10 - 4 = 6$	because	$6 + 4 = 10$
$15 - 8 = 7$	because	$7 + 8 = 15$

To subtract numbers with two or more digits, we align the numbers vertically and subtract in columns.

EXAMPLE 1   Subtract $376 - 241$.

SOLUTION   We write the problem vertically, aligning digits with the same place value. Then we subtract in columns.

$$
\begin{array}{r}
376 \\
-\ 241 \\
\hline
135
\end{array}
$$
⟵ *Subtract the bottom number in each column from the number above it*

We can visualize Example 1 using money.

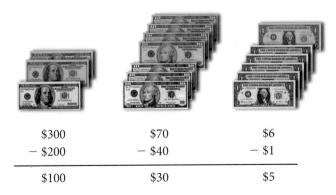

$$
\begin{array}{ccc}
\$300 & \$70 & \$6 \\
-\ \$200 & -\ \$40 & -\ \$1 \\
\hline
\$100 & \$30 & \$5
\end{array}
$$

EXAMPLE 2   Subtract 503 from 7,835.

SOLUTION   In symbols this statement is equivalent to

$$7,835 - 503$$

To subtract we write 503 below 7,835 and then subtract in columns.

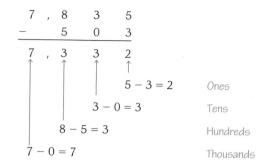

	7	,	8	3	5
−			5	0	3
	7	,	3	3	2

$5 - 3 = 2$ *Ones*

$3 - 0 = 3$ *Tens*

$8 - 5 = 3$ *Hundreds*

$7 - 0 = 7$ *Thousands*

As you can see, subtraction problems like the ones in Examples 1 and 2 are fairly simple. We write the problem vertically, lining up the digits with the same place value, and subtract in columns. We always subtract the bottom number from the top number.

Subtraction with Borrowing

Subtraction must involve *borrowing* when the bottom digit in any column is larger than the digit above it. In one sense borrowing is the reverse of the carrying we did in addition.

EXAMPLE 3   Subtract $92 - 45$.

SOLUTION   We write the problem vertically with the place values of the digits showing:

$$92 = 9 \text{ tens} + 2 \text{ ones}$$
$$- 45 = 4 \text{ tens} + 5 \text{ ones}$$

Note

The discussion here shows why borrowing is necessary and how we go about it. To understand borrowing you should pay close attention to this discussion.

Look at the ones column. We cannot subtract immediately, because 5 is larger than 2. Instead, we borrow 1 ten from the 9 tens in the tens column. We can rewrite the number 92 as

$$9 \text{ tens} + 2 \text{ ones}$$
$$= 8 \text{ tens} + 1 \text{ ten} + 2 \text{ ones}$$
$$= 8 \text{ tens} + 12 \text{ ones}$$

Now we are in a position to subtract.

$$92 = 9 \text{ tens} + 2 \text{ ones} = 8 \text{ tens} + 12 \text{ ones}$$
$$- 45 = 4 \text{ tens} + 5 \text{ ones} = 4 \text{ tens} + 5 \text{ ones}$$
$$\overline{\hspace{4cm} 4 \text{ tens} + 7 \text{ ones}}$$

The result is 4 tens + 7 ones, which can be written in standard form as 47.

Writing the problem out in this way is more trouble than is actually necessary. The shorthand form of the same problem looks like this:

$$\begin{array}{cc} \overset{8}{\cancel{9}} & \overset{12}{\cancel{2}} \\ - \ 4 & 5 \\ \hline 4 & 7 \end{array}$$ ←———— *This shows we have borrowed 1 ten to go with the 2 ones*

$12 - 5 = 7$ *Ones*

$8 - 4 = 4$ *Tens*

This shortcut form shows all the necessary work involved in subtraction with borrowing. We will use it from now on.

The borrowing that changed 9 tens + 2 ones into 8 tens + 12 ones can be visualized with money.

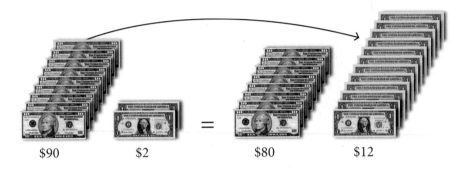

$90 $2   = $80 $12

![EXAMPLE 4] **EXAMPLE 4**   Find the difference of 549 and 187.

SOLUTION   In symbols the difference of 549 and 187 is written

$549 - 187$

Writing the problem vertically so that the digits with the same place value are aligned, we have

$$\begin{array}{r} 549 \\ - \ 187 \\ \hline \end{array}$$

The top number in the tens column is smaller than the number below it. This means that we will have to borrow from the next larger column.

$$\begin{array}{ccc} \overset{4}{\cancel{5}} & \overset{14}{\cancel{4}} & 9 \\ - \ 1 & 8 & 7 \\ \hline 3 & 6 & 2 \end{array}$$ ←———— *Borrow 1 hundred to go with the 4 tens*

$9 - 7 = 2$ *Ones*

$14 - 8 = 6$ *Tens*

$4 - 1 = 3$ *Hundreds*

The actual work we did in borrowing looks like this:

5 hundreds + 4 tens + 9 ones

= 4 hundreds + 1 hundred + 4 tens + 9 ones

= 4 hundreds + 14 tens + 9 ones

EXAMPLE 5 Jo Ann has $742 in her checking account. If she writes a check for $615 to pay the rent, how much is left in her checking account?

		RECORD ALL CHARGES OR CREDITS THAT AFFECT YOUR ACCOUNT			
			PAYMENT/DEBIT (-)	DEPOSIT/CREDIT (+)	BALANCE $742 00
NUMBER	DATE	DESCRIPTION OF TRANSACTION			?
1402	12/1	Rent	$615 00		

SOLUTION To find the amount left in the account after she has written the rent check, we subtract

$$\begin{array}{r} \$7 \overset{3}{\cancel{4}} \overset{12}{\cancel{2}} \\ -615 \\ \hline \$127 \end{array}$$

She has $127 left in her account after writing a check for the rent.

USING TECHNOLOGY

Calculators

Here is how we would work the problem shown in Example 5 on a calculator:

Scientific Calculator: 742 $\boxed{-}$ 615 $\boxed{=}$
Graphing Calculator: 742 $\boxed{-}$ 615 $\boxed{\text{ENT}}$

Estimating

One way to estimate the answer to the problem shown in Example 5 is to round 742 to 700 and 615 to 600 and then subtract 600 from 700 to obtain 100, which is an estimate of the difference. Making a mental estimate in this manner will help you catch some of the errors that will occur if you press the wrong buttons on your calculator.

GETTING READY FOR CLASS

After reading through the preceding section, respond in your own words and in complete sentences.

1. Which sentence below describes the problem in Example 1?
   a. The difference of 241 and 376 is 135.
   b. The difference of 376 and 241 is 135.
2. Write a subtraction problem using the number 234 that involves borrowing from the tens column to the ones column.
3. Write a subtraction problem using the number 234 in which the answer is 111.
4. Describe how you would subtract the number 56 from the number 93.

Problem Set 1.4

Perform the indicated operation.

1. Subtract 24 from 56.

2. Subtract 71 from 89.

3. Subtract 23 from 45.

4. Subtract 97 from 98.

5. Find the difference of 29 and 19.

6. Find the difference of 37 and 27.

7. Find the difference of 126 and 15.

8. Find the difference of 348 and 32.

Work each of the following subtraction problems.

9.	975	**10.**	480	**11.**	904	**12.**	657
	− 663		− 260		− 501		− 507

13.	9,876	**14.**	5,008	**15.**	7,976	**16.**	6,980
	− 8,765		− 3,002		− 3,432		− 470

Find the difference in each case. (These problems all involve borrowing.)

17. 52 − 37

18. 65 − 48

19. 70 − 37

20. 90 − 21

21. 74 − 69

22. 31 − 28

23. 51 − 18

24. 64 − 58

25. 329 − 234

26. 518 − 492

27. 348 − 196

28. 759 − 661

29.	932	**30.**	895	**31.**	647	**32.**	842
	− 658		− 597		− 159		− 199

33.	905	**34.**	804	**35.**	600	**36.**	800
	− 367		− 238		− 437		− 342

37.	4,583	**38.**	7,849	**39.**	79,040	**40.**	86,492
	− 2,973		− 2,957		− 32,957		− 78,506

Complete the following tables.

41.

First Number a	Second Number b	The Difference of a and b a − b
25	15	
24	16	
23	17	
22	18	

42.

First Number a	Second Number b	The Difference of a and b a − b
90	79	
80	69	
70	59	
60	49	

39

43.

First Number a	Second Number b	The Difference of a and b a − b
400	256	
400	144	
225	144	
225	81	

44.

First Number a	Second Number b	The Difference of a and b a − b
100	36	
100	64	
25	16	
25	9	

Write each of the following expressions in words. Use the word *difference* in each case.

45. $10 - 2$

46. $9 - 5$

47. $a - 6$

48. $7 - x$

49. $8 - 2 = 6$

50. $m - 1 = 4$

Write each of the following expressions in symbols.

51. The difference of 8 and 3

52. The difference of x and 2

53. 9 subtracted from y

54. a subtracted from b

55. The difference of 3 and 2 is 1.

56. The difference of 10 and y is 5.

Applying the Concepts

Not all of the following application problems involve only subtraction. Some involve addition as well. Be sure to read each problem carefully.

57. Checkbook Balance Diane has $504 in her checking account. If she writes five checks for a total of $249, how much does she have left in her account?

58. Checkbook Balance Larry has $763 in his checking account. If he writes a check for each of the three bills listed, how much will he have left in his account?

Item	Amount
Rent	$418
Phone	25
Car repair	117

59. Tallest Mountain The world's tallest mountain is Mount Everest. On May 5, 1999, it was found to be 7 feet taller than it was previously thought to be. Before this date, Everest was thought to be 29,028 feet high. That height was determined by B. L. Gulatee in 1954. The first measurement of Everest was in 1847. At that time the height was thought to be 29,002 feet. What is the difference between the current height of Everest and the height measured in 1847?

60. **Home Prices** In 1985, Mr. Hicks paid $137,500 for his home. He sold it in 2000 for $260,600. What is the difference between what he sold it for and what he bought it for?

61. **Enrollment** Six years ago, there were 567 students attending Smith Elementary School. Today the same school has an enrollment of 399 students. How much of a decrease in enrollment has there been in the last six years at Smith School?

62. **Oil Spills** In March 1977, an oil tanker hit a reef off Taiwan and spilled 3,134,500 gallons of oil. In March 1989, an oil tanker hit a reef off Alaska and spilled 10,080,000 gallons of oil. How much more oil was spilled in the 1989 disaster?

Checkbook Balance On Monday Gil has a balance of $425 in his checkbook. On Tuesday he deposits $149 into the account. On Wednesday he writes a check for $37, and on Friday he writes a check for $188. Use this information to answer Problems 63–66.

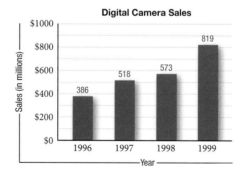

63. Find Gil's balance after he makes the deposit on Tuesday.

64. What is his balance after he writes the check on Wednesday?

65. To the nearest ten dollars, what is his balance at the end of the week?

66. To the nearest ten dollars, what is his balance before he writes the check on Friday?

67. **Digital Camera Sales** The bar chart below shows the sales of digital cameras from 1996–1999.

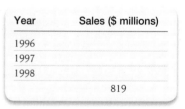

Year	Sales ($ millions)
1996	
1997	
1998	
	819

Data from *USA Today*. Copyright 2000.

a. Use the information in the bar chart to fill in the missing entries in the table.

b. What is the difference in camera sales between 1999 and 1997?

68. Wireless Phone Costs The bar chart below shows the costs of wireless phone use through 2003.

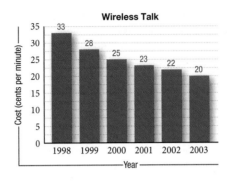

Year	Cents/Minute
	33
1999	
2000	
2001	
2002	
	20

Data from *USA Today*. Copyright 2000.

a. Use the chart to fill in the missing entries in the table.

b. What is the difference in cost between 1998 and 1999?

Multiplication with Whole Numbers, and Area

A supermarket orders 35 cases of a certain soft drink. If each case contains 12 cans of the drink, how many cans were ordered?

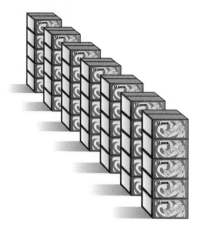

To solve this problem and others like it, we must use multiplication. Multiplication is what we will cover in this section.

To begin we can think of multiplication as shorthand for repeated addition. That is, multiplying 3 times 4 can be thought of this way:

$$3 \text{ times } 4 = 4 + 4 + 4 = 12$$

Multiplying 3 times 4 means to add three 4's. We can write 3 times 4 as 3×4, or $3 \cdot 4$.

EXAMPLE 1   Multiply $3 \cdot 4,000$.

SOLUTION   Using the definition of multiplication as repeated addition, we have

$$3 \cdot 4,000 = 4,000 + 4,000 + 4,000$$
$$= 12,000$$

Here is one way to visualize this process.

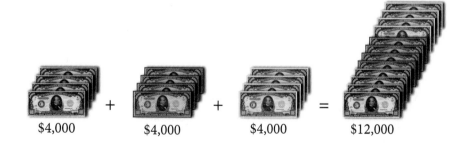

$4,000 \quad + \quad 4,000 \quad + \quad 4,000 \quad = \quad 12,000$

Notice that if we had multiplied 3 and 4 to get 12 and then attached three zeros on the right, the result would have been the same.

Notation

There are many ways to indicate multiplication. All the following statements are equivalent. They all indicate multiplication with the numbers 3 and 4.

$$3 \cdot 4, \quad 3 \times 4, \quad 3(4), \quad (3)4, \quad (3)(4), \qquad \begin{array}{r} 4 \\ \times\,3 \\ \hline \end{array}$$

If one or both of the numbers we are multiplying are represented by letters, we may also use the following notation:

$5n$	means	5 times n
ab	means	a times b

Vocabulary

We use the word *product* to indicate multiplication. If we say "The product of 3 and 4 is 12," then we mean

$$3 \cdot 4 = 12$$

Both $3 \cdot 4$ and 12 are called the product of 3 and 4. The 3 and 4 are called *factors*.

> **TABLE 1**
>
> | In English | In Symbols |
> |---|---|
> | The product of 2 and 5 | $2 \cdot 5$ |
> | The product of 5 and 2 | $5 \cdot 2$ |
> | The product of 4 and n | $4n$ |
> | The product of x and y | xy |
> | The product of 9 and 6 is 54 | $9 \cdot 6 = 54$ |
> | The product of 2 and 8 is 16 | $2 \cdot 8 = 16$ |

EXAMPLE 2   Identify the products and factors in the statement

$$9 \cdot 8 = 72$$

SOLUTION   The factors are 9 and 8, and the products are $9 \cdot 8$ and 72.

EXAMPLE 3   Identify the products and factors in the statement

$$30 = 2 \cdot 3 \cdot 5$$

SOLUTION   The factors are 2, 3, and 5. The products are $2 \cdot 3 \cdot 5$ and 30.

Distributive Property

To develop an efficient method of multiplication, we need to use what is called the *distributive property*. To begin, consider the following two problems:

Basic Multiplication Facts

×	1	2	3	4	5	6	7	8	9
1	1	2	3	4	5	6	7	8	9
2	2	4	6	8	10	12	14	16	18
3	3	6	9	12	15	18	21	24	27
4	4	8	12	16	20	24	28	32	36
5	5	10	15	20	25	30	35	40	45
6	6	12	18	24	30	36	42	48	54
7	7	14	21	28	35	42	49	56	63
8	8	16	24	32	40	48	56	64	72
9	9	18	27	36	45	54	63	72	81

Problem 1	*Problem 2*
$3(4 + 5)$	$3(4) + 3(5)$
$= 3(9)$	$= 12 + 15$
$= 27$	$= 27$

The result in both cases is the same number, 27. This indicates that the original two expressions must have been equal also. That is,

$$3(4 + 5) = 3(4) + 3(5)$$

This is an example of the distributive property. We say that multiplication *distributes* over addition.

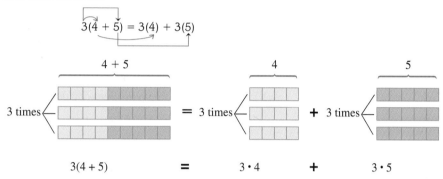

$$3(4 + 5) = 3(4) + 3(5)$$

We can write this property in symbols using the letters a, b, and c to represent any three whole numbers.

Distributive Property

If a, b, and c represent any three whole numbers, then

$$a(b + c) = a(b) + a(c)$$

Multiplication with Whole Numbers

Suppose we want to find the product $7(65)$. By writing 65 as $60 + 5$ and applying the distributive property, we have:

$$
\begin{aligned}
7(65) &= 7(60 + 5) && 65 = 60 + 5 \\
&= 7(60) + 7(5) && \text{Distributive property} \\
&= 420 + 35 && \text{Multiplication} \\
&= 455 && \text{Addition}
\end{aligned}
$$

We can write the same problem vertically like this:

$$
\begin{array}{r}
60 + 5 \\
\times \quad 7 \\
\hline
35 \quad \leftarrow \quad 7(5) = 35 \\
+ \quad 420 \quad \leftarrow \quad 7(60) = 420 \\
\hline
455
\end{array}
$$

This saves some space in writing. But notice that we can cut down on the amount of writing even more if we write the problem this way:

Step 2 $7(6) = 42$; add the 3 we carried to 42 to get 45

$$
\begin{array}{r}
\overset{3}{65} \\
\times \ 7 \\
\hline
455
\end{array}
$$

Step 1 $7(5) = 35$; write the 5 in the ones column, and then carry the 3 to the tens column

This shortcut notation takes some practice.

EXAMPLE 4   Multiply: 9(43)

Step 2   9(4) = 36; add the   2
2 we carried to 36 to get 38   →43   Step 1   9(3) = 27; write the 7
                                 × 9   in the ones column, and then carry
                                 ―――   the 2 to the tens column
                                 387 ←

EXAMPLE 5   Multiply: 52(37)

SOLUTION   This is the same as 52(30 + 7) or by the distributive property

52(30) + 52(7)

We can find each of these products by using the shortcut method:

```
                1
  52           52
× 30          × 7
―――          ―――
1,560         364
```

The sum of these two numbers is 1,560 + 364 = 1,924. Here is a summary of what we have so far:

52(37) = 52(30 + 7)   37 = 30 + 7
       = 52(30) + 52(7)   Distributive property
       = 1,560 + 364   Multiplication
       = 1,924   Addition

The shortcut form for this problem is

```
    52
×   37
―――――
   364    ←―――  7(52) = 364
+ 1,560   ←―――  30(52) = 1,560
―――――
 1,924
```

In this case we have not shown any of the numbers we carried, simply because it becomes very messy.

Note

This discussion is to show why we multiply the way we do. You should go over it in detail, so you will understand the reasons behind the process of multiplication. Besides being able to do multiplication, you should understand it.

EXAMPLE 6   Multiply: 279(428)

SOLUTION
```
      279
×     428
――――――――
    2,232    ←――― 8(279) = 2,232
    5,580    ←――― 20(279) = 5,580
+ 111,600    ←――― 400(279) = 111,600
――――――――
  119,412
```

USING TECHNOLOGY   *Calculators*

Here is how we would work the problem shown in Example 6 on a calculator:
Scientific Calculator: 279 × 428 =
Graphing Calculator: 279 × 428 ENT

Estimating

One way to estimate the answer to the problem shown in Example 6 is to round each number to the nearest hundred and then multiply the rounded numbers. Doing so would give us this:

$$300(400) = 120,000$$

Our estimate of the answer is 120,000, which is close to the actual answer, 119,412. Making estimates is important when we are using calculators; having an estimate of the answer will keep us from making major errors in multiplication.

Applications

EXAMPLE 7   A supermarket orders 35 cases of a certain soft drink. If each case contains 12 cans of the drink, how many cans were ordered?

SOLUTION   We have 35 cases and each case has 12 cans. The total number of cans is the product of 35 and 12, which is 35(12):

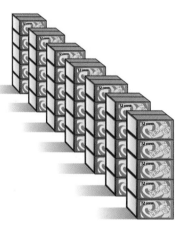

$$
\begin{array}{r}
12 \\
\times\ 35 \\
\hline
60 \quad\longleftarrow\quad 5(12) = 60 \\
+\ 360 \quad\longleftarrow\quad 30(12) = 360 \\
\hline
420
\end{array}
$$

There is a total of 420 cans of the soft drink.

EXAMPLE 8   Shirley earns $12 an hour for the first 40 hours she works each week. If she has $109 deducted from her weekly check for taxes and retirement, how much money will she take home if she works 38 hours this week?

SOLUTION   To find the amount of money she earned for the week, we multiply 12 and 38. From that total we subtract 109. The result is her take-home pay. Without showing all the work involved in the calculations, here is the solution:

$$38(\$12) = \$456 \qquad \text{Her total weekly earnings}$$
$$\$456 - \$109 = \$347 \qquad \text{Her take-home pay}$$

Note

The letter g that is shown after some of the numbers in the nutrition label in Figure 1 stands for grams, a unit used to measure weight. The unit mg stands for milligrams, another, smaller unit of weight. We will have more to say about these units later in the book.

EXAMPLE 9   In 1993, the government standardized the way in which nutrition information is presented on the labels of most packaged food products. Figure 1 shows one of these standardized food labels. It is from a package of Fritos Corn Chips that I ate the day I was writing this example. Approximately how many chips are in the bag, and what is the total number of calories consumed if all the chips in the bag are eaten?

SOLUTION   Reading toward the top of the label, we see that there are about 32 chips in one serving, and approximately 3 servings in the bag. Therefore, the total number of chips in the bag is

Nutrition Facts
Serving Size 1 oz. (28g/About 32 chips)
Servings Per Container: 3

Amount Per Serving	
Calories 160	Calories from fat 90
	% Daily Value*
Total Fat 10 g	16%
Saturated Fat 1.5g	8%
Cholesterol 0mg	0%
Sodium 160mg	7%
Total Carbohydrate 15g	5%
Dietary Fiber 1g	4%
Sugars 0g	
Protein 2g	

Vitamin A 0%	•	Vitamin C 0%
Calcium 2%	•	Iron 0%

*Percent Daily Values are based on a 2,000 calorie diet

FIGURE 1

$$3(32) = 96 \text{ chips}$$

This is an approximate number, because each serving is approximately 32 chips. Reading further we find that each serving contains 160 calories. Therefore, the total number of calories consumed by eating all the chips in the bag is

$$3(160) = 480 \text{ calories}$$

As we progress through the book, we will study more of the information in nutrition labels.

EXAMPLE 10   The table below lists the number of calories burned in 1 hour of exercise by a person who weighs 150 pounds. Suppose a 150-pound person goes bowling for 2 hours after having eaten the bag of chips mentioned in Example 9. Will he or she burn all the calories consumed from the chips?

Activity	Calories Burned in 1 Hour by a 150-Pound Person
Bicycling	374
Bowling	265
Handball	680
Jazzercize	340
Jogging	680
Skiing	544

SOLUTION   Each hour of bowling burns 265 calories. If the person bowls for 2 hours, a total of

$$2(265) = 530 \text{ calories}$$

will have been burned. Because the bag of chips contained only 480 calories, all of them have been burned with 2 hours of bowling.

Area

Note

To understand some of the notation we use for area, we need to talk about exponents. The 2 in the expression 3^2 is an exponent. The expression 3^2 is read "3 to the second power," or "3 squared," and it is defined this way:

$$3^2 = 3 \cdot 3 = 9$$

As you can see, the exponent 2 in the expression 3^2 tells us to multiply two 3s together. Here are some additional expressions containing the exponent 2.

$$4^2 = 4 \cdot 4 = 16$$
$$5^2 = 5 \cdot 5 = 25$$
$$11^2 = 11 \cdot 11 = 121$$

We will cover exponents in more detail later in this chapter.

The *area* of a flat object is a measure of the amount of surface the object has. The rectangle in Figure 2 below has an area of 6 square inches, because that is the number of squares (each of which is 1 inch long and 1 inch wide) it takes to cover the rectangle.

one square inch	one square inch	one square inch
one square inch	one square inch	one square inch

2 inches

3 inches

FIGURE 2   A rectangle with an area of 6 square inches

It is no coincidence that the area of the rectangle in Figure 2 and the product of the length and the width are the same number. We can calculate the area of the rectangle in Figure 2 by simply multiplying the length and the width together:

$$\begin{aligned}
\text{Area} &= (\text{length}) \cdot (\text{width}) \\
&= (3 \text{ inches}) \cdot (2 \text{ inches}) \\
&= (3 \cdot 2) \cdot (\text{inches} \cdot \text{inches}) \\
&= 6 \text{ square inches}
\end{aligned}$$

The unit *square inches* can be abbreviated as *sq. in.* or *in²*.

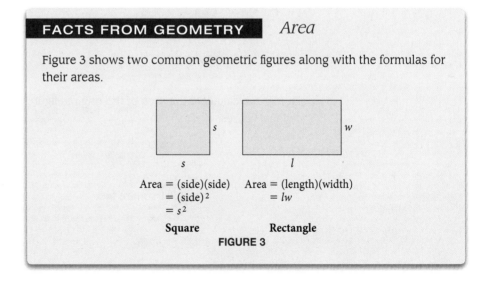

FACTS FROM GEOMETRY   *Area*

Figure 3 shows two common geometric figures along with the formulas for their areas.

s

s

w

l

$$\begin{aligned}
\text{Area} &= (\text{side})(\text{side}) \\
&= (\text{side})^2 \\
&= s^2
\end{aligned}$$

$$\begin{aligned}
\text{Area} &= (\text{length})(\text{width}) \\
&= lw
\end{aligned}$$

Square   **Rectangle**

FIGURE 3

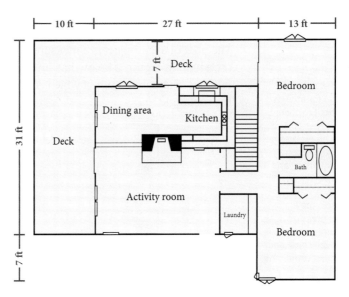

FIGURE 4   *Source:* **Image courtesy of COOLhouseplans.com.**

![EXAMPLE 11] Find the total area of the house and deck shown in Figure 4.

SOLUTION   We begin by drawing an additional line (shown as a broken line in Figure 5) so that the original figure is now composed of two rectangles. Next, we fill in the missing dimensions on the two rectangles (Figure 6).

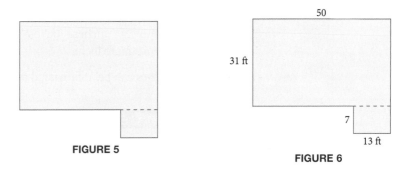

FIGURE 5

FIGURE 6

Finally, we calculate the area of the original figure by adding the areas of the individual figures:

$$
\begin{aligned}
\text{Area} &= \text{Area small rectangle} + \text{Area large rectangle} \\
&= \quad\quad 13 \cdot 7 \quad\quad\quad + \quad\quad 50 \cdot 31 \\
&= \quad\quad\quad 91 \quad\quad\quad\quad + \quad\quad\; 1550 \\
&= 1641 \text{ square feet}
\end{aligned}
$$

More Properties of Multiplication

Multiplication Property of 0

If a represents any number, then

$$a \cdot 0 = 0 \quad \text{and} \quad 0 \cdot a = 0$$

In words: Multiplication by 0 always results in 0.

Multiplication Property of 1

If a represents any number, then

$$a \cdot 1 = a \quad \text{and} \quad 1 \cdot a = a$$

In words: Multiplying any number by 1 leaves that number unchanged.

Commutative Property of Multiplication

If a and b are any two numbers, then

$$ab = ba$$

In words: The order of the numbers in a product doesn't affect the result.

Associative Property of Multiplication

If a, b, and c represent any three numbers, then

$$(ab)c = a(bc)$$

In words: We can change the grouping of the numbers in a product without changing the result.

To visualize the commutative property, we can think of an instructor with 12 students.

3 chairs across, 4 chairs back   =   4 chairs across, 3 chairs back

EXAMPLE 12   Use the commutative property of multiplication to rewrite each of the following products:

    a. $7 \cdot 9$ **b.** $4(6)$

SOLUTION   Applying the commutative property to each expression, we have:

    a. $7 \cdot 9 = 9 \cdot 7$ **b.** $4(6) = 6(4)$

EXAMPLE 13   Use the associative property of multiplication to rewrite each of the following products:

      a. $(2 \cdot 7) \cdot 9$ **b.** $3 \cdot (8 \cdot 2)$

SOLUTION   Applying the associative property of multiplication, we regroup as follows:

      a. $(2 \cdot 7) \cdot 9 = 2 \cdot (7 \cdot 9)$ **b.** $3 \cdot (8 \cdot 2) = (3 \cdot 8) \cdot 2$ ■

Solving Equations

If n is used to represent a number, then the equation

$$4 \cdot n = 12$$

is read "4 times n is 12," or "The product of 4 and n is 12." This means that we are looking for the number we multiply by 4 to get 12. The number is 3. Because the equation becomes a true statement if n is 3, we say that 3 is the solution to the equation.

EXAMPLE 14   Find the solution to each of the following equations:

      a. $6 \cdot n = 24$ **b.** $4 \cdot n = 36$ **c.** $15 = 3 \cdot n$ **d.** $21 = 3 \cdot n$

SOLUTION   **a.** The solution to $6 \cdot n = 24$ is 4, because $6 \cdot 4 = 24$.

            b. The solution to $4 \cdot n = 36$ is 9, because $4 \cdot 9 = 36$.

            c. The solution to $15 = 3 \cdot n$ is 5, because $15 = 3 \cdot 5$.

            d. The solution to $21 = 3 \cdot n$ is 7, because $21 = 3 \cdot 7$.   ■

GETTING READY FOR CLASS

After reading through the preceding section, respond in your own words and in complete sentences.

1. Use the numbers 7, 8, and 9 to give an example of the distributive property.

2. When we write the distributive property in words, we say "multiplication distributes over addition." It is also true that multiplication distributes over subtraction. Use the variables a, b, and c to write the distributive property using multiplication and subtraction.

3. We can multiply 8 and 487 by writing 487 in expanded form as $400 + 80 + 7$ and then applying the distributive property. Apply the distributive property to the expression below and then simplify.

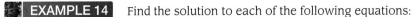

$$8(400 + 80 + 7) =$$

4. Find the mistake in the following multiplication problem. Then work the problem correctly.

$$
\begin{array}{r}
43 \\
\times\ 68 \\
\hline
344 \\
+\ 258 \\
\hline
602 \\
\end{array}
$$

Problem Set 1.5

Multiply each of the following.

1. $3 \cdot 100$ **2.** $7 \cdot 100$ **3.** $3 \cdot 200$ **4.** $4 \cdot 200$

5. $6 \cdot 500$ **6.** $8 \cdot 400$ **7.** $5 \cdot 1,000$ **8.** $8 \cdot 1,000$

9. $3 \cdot 7,000$ **10.** $6 \cdot 7,000$ **11.** $9 \cdot 9,000$ **12.** $7 \cdot 7,000$

Find each of the following products (Multiply). In each case use the shortcut method.

13. 25 ×4   **14.** 43 ×9   **15.** 38 ×6   **16.** 45 ×7

17. 18 ×2   **18.** 29 ×3   **19.** 72 ×20   **20.** 68 ×30

21. 19 ×50   **22.** 24 ×40   **23.** 69 ×25   **24.** 27 ×36

25. 11 ×11   **26.** 12 ×21   **27.** 97 ×16   **28.** 24 ×39

29. 168 ×25   **30.** 452 ×34   **31.** 728 ×91   **32.** 680 ×76

33. 698 ×400   **34.** 879 ×600   **35.** 111 ×111   **36.** 123 ×321

37. 532 ×200   **38.** 277 ×900   **39.** 856 ×232   **40.** 455 ×248

41. 976 ×628   **42.** 432 ×555   **43.** 2,468 ×135   **44.** 2,725 ×324

45. 24,563 ×735   **46.** 56,728 ×852   **47.** 44,777 ×5,888   **48.** 33,999 ×2,555

Complete the following tables.

49.

First Number a	Second Number b	Their Product ab
11	11	
11	22	
22	22	
22	44	

50.

First Number a	Second Number b	Their Product ab
25	15	
25	30	
50	15	
50	30	

51.

First Number a	Second Number b	Their Product ab
25	10	
25	100	
25	1,000	
25	10,000	

52.

First Number a	Second Number b	Their Product ab
11	111	
11	222	
22	111	
22	222	

53.

First Number a	Second Number b	Their Product ab
12	20	
36	20	
12	40	
36	40	

54.

First Number a	Second Number b	Their Product ab
10	12	
100	12	
1,000	12	
10,000	12	

Write each of the following expressions in words, using the word *product.*

55. $6 \cdot 7$

56. $9(4)$

57. $2 \cdot n$

58. $5 \cdot x$

59. $9 \cdot 7 = 63$

60. $(5)(6) = 30$

Write each of the following in symbols.

61. The product of 7 and n.

62. The product of 9 and x.

63. The product of 6 and 7 is 42.

64. The product of 8 and 9 is 72.

65. The product of 0 and 6 is 0.

66. The product of 1 and 6 is 6.

Identify the products in each statement.

67. $9 \cdot 7 = 63$ **68.** $2(6) = 12$ **69.** $4(4) = 16$ **70.** $5 \cdot 5 = 25$

Identify the factors in each statement.

71. $2 \cdot 3 \cdot 4 = 24$ **72.** $6 \cdot 1 \cdot 5 = 30$ **73.** $12 = 2 \cdot 2 \cdot 3$ **74.** $42 = 2 \cdot 3 \cdot 7$

Rewrite each of the following using the commutative property of multiplication.

75. $5(9)$ **76.** $4(3)$ **77.** $6 \cdot 7$ **78.** $8 \cdot 3$

Rewrite each of the following using the associative property of multiplication.

79. $2 \cdot (7 \cdot 6)$ **80.** $4 \cdot (8 \cdot 5)$ **81.** $3 \times (9 \times 1)$ **82.** $5 \times (8 \times 2)$

Use the distributive property to rewrite each expression, then simplify.

83. $7(2 + 3)$ **84.** $4(5 + 8)$ **85.** $9(4 + 7)$ **86.** $6(9 + 5)$

87. $3(x + 1)$ **88.** $5(x + 8)$ **89.** $2(x + 5)$ **90.** $4(x + 3)$

Find a solution for each equation.

91. $4 \cdot n = 12$ **92.** $3 \cdot n = 12$ **93.** $9 \cdot n = 81$ **94.** $6 \cdot n = 36$

95. $0 = n \cdot 5$ **96.** $6 = 1 \cdot n$

Find the area enclosed by each figure. (Note that some of the units on the figures come from the metric system. The abbreviations are as follows: Meter is abbreviated m, centimeter is cm, and millimeter is abbreviated mm. A meter is about 3 inches longer than a yard.)

97.
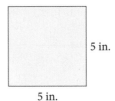
5 in.

5 in.

98.

10 in.

10 in.

99.

12 m

7 m

100.
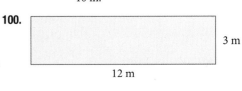
3 m

12 m

101.

102.

103. Planning a Trip A family decides to drive their compact car on their vacation. They figure it will require a total of about 130 gallons of gas for the vacation. If each gallon of gas will take them 22 miles, how long is the trip they are planning?

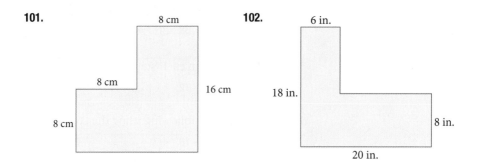

104. Rent A student pays $475 rent each month. How much money does she spend on rent in 2 years?

105. Reading House Plans Find the area of the floor of the house shown here if the garage is not included with the house and if the garage is included with the house. (The symbol ' represents feet.)

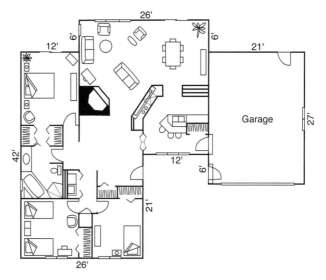

Source: Image courtesy of COOLhouseplans.com

106. The Alphabet and Area Have you ever wondered which of the letters in the alphabet is the most popular? No? The box shown here is called a *typecase,* or *printer's tray.* It was used in early typesetting to store the letters that would be used to layout a page of type. Use the box to answer the questions below.

a. What letter is used most often in printed material?

b. Which letter is printed more often, the letter *i* or the letter *f*?

c. What is the relationship between area and how often a letter in the alphabet is printed?

Exercise and Calories The table below is an extension of the table we used in Example 10 of this section. It gives the amount of energy expended during 1 hour of various activities for people of different weights. The accompanying figure is a nutrition label from a bag of Doritos tortilla chips. Use the information from the table and the nutrition label to answer Problems 107–112.

Nutrition Facts		
Serving Size 1 oz. (28g/About 12 chips)		
Servings Per Container About 2		
Amount Per Serving		
Calories 140	Calories from fat 60	
		% Daily Value*
Total Fat 7g		11%
Saturated Fat 1g		6%
Cholesterol 0mg		0%
Sodium 170mg		7%
Total Carbohydrate 18g		6%
Dietary Fiber 1g		4%
Sugars less than 1g		
Protein 2g		
Vitamin A 0%	•	Vitamin C 0%
Calcium 4%	•	Iron 2%
*Percent Daily Values are based on a 2,000 calorie diet		

CALORIES BURNED THROUGH EXERCISE

Activity	Calories Per Hour 120 Pounds	150 Pounds	180 Pounds
Bicycling	299	374	449
Bowling	212	265	318
Handball	544	680	816
Jazzercise	272	340	408
Jogging	544	680	816
Skiing	435	544	653

107. Suppose you weigh 180 pounds. How many calories would you burn if you play handball for 2 hours and then ride your bicycle for 1 hour?

108. How many calories are burned by a 120-lb person who jogs for 1 hour and then goes bike riding for 2 hours?

109. How many calories would you consume if you ate the entire bag of chips?

110. Approximately how many chips are in the bag?

111. If you weigh 180 pounds, will you burn off the calories consumed by eating 3 servings of tortilla chips if you ride your bike 1 hour?

112. If you weigh 120 pounds, will you burn off the calories consumed by eating 3 servings of tortilla chips if you ride your bike for 1 hour?

Estimating

Mentally estimate the answer to each of the following problems by rounding each number to the indicated place and then multiplying.

113.   750   hundred
       × 12   ten

114.   591   hundred
       × 323   hundred

115.   3,472   thousand
       ×   511   hundred

116.   399   hundred
       × 298   hundred

117.   2,399   thousand
       ×   698   hundred

118.   9,999   thousand
       ×   666   hundred

Darlene is planning a party and would like to serve 8-ounce glasses of soda. The glasses will be filled from 32-ounce bottles of soda. In order to know how many bottles of soda to buy, she needs to find out how many of the 8-ounce glasses can be filled by one of the 32-ounce bottles. One way to solve this problem is with division: dividing 32 by 8. A diagram of the problem is shown in Figure 1.

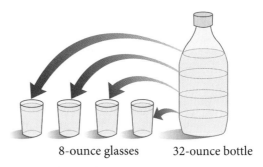

8-ounce glasses   32-ounce bottle

FIGURE 1

As a division problem: As a multiplication problem:

$32 \div 8 = 4$ $4 \cdot 8 = 32$

Notation

As was the case with multiplication, there are many ways to indicate division. All the following statements are equivalent. They all mean 10 divided by 5.

$$10 \div 5, \quad \frac{10}{5}, \quad 10/5, \quad 5\overline{)10}$$

The kind of notation we use to write division problems will depend on the situation. We will use the notation $5\overline{)10}$ mostly with the long division problems found in this chapter. The notation $\frac{10}{5}$ will be used in the chapter on fractions and in later chapters. The horizontal line used with the notation $\frac{10}{5}$ is called the *fraction bar.*

Vocabulary

The word *quotient* is used to indicate division. If we say "The quotient of 10 and 5 is 2," then we mean

$$10 \div 5 = 2 \quad \text{or} \quad \frac{10}{5} = 2$$

The 10 is called the *dividend,* and the 5 is called the *divisor.* All the expressions, $10 \div 5, \frac{10}{5}$, and 2, are called the *quotient* of 10 and 5.

TABLE 1

In English	In Symbols
The quotient of 15 and 3	$15 \div 3$, or $\frac{15}{3}$, or $15/3$
The quotient of 3 and 15	$3 \div 15$, or $\frac{3}{15}$, or $3/15$
The quotient of 8 and n	$8 \div n$, or $\frac{8}{n}$, or $8/n$
x divided by 2	$x \div 2$, or $\frac{x}{2}$, or $x/2$
The quotient of 21 and 3 is 7	$21 \div 3 = 7$, or $\frac{21}{3} = 7$

The Meaning of Division

One way to arrive at an answer to a division problem is by thinking in terms of multiplication. For example, if we want to find the quotient of 32 and 8, we may ask, "What do we multiply by 8 to get 32?"

$$32 \div 8 = ? \qquad \text{means} \qquad 8 \cdot ? = 32$$

Because we know from our work with multiplication that $8 \cdot 4 = 32$, it must be true that

$$32 \div 8 = 4$$

Table 2 lists some additional examples.

TABLE 2

Division		Multiplication
$18 \div 6 = 3$	because	$6 \cdot 3 = 18$
$32 \div 8 = 4$	because	$8 \cdot 4 = 32$
$10 \div 2 = 5$	because	$2 \cdot 5 = 10$
$72 \div 9 = 8$	because	$9 \cdot 8 = 72$

Division by One-Digit Numbers

Consider the following division problem:

$$465 \div 5$$

We can think of this problem as asking the question, "How many fives can we subtract from 465?" To answer the question we begin subtracting multiples of 5. One way to organize this process is shown below:

$$
\begin{array}{r}
90 \\
5\overline{)465} \\
-450 \\
\hline
15
\end{array}
$$

← We first guess that there are at least 90 fives in 465

← $90(5) = 450$

← 15 is left after we subtract 90 fives from 465

What we have done so far is subtract 90 fives from 465 and found that 15 is still left. Because there are 3 fives in 15, we continue the process.

$$
\begin{array}{r}
3 \quad\leftarrow \text{There are 3 fives in 15} \\
90 \\
5\overline{)465} \\
-\ 450 \\
\hline
15 \\
-\ 15 \quad\leftarrow\ 3 \cdot 5 = 15 \\
\hline
0 \quad\leftarrow \text{The difference is 0}
\end{array}
$$

The total number of fives we have subtracted from 465 is

$$90 + 3 = 93 \text{ the number of fives subtracted from 465}$$

We now summarize the results of our work.

$$465 \div 5 = 93$$ which we check with multiplication $\longrightarrow$
$$
\begin{array}{r}
\overset{1}{9}3 \\
\times\ 5 \\
\hline
465
\end{array}
$$

Notation

The division problem just shown can be shortened by eliminating the subtraction signs, eliminating the zeros in each estimate, and eliminating some of the numbers that are repeated in the problem.

The shorthand form for this problem

$$
\begin{array}{r}
3 \\
90 \\
5\overline{)465} \\
-\ 450 \\
\hline
15 \\
-15 \\
\hline
0
\end{array}
$$

looks like this.

$$
\begin{array}{r}
93 \\
5\overline{)465} \\
-45\downarrow \\
\hline
15 \\
-15 \\
\hline
0
\end{array}
$$

The arrow indicates that we bring down the 5 after we subtract.

The problem shown above on the right is the shortcut form of what is called *long division*. Here is an example showing this shortcut form of long division from start to finish.

EXAMPLE 1   Divide $595 \div 7$.

SOLUTION   Because $7(8) = 56$, our first estimate of the number of sevens that can be subtracted from 595 is 80.

$$
\begin{array}{r}
8 \quad\leftarrow \text{The 8 is placed above the tens column} \\
7\overline{)595} \qquad \text{so we know our first estimate is 80} \\
-56\downarrow \quad\leftarrow 8(7) = 56 \\
\hline
35 \quad\leftarrow 59 - 56 = 3; \text{ then bring down the 5}
\end{array}
$$

Since $7(5) = 35$, we have

$$
\begin{array}{r}
85 \quad\leftarrow \text{There are 5 sevens in 35} \\
7\overline{)595} \\
-56\downarrow \\
\hline
35 \\
-35 \quad\leftarrow 5(7) = 35 \\
\hline
0 \quad\leftarrow 35 - 35 = 0
\end{array}
$$

Our result is $595 \div 7 = 85$, which we can check with multiplication:

$$\begin{array}{r} \overset{3}{85} \\ \times\, 7 \\ \hline 595 \end{array}$$

Division by Two-Digit Numbers

EXAMPLE 2   Divide: $9{,}380 \div 35$

SOLUTION   In this case our divisor, 35, is a two-digit number. The process of division is the same. We still want to find the number of thirty-fives we can subtract from 9,380.

$$\begin{array}{r} 2 \\ 35\overline{)9{,}380} \\ -7\,0\downarrow \\ \hline 2\,38 \end{array}$$

⟵ The 2 is placed above the hundreds column
⟵ $2(35) = 70$
⟵ $93 - 70 = 23$; then bring down the 8

We can make a few preliminary calculations to help estimate how many thirty-fives are in 238:

$$5 \times 35 = 175 \qquad 6 \times 35 = 210 \qquad 7 \times 35 = 245$$

Because 210 is the closest to 238 without being larger than 238, we use 6 as our next estimate:

$$\begin{array}{r} 26 \\ 35\overline{)9{,}380} \\ -7\,0 \\ \hline 2\,38 \\ -2\,10\downarrow \\ \hline 280 \end{array}$$

⟵ 6 in the tens column means this estimate is 60
⟵ $6(35) = 210$
⟵ $238 - 210 = 28$; bring down the 0

Because $35(8) = 280$, we have

$$\begin{array}{r} 268 \\ 35\overline{)9{,}380} \\ -7\,0 \\ \hline 2\,38 \\ -2\,10 \\ \hline 280 \\ -280 \\ \hline 0 \end{array}$$

⟵ $8(35) = 280$
⟵ $280 - 280 = 0$

We can check our result with multiplication:

$$\begin{array}{r} 268 \\ \times\ \ 35 \\ \hline 1{,}340 \\ +8{,}040 \\ \hline 9{,}380 \end{array}$$

EXAMPLE 3   Divide: 1,872 by 18.

SOLUTION   Here is the first step.

$$
\begin{array}{r}
1 \\
18\overline{)1{,}872} \\
-18 \\
\hline
0
\end{array}
$$

⟵ 1 is placed above hundred column

⟵ Multiply 1(18) to get 18

⟵ Subtract to get 0

The next step is to bring down the 7 and divide again.

$$
\begin{array}{r}
10 \\
18\overline{)1{,}872} \\
-18\downarrow \\
\hline
07 \\
-\ 0 \\
\hline
7
\end{array}
$$

⟵ 0 is placed above tens column. 0 is the largest number
we can multiply by 18 and not go over 7

⟵ Multiply 0(18) to get 0

⟵ Subtract to get 7

Here is the complete problem.

$$
\begin{array}{r}
104 \\
18\overline{)1{,}872} \\
-18\downarrow\ \ \\
\hline
07\ \ \\
-\ 0\downarrow \\
\hline
72 \\
-\ 72 \\
\hline
0
\end{array}
$$

To show our answer is correct, we multiply.

$$18(104) = 1{,}872$$

Division with Remainders

Suppose Darlene were planning to use 6-ounce glasses instead of 8-ounce glasses for her party. To see how many glasses she could fill from the 32-ounce bottle, she would divide 32 by 6. If she did so, she would find that she could fill 5 glasses, but after doing so she would have 2 ounces of soda left in the bottle. A diagram of this problem is shown in Figure 2.

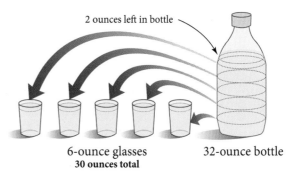

2 ounces left in bottle

6-ounce glasses
30 ounces total

32-ounce bottle

FIGURE 2

Writing the results in the diagram as a division problem looks like this:

$$\begin{array}{r} 5 \quad\longleftarrow \text{ Quotient} \\ \text{Divisor} \longrightarrow 6\overline{)32} \quad\longleftarrow \text{ Dividend} \\ -30 \\ \hline 2 \quad\longleftarrow \text{ Remainder} \end{array}$$

EXAMPLE 4   Divide: $1{,}690 \div 67$

SOLUTION   Dividing as we have previously, we get

$$\begin{array}{r} 25 \\ 67\overline{)1{,}690} \\ -1\ 34\downarrow \\ \hline 350 \\ -335 \\ \hline 15 \quad\longleftarrow \text{ 15 is left over} \end{array}$$

We have 15 left, and because 15 is less than 67, no more sixty-sevens can be subtracted. In a situation like this we call 15 the *remainder* and write

These indicate that the remainder is 15

$$\begin{array}{r} 25 \ \text{R } 15 \\ 67\overline{)1{,}690} \\ -1\ 34\downarrow \\ \hline 350 \\ -335 \\ \hline 15 \end{array} \qquad \text{or} \qquad \begin{array}{r} 25\tfrac{15}{67} \\ 67\overline{)1{,}690} \\ -1\ 34\downarrow \\ \hline 350 \\ -335 \\ \hline 15 \end{array}$$

Both forms of notation shown above indicate that 15 is the remainder. The notation R 15 is the notation we will use in this chapter. The notation $\tfrac{15}{67}$ will be useful in the chapter on fractions.

To check a problem like this, we multiply the divisor and the quotient as usual, and then add the remainder to this result:

$$\begin{array}{r} 67 \\ \times 25 \\ \hline 335 \\ +1{,}340 \\ \hline 1{,}675 \quad\longleftarrow \text{ Product of divisor and quotient} \end{array}$$

$$1{,}675 + 15 = 1{,}690$$

Remainder   Dividend

EXAMPLE 5   A family has an annual income of $35,880. How much is their average monthly income?

SOLUTION   Because there are 12 months in a year and the yearly (annual) income is $35,880, we want to know what $35,880 divided into 12 equal parts is. Therefore we have

$$
\begin{array}{r}
2\,990 \\
12\overline{)35,880} \\
-\underline{24} \\
11\,8 \\
-\underline{10\,8} \\
1\,08 \\
-\underline{1\,08} \\
00
\end{array}
$$

Because $35,880 \div 12 = 2,990$, the monthly income for this family is $2,990.

Division by Zero

We cannot divide by 0. That is, we cannot use 0 as a divisor in any division problem. Here's why.

Suppose there was an answer to the problem

$$\frac{8}{0} = ?$$

That would mean that

$$0 \cdot ? = 8$$

But we already know that multiplication by 0 always produces 0. There is no number we can use for the ? to make a true statement out of

$$0 \cdot ? = 8$$

Because this was equivalent to the original division problem

$$\frac{8}{0} = ?$$

we have no number to associate with the expression $\frac{8}{0}$. It is undefined.

Rule

Division by 0 is undefined. Any expression with a divisor of 0 is undefined. We cannot divide by 0.

GETTING READY FOR CLASS

After reading through the preceding section, respond in your own words and in complete sentences.

1. Which sentence below describes the problem shown in Example 1?
   a. The quotient of 7 and 595 is 85.
   b. Seven divided by 595 is 85.
   c. The quotient of 595 and 7 is 85.

2. In Example 2, we divide 9,380 by 35 to obtain 268. Suppose we add 35 to 9,380, making it 9,415. What will our answer be if we divide 9,415 by 35?

3. Example 4 shows that $1,690 \div 67$ gives a quotient of 25 with a remainder of 15. If we were to divide 1,692 by 67, what would the remainder be?

4. Explain why division by 0 is undefined in mathematics.

Problem Set 1.6

Write each of the following in symbols.

1. The quotient of 6 and 3

2. The quotient of 3 and 6

3. The quotient of 45 and 9

4. The quotient of 12 and 4

5. The quotient of r and s

6. The quotient of s and r

7. The quotient of 20 and 4 is 5.

8. The quotient of 20 and 5 is 4.

Write a multiplication statement that is equivalent to each of the following division statements.

9. $6 \div 2 = 3$

10. $6 \div 3 = 2$

11. $\dfrac{36}{9} = 4$

12. $\dfrac{36}{4} = 9$

13. $\dfrac{48}{6} = 8$

14. $\dfrac{35}{7} = 5$

15. $28 \div 7 = 4$

16. $81 \div 9 = 9$

Find each of the following quotients. (Divide.)

17. $25 \div 5$

18. $72 \div 8$

19. $40 \div 5$

20. $12 \div 2$

21. $9 \div 0$

22. $7 \div 1$

23. $360 \div 8$

24. $285 \div 5$

25. $\dfrac{138}{6}$

26. $\dfrac{267}{3}$

27. $5\overline{)7,650}$

28. $5\overline{)5,670}$

29. $5\overline{)6,750}$

30. $5\overline{)6,570}$

31. $3\overline{)54,000}$

32. $3\overline{)50,400}$

33. $3\overline{)50,040}$

34. $3\overline{)50,004}$

Estimating

Work Problems 35 through 38 mentally, without using a calculator.

35. The quotient $845 \div 93$ is closest to which of the following numbers.
    a. 10   **b.** 100   **c.** 1,000   **d.** 10,000

36. The quotient $762 \div 43$ is closest to which of the following numbers?
    a. 2   **b.** 20   **c.** 200   **d.** 2,000

37. The quotient $15,208 \div 771$ is closest to which of the following numbers?
    a. 2   **b.** 20   **c.** 200   **d.** 2,000

38. The quotient $24,471 \div 523$ is closest to which of the following numbers?
    a. 5   **b.** 50   **c.** 500   **d.** 5,000

Without a calculator give a one-digit estimate for each of the following quotients. That is, for each quotient, mentally estimate the answer using one of the digits 1, 2, 3, 4, 5, 6, 7, 8, or 9.

39. $316 \div 289$

40. $662 \div 289$

41. $728 \div 355$

42. $728 \div 177$

43. $921 \div 243$

44. $921 \div 442$

45. $673 \div 109$

46. $673 \div 218$

Divide. You shouldn't have any wrong answers because you can always check your results with multiplication.

47. $1,440 \div 32$ **48.** $1,206 \div 67$ **49.** $\dfrac{2,401}{49}$ **50.** $\dfrac{4,606}{49}$

51. $28\overline{)12,096}$ **52.** $28\overline{)96,012}$ **53.** $63\overline{)90,594}$ **54.** $45\overline{)17,595}$

55. $87\overline{)61,335}$ **56.** $79\overline{)48,032}$ **57.** $45\overline{)135,900}$ **58.** $56\overline{)227,920}$

Complete the following tables.

59.

First Number a	Second Number b	The Quotient of a and b $\dfrac{a}{b}$
100	25	
100	26	
100	27	
100	28	

60.

First Number a	Second Number b	The Quotient of a and b $\dfrac{a}{b}$
100	25	
101	25	
102	25	
103	25	

Divide. The following division problems all have remainders.

61. $6\overline{)370}$ **62.** $8\overline{)390}$ **63.** $3\overline{)271}$ **64.** $3\overline{)172}$

65. $26\overline{)345}$ **66.** $26\overline{)543}$ **67.** $71\overline{)16,620}$ **68.** $71\overline{)33,240}$

69. $23\overline{)9,250}$ **70.** $23\overline{)20,800}$ **71.** $169\overline{)5,950}$ **72.** $391\overline{)34,450}$

Applying the Concepts

The application problems that follow may involve more than merely division. Some may require addition, subtraction, or multiplication, whereas others may use a combination of two or more operations.

73. Monthly Income A family has an annual income of $22,200. How much is their monthly income?

74. Hourly Wages If a man works an 8-hour shift and is paid $96, how much does he make for 1 hour?

75. Price per Pound If 6 pounds of a certain kind of fruit cost 96¢, how much does 1 pound cost?

76. Cost of a Dress A dress shop orders 45 dresses for a total of $675. If they paid the same amount for each dress, how much was each dress?

77. Fitness Walking The guidelines for fitness now indicate that a person who walks 10,000 steps daily is physically fit. According to *The Walking Site* on the Internet, it takes just over 2,000 steps to walk one mile. If that is the case, how many miles do you need to walk in order to take 10,000 steps?

78. Filling Glasses How many 8-ounce glasses can be filled from three 32-ounce bottles of soda?

79. Filling Glasses How many 5-ounce glasses can be filled from a 32-ounce bottle of milk? How many ounces of milk will be left in the bottle when all the glasses are full?

80. Filling Glasses How many 3-ounce glasses can be filled from a 28-ounce bottle of milk? How many ounces of milk will be left in the bottle when all the glasses are filled?

81. Filling Glasses How many 32-ounce bottles of Coke will be needed to fill sixteen 6-ounce glasses?

82. Filling Glasses How many 28-ounce bottles of 7-Up will be needed to fill fourteen 6-ounce glasses?

83. Cost of Wine If a person paid $192 for 16 bottles of wine, how much did each bottle cost?

84. Miles per Gallon A traveling salesman kept track of his mileage for 1 month. He found that he traveled 1,104 miles and used 48 gallons of gas. How many miles did he travel on each gallon of gas?

85. Milligrams of Calcium Suppose one egg contains 25 milligrams of calcium, a piece of toast contains 40 milligrams of calcium, and a glass of milk contains 215 milligrams of calcium. How many milligrams of calcium are contained in a breakfast that consists of three eggs, two glasses of milk, and four pieces of toast?

86. Milligrams of Iron Suppose a glass of juice contains 3 milligrams of iron and a piece of toast contains 2 milligrams of iron. If Diane drinks two glasses of juice and has three pieces of toast for breakfast, how much iron is contained in the meal?

Calculator Problems

Find each of the following quotients using a calculator.

87. $305,026 \div 698$

88. $771,537 \div 949$

89. 18,436,466 divided by 5,678

90. 2,492,735 divided by 2,345

91. The quotient of 603,955 and 695.

92. The quotient of 875,124 and 876.

93. $4,903\overline{)27,868,652}$

94. $3,090\overline{)2,308,230}$

95. Gallons per Minute If a 79,768-gallon tank can be filled in 472 minutes, how many gallons enter the tank each minute?

96. Weight per Case A truckload of 632 crates of motorcycle parts weighs 30,968 pounds. How much does each of the crates weigh, if they each weigh the same amount?

Exponents, Order of Operations, and Averages

Exponents are a shorthand way of writing repeated multiplication. In the expression 2^3, 2 is called the *base* and 3 is called the *exponent.* The expression 2^3 is read "2 to the third power" or "2 cubed." The exponent 3 tells us to use the base 2 as a multiplication factor three times.

$$2^3 = 2 \cdot 2 \cdot 2 \quad \text{2 is used as a factor three times}$$

We can simplify the expression by multiplication:

$$2^3 = 2 \cdot 2 \cdot 2$$
$$= 4 \cdot 2$$
$$= 8$$

The expression 2^3 is equal to the number 8. We can summarize this discussion with the following definition:

DEFINITION

An **exponent** is a whole number that indicates how many times the base is to be used as a factor. Exponents indicate repeated multiplication.

For example, in the expression 5^2, 5 is the base and 2 is the exponent. The meaning of the expression is

$$5^2 = 5 \cdot 5 \quad \text{5 is used as a factor two times}$$
$$= 25$$

The expression 5^2 is read "5 to the second power" or "5 squared."

Here are some more examples.

EXAMPLE 1 Name the base and the exponent:

    a. 3^2

    b. 3^3

    c. 2^4

SOLUTION **a.** The base is 3, and the exponent is 2. The expression is read "3 to the second power" or "3 squared."

    b. The base is 3, and the exponent is 3. The expression is read "3 to the third power" or "3 cubed."

    c. The base is 2, and the exponent is 4. The expression is read "2 to the fourth power."

As you can see from this example, a base raised to the second power is also said to be *squared,* and a base raised to the third power is also said to be *cubed.* These are the only two exponents (^{2} and ^{3}) that have special names. All other exponents are referred to only as "fourth powers," "fifth powers," "sixth powers," and so on.

The next example shows how we can simplify expressions involving exponents by using repeated multiplication.

EXAMPLE 2   Expand and multiply:

a. 3^2 b. 4^2 c. 3^3 d. 2^4

SOLUTION   a. $3^2 = 3 \cdot 3 = 9$

b. $4^2 = 4 \cdot 4 = 16$

c. $3^3 = 3 \cdot 3 \cdot 3 = 9 \cdot 3 = 27$

d. $2^4 = 2 \cdot 2 \cdot 2 \cdot 2 = 4 \cdot 4 = 16$

USING TECHNOLOGY

Calculators

Here is how we use a calculator to evaluate exponents, as we did in Example 1c:

Scientific Calculator: 2 $\boxed{x^y}$ 4 $\boxed{=}$

Graphing Calculator: 2 $\boxed{\wedge}$ 4 $\boxed{\text{ENT}}$ or   2 $\boxed{x^y}$ 4 $\boxed{\text{ENT}}$
(depending on the calculator)

Finally, we should consider what happens when the numbers 0 and 1 are used as exponents. First of all, any number raised to the first power is itself. That is, if we let the letter a represent any number, then

$$a^1 = a$$

To take care of the cases when 0 is used as an exponent, we must use the following definition:

DEFINITION

Any number other than 0 raised to the 0 power is 1. That is, if a represents any nonzero number, then it is always true that

$$a^0 = 1$$

EXAMPLE 3   Simplify:

a. 5^1 b. 9^1 c. 4^0 d. 8^0

SOLUTION   a. $5^1 = 5$

b. $9^1 = 9$

c. $4^0 = 1$

d. $8^0 = 1$

Order of Operations

The symbols we use to specify operations, $+, -, \cdot, \div$, along with the symbols we use for grouping, ( ) and [ ], serve the same purpose in mathematics as punctuation marks in English. They may be called the punctuation marks of mathematics.

Consider the following sentence:

> Bob said John is tall.

It can have two different meanings, depending on how we punctuate it:

> 1 "Bob," said John, "is tall."
> 2 Bob said, "John is tall."

Without the punctuation marks we don't know which meaning the sentence has.

Now, consider the following mathematical expression:

$$4 + 5 \cdot 2$$

What should we do? Should we add 4 and 5 first, or should we multiply 5 and 2 first? There seem to be two different answers. In mathematics we want to avoid situations in which two different results are possible. Therefore we follow the rule for order of operations.

DEFINITION

Order of Operations   When evaluating mathematical expressions, we will perform the operations in the following order:

1. If the expression contains grouping symbols, such as parentheses ( ), brackets [ ], or a fraction bar, then we perform the operations inside the grouping symbols, or above and below the fraction bar, first.

2. Then we evaluate, or simplify, any numbers with exponents.

3. Then we do all multiplications and divisions in order, starting at the left and moving right.

4. Finally, we do all additions and subtractions, from left to right.

According to our rule, the expression $4 + 5 \cdot 2$ would have to be evaluated by multiplying 5 and 2 first, and then adding 4. The correct answer—and the only answer—to this problem is 14.

$$4 + 5 \cdot 2 = 4 + 10 \quad \text{Multiply first}$$
$$= 14 \quad \text{Then add}$$

Here are some more examples that illustrate how we apply the rule for order of operations to simplify (or evaluate) expressions.

EXAMPLE 4   Simplify: $4 \cdot 8 - 2 \cdot 6$

SOLUTION   We multiply first and then subtract:

$$4 \cdot 8 - 2 \cdot 6 = 32 - 12 \quad \text{Multiply first}$$
$$= 20 \quad \text{Then subtract}$$

EXAMPLE 5   Simplify: $5 + 2(7 - 1)$

SOLUTION   According to the rule for the order of operations, we must do what is inside the parentheses first:

$$\begin{aligned}
5 + 2(7 - 1) &= 5 + 2(6) &&\text{Inside parentheses first}\\
&= 5 + 12 &&\text{Then multiply}\\
&= 17 &&\text{Then add}
\end{aligned}$$

EXAMPLE 6   Simplify: $9 \cdot 2^3 + 36 \div 3^2 - 8$

SOLUTION

$$\begin{aligned}
9 \cdot 2^3 + 36 \div 3^2 - 8 &= 9 \cdot 8 + 36 \div 9 - 8 &&\text{Exponents first}\\
&= 72 + 4 - 8 &&\text{Then multiply and divide, left to right}\\
&= 76 - 8 &&\text{Add and subtract,}\\
&= 68 &&\text{left to right}
\end{aligned}$$

USING TECHNOLOGY   *Calculators*

Here is how we use a calculator to work the problem shown in Example 5:

Scientific Calculator: 5 [+] 2 [×] [(] 7 [−] 1 [)] [=]

Graphing Calculator: 5 [+] 2 [(] 7 [−] 1 [)] [ENT]

Example 6 on a calculator looks like this:

Scientific Calculator: 9 [×] 2 [x^y] 3 [+] 36 [÷] 3 [x^y] 2 [−] 8 [=]

Graphing Calculator: 9 [×] 2 [^] 3 [+] 36 [÷] 3 [^] 2 [−] 8 [ENT]

EXAMPLE 7   Simplify: $3 + 2[10 - 3(5 - 2)]$

SOLUTION   The brackets, [ ], are used in the same way as parentheses. In a case like this we move to the innermost grouping symbols first and begin simplifying:

$$\begin{aligned}
3 + 2[10 - 3(5 - 2)] &= 3 + 2[10 - 3(3)]\\
&= 3 + 2[10 - 9]\\
&= 3 + 2[1]\\
&= 3 + 2\\
&= 5
\end{aligned}$$

Table 1 lists some English expressions and their corresponding mathematical expressions written in symbols.

TABLE 1	
In English	**Mathematical Equivalent**
5 times the sum of 3 and 8	$5(3 + 8)$
Twice the difference of 4 and 3	$2(4 - 3)$
6 added to 7 times the sum of 5 and 6	$6 + 7(5 + 6)$
The sum of 4 times 5 and 8 times 9	$4 \cdot 5 + 8 \cdot 9$
3 subtracted from the quotient of 10 and 2	$10 \div 2 - 3$

DESCRIPTIVE STATISTICS *Average*

Next we turn our attention to averages. If we go online to the Merriam-Webster dictionary at www.m-w.com, we find the following definition for the word *average* when it is used as a noun:

av · er · age *noun*: a single value (as a mean, mode, or median) that summarizes or represents the general significance of a set of unequal values . . .

In everyday language, the word *average* can refer to the mean, the median, or the mode. The mean is probably the most common average.

Mean

DEFINITION

To find the **mean** for a set of numbers, we add all the numbers and then divide the sum by the number of numbers in the set. The mean is sometimes called the *arithmetic mean.*

EXAMPLE 8 An instructor at a community college earned the following salaries for the first five years of teaching. Find the mean of these salaries.

$35,344 $38,290 $39,199 $40,346 $42,866

SOLUTION We add the five numbers and then divide by 5, the number of numbers in the set.

$$\text{Mean} = \frac{35{,}344 + 38{,}290 + 39{,}199 + 40{,}346 + 42{,}866}{5} = \frac{196{,}045}{5} = 39{,}209$$

The instructor's mean salary for the first five years of work is $39,209 per year.

Median

The table below appeared in *Parade* magazine in February 2000. It shows the weekly wages for a number of professions.

WEEKLY WAGES

All Americans $549
Butchers $400
Dietitians...$577
Social workers $601
Electricians $645
Clergy...$657
Special ed teachers $677
Lawyers..$1168

Source: U.S. Bureau of Labor Statistics (all wages are median figures for 1999)

If you look at the type at the bottom of the table, you can see that the numbers are the *median* figures for 1999. The median for a set of numbers is the number such that half of the numbers in the set are above it and half are below it. Here is the exact definition.

> **DEFINITION**
>
> To find the **median** for a set of numbers, we write the numbers in order from smallest to largest. If there is an odd number of numbers, the median is the middle number. If there is an even number of numbers, then the median is the mean of the two numbers in the middle.

EXAMPLE 9   Find the median of the numbers given in Example 8.

SOLUTION   The numbers in Example 8, written from smallest to largest, are shown below. Because there are an odd number of numbers in the set, the median is the middle number.

$$35{,}344 \quad 38{,}290 \quad 39{,}199 \quad 40{,}346 \quad 42{,}866$$
$$\uparrow$$
$$\text{median}$$

The instructor's median salary for the first five years of teaching is $39,199.

EXAMPLE 10   A teacher at a community college in California will make the following salaries for the first four years she teaches.

$$\$51{,}890 \quad \$53{,}745 \quad \$55{,}601 \quad \$57{,}412$$

Find the mean and the median for the four salaries.

SOLUTION   To find the mean, we add the four numbers and then divide by 4:

$$\frac{51{,}890 + 53{,}745 + 55{,}601 + 57{,}412}{4} = \frac{218{,}648}{4} = 54{,}662$$

To find the median, we write the numbers in order from smallest to largest. Then, because there is an even number of numbers, we average the middle two numbers to obtain the median.

51,890   53,745   55,601   57,412

median
$$\frac{53,745 + 55,601}{2} = 54,673$$

The mean is $54,662, and the median is $54,673.

Mode

The mode is best used when we are looking for the most common eye color in a group of people, the most popular breed of dog in the United States, and the movie that was seen the most often. When we have a set of numbers in which one number occurs more often than the rest, that number is the mode.

> **DEFINITION**
>
> The *mode* for a set of numbers is the number that occurs most frequently. If all the numbers in the set occur the same number of times, there is no mode.

EXAMPLE 11   A math class with 18 students had the grades shown below on their first test. Find the mean, the median, and the mode.

77   87   100   65   79   87

79   85   87   95   56   87

56   75   79   93   97   92

SOLUTION   To find the mean, we add all the numbers and divide by 18:

$$\text{mean} = \frac{77+87+100+65+79+87+79+85+87+95+56+87+56+75+79+93+97+92}{18}$$

$$= \frac{1,476}{18} = 82$$

To find the median, we must put the test scores in order from smallest to largest; then, because there are an even number of test scores, we must find the mean of the middle two scores.

56 56 65 75 77 79 79 79 85 87 87 87 87 92 93 95 97 100

$$\text{Median} = \frac{85 + 87}{2} = 86$$

The mode is the most frequently occurring score. Because 87 occurs 4 times, and no other scores occur that many times, 87 is the mode.

The mean is 82, the median is 86, and the mode is 87.

More Vocabulary

When we used the word *average* for the first time in this section, we used it as a noun. It can also be used as an adjective and a verb. Below is the definition of the word *average* when it is used as a verb.

 av · er · age *verb*: to find the arithmetic mean of (a series of unequal quantities) . . .

In everyday language, if you are asked for, or given, the *average* of a set of numbers, the word *average* can represent the mean, the median, or the mode. When used in this way, the word *average* is a noun. However, if you are asked to *average* a set of numbers, then the word *average* is a verb, and you are being asked to find the mean of the numbers.

Before we leave this section, there is one more statistic we need. It is called the *range,* and it is used to give us an idea of how spread out the numbers in our samples are. Here is the definition.

> **DEFINITION**
>
> The *range* for a set of numbers is the difference between the largest number and the smallest number in the sample.

If we look back to Example 8, we find the range of salaries is

$$\text{Range} = 42{,}866 - 35{,}344 = \$7{,}522$$

Likewise, the range of test scores in Example 11 is

$$\text{Range} = 100 - 56 = 44$$

GETTING READY FOR CLASS

After reading through the preceding section, respond in your own words and in complete sentences.

1. In the expression 5^3, which number is the base?

2. Give a written description of the process you would use to simplify the expression below.

$$3 + 4(5 + 6)$$

3. What is the first step in simplifying the expression below?

$$8 + 6 \div 3 - 1$$

4. What number must we use for x, if the mean of 6, 8, and x is to be 8?

Problem Set 1.7

For each of the following expressions, name the base and the exponent.

1. 4^5 **2.** 5^4 **3.** 3^6 **4.** 6^3 **5.** 8^2

6. 2^8 **7.** 9^1 **8.** 1^9 **9.** 4^0 **10.** 0^4

Use the definition of exponents as indicating repeated multiplication to simplify each of the following expressions.

11. 6^2 **12.** 7^2 **13.** 2^3 **14.** 2^4 **15.** 1^4 **16.** 5^1 **17.** 9^0 **18.** 27^0

19. 9^2 **20.** 8^2 **21.** 10^1 **22.** 8^1 **23.** 12^1 **24.** 16^0 **25.** 45^0 **26.** 3^4

Use the rule for the order of operations to simplify each expression.

27. $16 - 8 + 4$ **28.** $16 - 4 + 8$ **29.** $20 \div 2 \cdot 10$

30. $40 \div 4 \cdot 5$ **31.** $20 - 4 \cdot 4$ **32.** $30 - 10 \cdot 2$

33. $3 + 5 \cdot 8$ **34.** $7 + 4 \cdot 9$ **35.** $3 \cdot 6 - 2$

36. $5 \cdot 1 + 6$ **37.** $6 \cdot 2 + 9 \cdot 8$ **38.** $4 \cdot 5 + 9 \cdot 7$

39. $4 \cdot 5 - 3 \cdot 2$ **40.** $5 \cdot 6 - 4 \cdot 3$ **41.** $5^2 + 7^2$

42. $4^2 + 9^2$ **43.** $480 + 12(32)^2$ **44.** $360 + 14(27)^2$

45. $3 \cdot 2^3 + 5 \cdot 4^2$ **46.** $4 \cdot 3^2 + 5 \cdot 2^3$ **47.** $8 \cdot 10^2 - 6 \cdot 4^3$

48. $5 \cdot 11^2 - 3 \cdot 2^3$ **49.** $2(3 + 6 \cdot 5)$ **50.** $8(1 + 4 \cdot 2)$

51. $19 + 50 \div 5^2$ **52.** $9 + 8 \div 2^2$ **53.** $9 - 2(4 - 3)$

54. $15 - 6(9 - 7)$ **55.** $4 \cdot 3 + 2(5 - 3)$ **56.** $6 \cdot 8 + 3(4 - 1)$

57. $4[2(3) + 3(5)]$ **58.** $3[2(5) + 3(4)]$ **59.** $(7 - 3)(8 + 2)$

60. $(9 - 5)(9 + 5)$ **61.** $3(9 - 2) + 4(7 - 2)$ **62.** $7(4 - 2) - 2(5 - 3)$

63. $18 + 12 \div 4 - 3$ **64.** $20 + 16 \div 2 - 5$ **65.** $4(10^2) + 20 \div 4$

66. $3(4^2) + 10 \div 5$ **67.** $8 \cdot 2^4 + 25 \div 5 - 3^2$ **68.** $5 \cdot 3^4 + 16 \div 8 - 2^2$

69. $5 + 2[9 - 2(4 - 1)]$ **70.** $6 + 3[8 - 3(1 + 1)]$ **71.** $3 + 4[6 + 8(2 - 0)]$

72. $2 + 5[9 + 3(4 - 1)]$ **73.** $\dfrac{15 + 5(4)}{17 - 12}$ **74.** $\dfrac{20 + 6(2)}{11 - 7}$

Translate each English expression into an equivalent mathematical expression written in symbols. Then simplify.

75. 8 times the sum of 4 and 2   **76.** 3 times the difference of 6 and 1

77. Twice the sum of 10 and 3   **78.** 5 times the difference of 12 and 6

79. 4 added to 3 times the sum of 3 and 4

80. 25 added to 4 times the difference of 7 and 5

81. 9 subtracted from the quotient of 20 and 2

82. 7 added to the quotient of 6 and 2

83. The sum of 8 times 5 and 5 times 4

84. The difference of 10 times 5 and 6 times 2

79

Find the mean and the range for each set of numbers.

85. 1, 2, 3, 4, 5   **86.** 2, 4, 6, 8, 10   **87.** 1, 3, 9, 11   **88.** 5, 7, 9, 12, 12

Find the median and the range for each set of numbers.

89. 5, 9, 11, 13, 15   **90.** 42, 48, 50, 64

91. 10, 20, 50, 90, 100   **92.** 700, 900, 1100

Find the mode and the range for each set of numbers.

93. 14, 18, 27, 36, 18, 73   **94.** 11, 27, 18, 11, 72, 11

Applying the Concepts

Nutrition Labels Use the three nutrition labels below to work Problems 95–100.

Spaghetti

Nutrition Facts
Serving Size 2 oz. (56g/l/8 of pkg) dry
Servings Per Container: 8
Amount Per Serving
Calories 210   Calories from fat 10
% Daily Value*
Total Fat 1g   2%
Saturated Fat 0g   0%
Poly unsaturated Fat 0.5g
Monounsaturated Fat 0g
Cholesterol 0mg   0%
Sodium 0mg   0%
Total Carbohydrate 42g   14%
Dietary Fiber 2g   7%
Sugars 3g
Protein 7g
Vitamin A 0%   •   Vitamin C 0%
Calcium 0%   •   Iron 10%
*Percent Daily Values are based on a 2,000 calorie diet

Canned Italian Tomatoes

Nutrition Facts
Serving Size 1/2 cup (121g)
Servings Per Container: about 3 1/2
Amount Per Serving
Calories 25   Calories from fat 0
% Daily Value*
Total Fat 0g   0%
Saturated Fat 0g   0%
Cholesterol 0mg   0%
Sodium 300mg   12%
Potassium 145mg   4%
Total Carbohydrate 4g   2%
Dietary Fiber 1g   4%
Sugars 4g
Protein 1g
Vitamin A 20%   •   Vitamin C 15%
Calcium 4%   •   Iron 15%
*Percent Daily Values are based on a 2,000 calorie diet. Your daily values may be higher or lower depending on your calorie needs.

Shredded Romano Cheese

Nutrition Facts
Serving Size 2 tsp (5g)
Servings Per Container: 34
Amount Per Serving
Calories 20   Calories from fat 10
% Daily Value*
Total Fat 1.5g   2%
Saturated Fat 1g   5%
Cholesterol 5mg   2%
Sodium 70mg   3%
Total Carbohydrate 0g   0%
Fiber 0g   0%
Sugars 0g
Protein 2g
Vitamin A 0%   •   Vitamin C 0%
Calcium 4%   •   Iron 0%
*Percent Daily Values (DV) are based on a 2,000 calorie diet

Find the total number of calories in each of the following meals.

95. Spaghetti   1 serving   **96.** Spaghetti   1 serving
    Tomatoes   1 serving   Tomatoes   2 servings
    Cheese   1 serving   Cheese   1 serving

97. Spaghetti   2 servings   **98.** Spaghetti   2 servings
    Tomatoes   1 serving   Tomatoes   1 serving
    Cheese   1 serving   Cheese   2 servings

Find the number of calories from fat in each of the following meals.

99. Spaghetti   2 servings   **100.**   Spaghetti   2 servings
    Tomatoes   1 serving   Tomatoes   1 serving
    Cheese   1 serving   Cheese   2 servings

The following table lists the number of calories consumed by eating some popular fast foods. Use the table to work Problems 101 and 102.

CALORIES IN FOOD	
Food	Calories
McDonald's hamburger	270
Burger King hamburger	260
Jack in the Box hamburger	280
McDonald's Big Mac	510
Burger King Whopper	630
Jack in the Box Colossus burger	940

101. Compare the total number of calories in the meal in Problem 95 with the number of calories in a McDonald's Big Mac.

102. Compare the total number of calories in the meal in Problem 98 with the number of calories in a Burger King hamburger.

103. Average If a basketball team has scores of 61, 76, 98, 55, 76, and 102 in their first six games, find
a. the mean score
b. the median score
c. the mode of the scores
d. the range of scores

104. Home Sales Below are listed the prices paid for 10 homes that sold during the month of February in a city in Texas.

$210,000 $139,000 $122,000 $145,000 $120,000
$540,000 $167,000 $125,000 $125,000 $950,000

a. Find the mean housing price for the month.
b. Find the median housing price for the month.
c. Find the mode of the housing prices for the month.
d. Which measure of "average" best describes the average housing price for the month? Explain your answer.

105. Average Enrollment The number of students enrolled in a community college during a 5-year period is shown in this table:

Year	Enrollment
1999	6,789
2000	6,970
2001	7,242
2002	6,981
2003	6,423

Find the mean enrollment and the range of enrollments for this 5-year period.

106. Car Prices The following prices were listed for Volkswagen Jettas on the ebay.com car auction site. Use the table to find each of the following:

a. the mean car price

b. the median car price

c. the mode for the car prices

d. the range of car prices

CAR PRICES	
Year	Price
1998	$10,000
1999	$14,500
1999	$10,500
1999	$11,700
1999	$15,500
2000	$10,500
2000	$18,200
2001	$19,900

Chapter 1 Summary

The numbers in brackets indicate the sections in which the topics were discussed.

The margins of the chapter summaries will be used for examples of the topics being reviewed, whenever it is convenient.

1. The number 42,103,045 written in words is "forty-two million, one hundred three thousand, forty-five." The number 5,745 written in expanded form is $5{,}000 + 700 + 40 + 5$

Place Values for Decimal Numbers [1.1]

The place values for the digits of any base 10 number are as follows:

TABLE 1

	Trillions			Billions			Millions			Thousands			Ones		
	Hundreds	Tens	Ones	Hundreds	Tens	Ones	Hundreds	Tens	Ones	Hundreds	Tens	Ones	Hundreds	Tens	Ones
			5	8	6	5	6	9	6	0	0	0	0	0	0

Vocabulary Associated with Addition, Subtraction, Multiplication, and Division [1.2, 1.4, 1.5, 1.6]

2. The sum of 5 and 2 is $5 + 2$. The difference of 5 and 2 is $5 - 2$. The product of 5 and 2 is $5 \cdot 2$. The quotient of 10 and 2 is $10 \div 2$.

The word *sum* indicates addition.

The word *difference* indicates subtraction.

The word *product* indicates multiplication.

The word *quotient* indicates division.

Properties of Addition and Multiplication [1.2, 1.5]

3. $3 + 2 = 2 + 3$
$3 \cdot 2 = 2 \cdot 3$
$(x + 3) + 5 = x + (3 + 5)$
$(4 \cdot 5) \cdot 6 = 4 \cdot (5 \cdot 6)$
$3(4 + 7) = 3(4) + 3(7)$

If a, b, and c represent any three numbers, then the properties of addition and multiplication used most often are

Commutative property of addition: $a + b = b + a$
Commutative property of multiplication: $a \cdot b = b \cdot a$
Associative property of addition: $(a + b) + c = a + (b + c)$
Associative property of multiplication: $(a \cdot b) \cdot c = a \cdot (b \cdot c)$
Distributive property: $a(b + c) = a(b) + a(c)$

Perimeter of a Polygon [1.2]

4. The perimeter of the rectangle below is
$P = 37 + 37 + 24 + 24$
$= 122$ feet

The *perimeter* of any polygon is the sum of the lengths of the sides, and it is denoted with the letter P.

24 ft

37 ft

Steps for Rounding Whole Numbers [1.3]

5. 5,482 to the nearest ten is 5,480

5,482 to the nearest hundred is 5,500

5,482 to the nearest thousand is 5,000

1. Locate the digit just to the right of the place you are to round to.

2. If that digit is less than 5, replace it and all digits to its right with zeros.

3. If that digit is 5 or more, replace it and all digits to its right with zeros, and add 1 to the digit to its left.

Formulas for Area [1.5]

Below are two common geometric figures, along with the formulas for their areas.

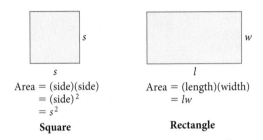

Area = (side)(side)   Area = (length)(width)
    = (side)^2 = lw
    = s^2
    Square   **Rectangle**

Order of Operations [1.7]

6. $4 + 6(8 - 2)$

$= 4 + 6(6)$ Inside parentheses first

$= 4 + 36$ Then multiply

$= 40$ Then add

To simplify a mathematical expression:

1. We simplify the expression inside the grouping symbols first. Grouping symbols are parentheses ( ), brackets [ ], or a fraction bar.

2. Then we evaluate any numbers with exponents.

3. We then perform all multiplications and divisions in order, starting at the left and moving right.

4. Finally, we do all the additions and subtractions, from left to right.

Average [1.7]

7. The mean of 4, 7, 9 and 12 is

$(4 + 7 + 9 + 12) \div 4 = 32 \div 4$

$= 8$

The *average* for a set of numbers can be the mean, the median, or the mode.

Division by 0 (Zero) [1.6]

8. Each expression below is undefined.

$5 \div 0 \quad \frac{7}{0} \quad 4/0$

Division by 0 is undefined. We cannot use 0 as a divisor in any division problem.

Exponents [1.7]

9. $2^3 = 2 \cdot 2 \cdot 2 = 8$

$5^0 = 1$

$3^1 = 3$

In the expression 2^3, 2 is the *base* and 3 is the *exponent*. An exponent is a shorthand notation for repeated multiplication. The exponent 0 is a special exponent. Any nonzero number to the 0 power is 1.

Chapter 1 Test Form A

These problems are all taken from examples in your text. Work each problem and check your answers. If you have made a mistake, work the problem again. If you cannot get the correct answer after two tries, look up the correct solution in your text.

1. Give the place value of each digit in 305, 965.

2. Write 56,094 in expanded form.

3. Write each number in words.

   a.   3,561   **b.** 53,662   **c.**   547,801

4. Add: $197 + 213 + 324$

5. Use the associative property to rewrite each sum.

   a. $(5 + 6) + 7$ **b.** $(3 + 9) + 1$ **c.** $6 + (8 + 2)$ **d.** $4 + (9 + n)$

6. Round 5,382 to the nearest hundred.

7. Round 47,256,344 to the nearest million.

8. Subtract: $92 - 45$

9. Find the difference of 549 and 187.

10. Multiply: $52(37)$

11. Find the solution to each of the following equations:

   a. $6 \cdot n = 24$ **b.** $4 \cdot n = 36$ **c.** $15 = 3 \cdot n$ **d.** $21 = 3 + n$

12. Divide: $9,380 \div 35$

13. Expand and multiply:

   a. 3^2 **b.** 4^2 **c.** 3^3 **d.** 2^4

14. Simplify:

   a. 5^1 **b.** 9^1 **c.** 4^0 **d.** 8^0

15. Simplify: $5 + 2(7 - 1)$

Chapter Test, Form B

For an alternate, more comprehensive, chapter test, go to MathTV.com and select the test and summary for this chapter of the textbook. Click the worksheet labeled Chapter 1 Test, Form B to download it.

Fractions 1: Multiplication and Division

2

iStockphoto.com © ZoneCreative

If you have had any problems with or testing of your thyroid gland, then you may have come in contact with radioactive Iodine-131. Like all radioactive elements, Iodine-131 decays naturally. The half-life of Iodine-131 is 8 days, which means that every 8 days a sample of Iodine-131 will decrease to half its original amount. The table and graph illustrate the radioactive decay of Iodine-131. The graph in Figure 1 is called a *line graph*. Like the bar charts in Chapter 1, line graphs give us a way of taking the information in a table and displaying it in a more visual form.

There are many radioactive materials in the world we inhabit. In each case, the simple fractions shown here are a straightforward, simple way to describe the way in which these materials decay.

Iodine-131 Decay	
Days Since Ingestion	Fraction of Dose Remaining
0	1
8	$\frac{1}{2}$
16	$\frac{1}{4}$
24	$\frac{1}{8}$
32	$\frac{1}{16}$
40	$\frac{1}{32}$

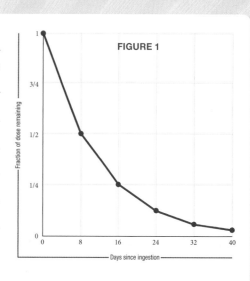

FIGURE 1

Study Skills

If you have successfully completed Chapter 1, then you have made a good start at developing the study skills necessary to succeed in all math classes. Some of the study skills for this chapter are a continuation of the skills from Chapter 1, while others are new to this chapter.

1. **Continue to Set and Keep a Schedule** Sometimes I find students do well in Chapter 1 and then become overconfident. They will begin to put in less time with their homework. Don't do it. Keep to the same schedule.

2. **Increase Effectiveness** You want to become more and more effective with the time you spend on your homework. Increase those activities that are the most beneficial and decrease those that have not given you the results you want.

3. **List Difficult Problems** Begin to make lists of problems that give you the most difficulty. These are the problems in which you are repeatedly making mistakes.

4. **Begin to Develop Confidence With Word Problems** It seems that the main difference between people who are good at working word problems and those who are not is confidence. People with confidence know that no matter how long it takes them, they will eventually be able to solve the problem. Those without confidence begin by saying to themselves, "I'll never be able to work this problem." If you are in this second category, then instead of telling yourself that you can't do word problems, decide to do whatever it takes to master them. The more word problems you work, the better you will become at them.

   Many of my students keep a notebook that contains everything that they need for the course: class notes, homework, quizzes, tests, and research projects. A three-ring binder with tabs is ideal. Organize your notebook so that you can easily get to any item you want to look at.

The Meaning and Properties of Fractions

The information in the table below was taken from the website for Cal Poly State University in California. The pie chart was created from the table. Both the table and pie chart use fractions to specify how the students at Cal Poly are distributed, among the different schools within the university.

CAL POLY STATE UNIVERSITY ENROLLMENT	
School	**Fraction Of Students**
Agriculture	$\frac{11}{50}$
Architecture and Environmental Design	$\frac{1}{10}$
Business	$\frac{3}{20}$
Engineering	$\frac{1}{4}$
Liberal Arts	$\frac{4}{25}$
Science and Mathematics	$\frac{3}{25}$

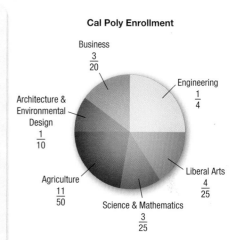

Cal Poly Enrollment

Business $\frac{3}{20}$
Engineering $\frac{1}{4}$
Architecture & Environmental Design $\frac{1}{10}$
Liberal Arts $\frac{4}{25}$
Agriculture $\frac{11}{50}$
Science & Mathematics $\frac{3}{25}$

Note

As we mentioned in Chapter 1, when we use a letter to represent a number, or a group of numbers, that letter is called a variable. In the definition below, we are restricting the numbers that the variable b can represent to numbers other than 0. As you will see later in the chapter, we do this to avoid writing an expression that would imply division by the number 0.

From the table, we see that $\frac{1}{4}$ (one-fourth) of the students are enrolled in the School of Engineering. This means that one out of every four students at Cal Poly is studying Engineering. The fraction $\frac{1}{4}$ tells us we have 1 part of 4 equal parts. That is, the students at Cal Poly could be divided into 4 equal groups, so that one of the groups contained all the engineering students and only engineering students.

Figure 1 at the left shows a rectangle that has been divided into equal parts, four different ways. The shaded area for each rectangle is $\frac{1}{2}$ the total area.

Now that we have an intuitive idea of the meaning of fractions, here are the more formal definitions and vocabulary associated with fractions.

a. $\frac{1}{2}$ is shaded

b. $\frac{2}{4}$ are shaded

c. $\frac{3}{6}$ are shaded

d. $\frac{4}{8}$ are shaded

FIGURE 1 Four Ways to Visualize $\frac{1}{2}$

DEFINITION

A **fraction** is any number that can be put in the form $\frac{a}{b}$ (also sometimes written a/b), where a and b are numbers and b is not 0.

Some examples of fractions are:

$$\frac{1}{2} \qquad \frac{3}{4} \qquad \frac{7}{8} \qquad \frac{9}{5}$$

One-half   *Three-fourths*   *Seven-eighths*   *Nine-fifths*

DEFINITION

For the fraction $\frac{a}{b}$, a and b are called the **terms** of the fraction. More specifically, a is called the **numerator,** and b is called the **denominator.**

EXAMPLE 1 Name the numerator and denominator for each fraction.

a. $\dfrac{3}{4}$ b. $\dfrac{a}{5}$ c. $\dfrac{7}{1}$

SOLUTION In each case we divide the numerator by the denominator:

a. The terms of the fraction $\dfrac{3}{4}$ are 3 and 4. The 3 is called the numerator, and the 4 is called the denominator.

b. The numerator of the fraction $\dfrac{a}{5}$ is a. The denominator is 5. Both a and 5 are called terms.

c. The number 7 may also be put in fraction form, because it can be written as $\frac{7}{1}$. In this case, 7 is the numerator and 1 is the denominator.

DEFINITION

A *proper fraction* is a fraction in which the numerator is less than the denominator. If the numerator is greater than or equal to the denominator, the fraction is called an *improper fraction.*

CLARIFICATION 1: The fractions $\frac{3}{4}, \frac{1}{8},$ and $\frac{9}{10}$ are all proper fractions, because in each case the numerator is less than the denominator.

CLARIFICATION 2: The numbers $\frac{9}{5}, \frac{10}{10},$ and 6 are all improper fractions, because in each case the numerator is greater than or equal to the denominator. (Remember that 6 can be written as $\frac{6}{1}$, in which case 6 is the numerator and 1 is the denominator.)

Fractions on the Number Line

Note

There are many ways to give meaning to fractions like $\frac{2}{3}$ other than by using the number line. One popular way is to think of cutting a pie into three equal pieces, as shown below. If you take two of the pieces, you have taken $\frac{2}{3}$ of the pie.

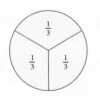

We can give meaning to the fraction $\frac{2}{3}$ by using a number line. If we take that part of the number line from 0 to 1 and divide it into *three equal parts,* we say that we have divided it into *thirds* (see Figure 2). Each of the three segments is $\frac{1}{3}$ (one third) of the whole segment from 0 to 1.

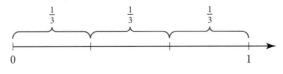

FIGURE 2

Two of these smaller segments together are $\frac{2}{3}$ (two thirds) of the whole segment. And three of them would be $\frac{3}{3}$ (three thirds), or the whole segment, as indicated in Figure 3.

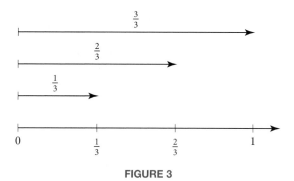

FIGURE 3

Let's do the same thing again with six and twelve equal divisions of the segment from 0 to 1 (see Figure 4).

The same point that we labeled with $\frac{1}{3}$ in Figure 3 is labeled with $\frac{2}{6}$ and with $\frac{4}{12}$ in Figure 4. It must be true then that

$$\frac{4}{12} = \frac{2}{6} = \frac{1}{3}$$

Although these three fractions look different, each names the same point on the number line, as shown in Figure 4. All three fractions have the same *value*, because they all represent the same number.

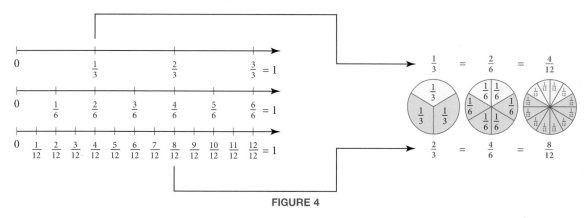

FIGURE 4

DEFINITION

Fractions that represent the same number are said to be **equivalent**. Equivalent fractions may look different, but they must have the same value.

It is apparent that every fraction has many different representations, each of which is equivalent to the original fraction. The next two properties give us a way of changing the terms of a fraction without changing its value.

> **Property 1 for Fractions**
>
> If a, b, and c are numbers and b and c are not 0, then it is always true that
>
> $$\frac{a}{b} = \frac{a \cdot c}{b \cdot c}$$
>
> *In words:* If the numerator and the denominator of a fraction are multiplied by the same nonzero number, the resulting fraction is equivalent to the original fraction.

EXAMPLE 2   Write $\frac{3}{4}$ as an equivalent fraction with denominator 20.

SOLUTION   The denominator of the original fraction is 4. The fraction we are trying to find must have a denominator of 20. We know that if we multiply 4 by 5, we get 20. Property 1 indicates that we are free to multiply the denominator by 5 so long as we do the same to the numerator.

$$\frac{3}{4} = \frac{3 \cdot 5}{4 \cdot 5} = \frac{15}{20}$$

The fraction $\frac{15}{20}$ is equivalent to the fraction $\frac{3}{4}$.

EXAMPLE 3   Write $\frac{3}{4}$ as an equivalent fraction with denominator $12x$.

SOLUTION   If we multiply 4 by $3x$, we will have $12x$:

$$\frac{3}{4} = \frac{3 \cdot 3x}{4 \cdot 3x} = \frac{9x}{12x}$$

> **Property 2 for Fractions**
>
> If a, b, and c are integers and b and c are not 0, then it is always true that
>
> $$\frac{a}{b} = \frac{a \div c}{b \div c}$$
>
> *In words*: If the numerator and the denominator of a fraction are divided by the same nonzero number, the resulting fraction is equivalent to the original fraction.

EXAMPLE 4   Write $\frac{10}{12}$ as an equivalent fraction with denominator 6.

SOLUTION   If we divide the original denominator 12 by 2, we obtain 6. Property 2 indicates that if we divide both the numerator and the denominator by 2, the resulting fraction will be equal to the original fraction:

$$\frac{10}{12} = \frac{10 \div 2}{12 \div 2} = \frac{5}{6}$$

The Number 1 and Fractions

There are two situations involving fractions and the number 1 that occur frequently in mathematics. The first is when the denominator of a fraction is 1. In this case, if we let a represent any number, then

$$\frac{a}{1} = a \qquad \text{for any number } a$$

The second situation occurs when the numerator and the denominator of a fraction are the same nonzero number:

$$\frac{a}{a} = 1 \qquad \text{for any nonzero number } a$$

EXAMPLE 5   Simplify each expression.

a. $\dfrac{24}{1}$ **b.** $\dfrac{24}{24}$ **c.** $\dfrac{48}{24}$ **d.** $\dfrac{72}{24}$

SOLUTION   In each case we divide the numerator by the denominator:

a. $\dfrac{24}{1} = 24$ **b.** $\dfrac{24}{24} = 1$ **c.** $\dfrac{48}{24} = 2$ **d.** $\dfrac{72}{24} = 3$

Comparing Fractions

We can compare fractions to see which is larger or smaller when they have the same denominator.

EXAMPLE 6   Write each fraction as an equivalent fraction with denominator 24. Then write them in order from smallest to largest.

$$\frac{5}{8} \qquad \frac{5}{6} \qquad \frac{3}{4} \qquad \frac{2}{3}$$

SOLUTION   We begin by writing each fraction as an equivalent fraction with denominator 24.

$$\frac{5}{8} = \frac{15}{24} \qquad \frac{5}{6} = \frac{20}{24} \qquad \frac{3}{4} = \frac{18}{24} \qquad \frac{2}{3} = \frac{16}{24}$$

Now that they all have the same denominator, the smallest fraction is the one with the smallest numerator and the largest fraction is the one with the largest numerator. Writing them in order from smallest to largest we have:

$$\frac{15}{24} \quad < \quad \frac{16}{24} \quad < \quad \frac{18}{24} \quad < \quad \frac{20}{24}$$

or

$$\frac{5}{8} \quad < \quad \frac{2}{3} \quad < \quad \frac{3}{4} \quad < \quad \frac{5}{6}$$

DESCRIPTIVE STATISTICS *Scatter Diagrams and Line Graphs*

The table and bar chart give the daily gain in the price of eCollege.com stock for one week in the year 2000, when stock prices were given in terms of fractions instead of decimals.

CHANGE IN STOCK PRICE	
Day	Gain
Monday	$\frac{3}{4}$
Tuesday	$\frac{9}{16}$
Wednesday	$\frac{3}{32}$
Thursday	$\frac{7}{32}$
Friday	$\frac{1}{16}$

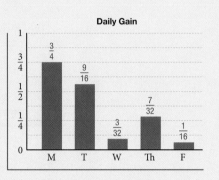

FIGURE 5 Bar Chart

Figure 6 below shows another way to visualize the information in the table. It is called a scatter diagram. In the *scatter diagram,* dots are used instead of the bars shown in Figure 5 to represent the gain in stock price for each day of the week. If we connect the dots in Figure 6 with straight lines, we produce the diagram in Figure 7, which is known as a *line graph.*

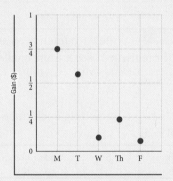

FIGURE 6 Scatter Diagram

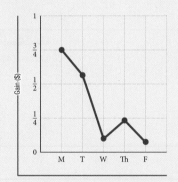

FIGURE 7 Line Graph

GETTING READY FOR CLASS

After reading through the preceding section, respond in your own words and in complete sentences.

1. Explain what a fraction is.
2. Which term in the fraction $\frac{7}{8}$ is the numerator?
3. Is the fraction $\frac{3}{9}$ a proper fraction?
4. What word do we use to describe fractions such as $\frac{1}{5}$ and $\frac{4}{20}$, which look different, but have the same value?

Problem Set 2.1

Name the numerator of each fraction.

1. $\dfrac{1}{3}$ **2.** $\dfrac{1}{4}$ **3.** $\dfrac{2}{3}$ **4.** $\dfrac{2}{4}$

5. $\dfrac{x}{8}$ **6.** $\dfrac{y}{10}$ **7.** $\dfrac{a}{b}$ **8.** $\dfrac{x}{y}$

Name the denominator of each fraction.

9. $\dfrac{2}{5}$ **10.** $\dfrac{3}{5}$ **11.** 6   **12.** 2

13. $\dfrac{a}{12}$ **14.** $\dfrac{b}{14}$

Complete the following tables.

15.

Numerator a	Denominator b	Fraction $\dfrac{a}{b}$
3	5	
1		$\dfrac{1}{7}$
	y	$\dfrac{x}{y}$
$x + 1$	x	

16.

Numerator a	Denominator b	Fraction $\dfrac{a}{b}$
2	9	
	3	$\dfrac{4}{3}$
1		$\dfrac{1}{x}$
x		$\dfrac{x}{x + 1}$

17. For the set of numbers $\left\{ \dfrac{3}{4}, \dfrac{6}{5}, \dfrac{12}{3}, \dfrac{1}{2}, \dfrac{9}{10}, \dfrac{20}{10} \right\}$, list all the proper fractions.

18. For the set of numbers $\left\{ \dfrac{1}{8}, \dfrac{7}{9}, \dfrac{6}{3}, \dfrac{18}{6}, \dfrac{3}{5}, \dfrac{9}{8} \right\}$, list all the improper fractions.

Indicate whether each of the following is *True* or *False*.

19. Every whole number greater than 1 can also be expressed as an improper fraction.

20. Some improper fractions are also proper fractions.

21. Adding the same number to the numerator and the denominator of a fraction will not change its value.

22. The fractions $\dfrac{3}{4}$ and $\dfrac{9}{16}$ are equivalent.

Divide the numerator and the denominator of each of the following fractions by 2.

23. $\dfrac{6}{8}$ **24.** $\dfrac{10}{12}$ **25.** $\dfrac{86}{94}$ **26.** $\dfrac{106}{142}$

Divide the numerator and the denominator of each of the following fractions by 3.

27. $\dfrac{12}{9}$ **28.** $\dfrac{33}{27}$ **29.** $\dfrac{39}{51}$ **30.** $\dfrac{57}{69}$

Write each of the following fractions as an equivalent fraction with denominator 6.

31. $\dfrac{2}{3}$ **32.** $\dfrac{1}{2}$ **33.** $\dfrac{55}{66}$ **34.** $\dfrac{65}{78}$

Write each of the following fractions as an equivalent fraction with denominator 12.

35. $\dfrac{2}{3}$ **36.** $\dfrac{5}{6}$ **37.** $\dfrac{56}{84}$ **38.** $\dfrac{143}{156}$

Write each fraction as an equivalent fraction with denominator 12x.

39. $\dfrac{1}{6}$ **40.** $\dfrac{3}{4}$

Write each number as an equivalent fraction with denominator 8.

41. 2   **42.** 1   **43.** 5   **44.** 8

45. One-fourth of the first circle below is shaded. Use the other three circles to show three other ways to shade one-fourth of the circle.

46. The six-sided figures below are hexagons. One-third of the first hexagon is shaded. Shade the other three hexagons to show three other ways to represent one-third.

Simplify by dividing the numerator by the denominator.

47. $\dfrac{3}{1}$ **48.** $\dfrac{3}{3}$ **49.** $\dfrac{6}{3}$ **50.** $\dfrac{12}{3}$ **51.** $\dfrac{37}{1}$ **52.** $\dfrac{37}{37}$

53. For each square below, what fraction of the area is given by the shaded region?

  a.   **b.**   **c.**   **d.**

54. For each square below, what fraction of the area is given by the shaded region?

a. 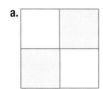 b. 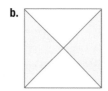 c. 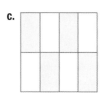 d.

The number line below extends from 0 to 2, with the segment from 0 to 1 and the segment from 1 to 2 each divided into 8 equal parts. Locate each of the following numbers on this number line.

55. $\dfrac{1}{4}$ **56.** $\dfrac{1}{8}$ **57.** $\dfrac{1}{16}$ **58.** $\dfrac{5}{8}$ **59.** $\dfrac{3}{4}$

60. $\dfrac{15}{16}$ **61.** $\dfrac{3}{2}$ **62.** $\dfrac{5}{4}$ **63.** $\dfrac{31}{16}$ **64.** $\dfrac{15}{8}$

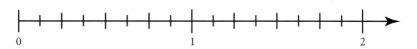

65. Write each fraction as an equivalent fraction with denominator 100. Then write them in order from smallest to largest.

$$\frac{3}{10} \qquad \frac{1}{20} \qquad \frac{4}{25} \qquad \frac{2}{5}$$

66. Write each fraction as an equivalent fraction with denominator 30. Then write them in order from smallest to largest.

$$\frac{1}{15} \qquad \frac{5}{6} \qquad \frac{7}{10} \qquad \frac{1}{2}$$

Applying the Concepts

67. Sending E-mail The pie chart below shows the fraction of workers who responded to a survey about sending non-work-related e-mail from the office. Use the pie chart to fill in the table.

Workers sending personal e-mail from the office

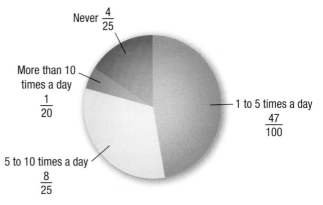

Never $\dfrac{4}{25}$

More than 10 times a day $\dfrac{1}{20}$

1 to 5 times a day $\dfrac{47}{100}$

5 to 10 times a day $\dfrac{8}{25}$

How Often Workers Send Non-Work-Related E-Mail From the Office	Fraction of Respondents Saying Yes
never	
1 to 5 times a day	
5 to 10 times a day	
more than 10 times a day	

68. **Surfing the Internet** The pie chart below shows the fraction of workers who responded to a survey about viewing non-work-related sites during working hours. Use the pie chart to fill in the table.

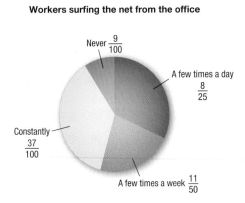

Workers surfing the net from the office

Never $\frac{9}{100}$

A few times a day $\frac{8}{25}$

Constantly $\frac{37}{100}$

A few times a week $\frac{11}{50}$

How Often Workers View Non-Work-Related Sites From the Office	Fraction of Respondents Saying Yes
never	
a few times a week	
a few times a day	
constantly	

69. **Number of Children** If there are 3 girls in a family with 5 children, then we say that $\frac{3}{5}$ of the children are girls. If there are 4 girls in a family with 5 children, what fraction of the children are girls?

70. **Medical School** If 3 out of every 7 people who apply to medical school actually get accepted, what fraction of the people who apply get accepted?

71. **Number of Students** Of the 43 people who started a math class meeting at 10:00 each morning, only 29 finished the class. What fraction of the people finished the class?

72. **Number of Students** In a class of 51 students, 23 are freshmen and 28 are juniors. What fraction of the students are freshmen?

73. **Expenses** If your monthly income is $1,791 and your house payment is $1,121, what fraction of your monthly income must go to pay your house payment?

74. **Expenses** If you spend $623 on food each month and your monthly income is $2,599, what fraction of your monthly income do you spend on food?

75. Half-life of an Antidepressant The half-life of a medication tells how quickly the medication is eliminated from a person's system. The line graph below shows the fraction of an antidepressant that remains in a patient's system once the patient stops taking the antidepressant. The half-life of the antidepressant is 5 days. Use the line graph to complete the table.

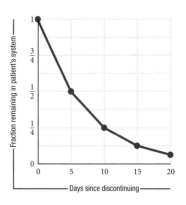

CONCENTRATION OF ANTIDEPRESSANT	
Days Since Discontinuing	Fraction Remaining in Patient's System
0	1
5	
	$\frac{1}{4}$
	$\frac{1}{16}$

76. Carbon Dating All living things contain a small amount of carbon-14, which is radioactive and decays. The half-life of carbon-14 is 5,600 years. During the lifetime of an organism, the carbon-14 is replenished, but after its death the carbon-14 begins to disappear. By measuring the amount left, the age of the organism can be determined with surprising accuracy. The line graph below shows the fraction of carbon-14 remaining after the death of an organism. Use the line graph to complete the table.

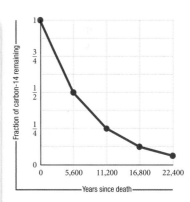

CONCENTRATION OF CARBON-14	
Years Since Death of Organism	Fraction of Carbon-14 Remaining
0	1
11,200	$\frac{1}{4}$
16,800	$\frac{1}{8}$
	$\frac{1}{16}$

Estimating

77. Which of the following fractions is closest to the number 0?

    a. $\dfrac{1}{2}$ **b.** $\dfrac{1}{3}$ **c.** $\dfrac{1}{4}$ **d.** $\dfrac{1}{5}$

78. Which of the following fractions is closest to the number 1?

    a. $\dfrac{1}{2}$ **b.** $\dfrac{1}{3}$ **c.** $\dfrac{1}{4}$ **d.** $\dfrac{1}{5}$

79. Which of the following fractions is closest to the number 0?

    a. $\dfrac{1}{8}$ **b.** $\dfrac{3}{8}$ **c.** $\dfrac{5}{8}$ **d.** $\dfrac{7}{8}$

80. Which of the following fractions is closest to the number 1?

    a. $\dfrac{1}{8}$ **b.** $\dfrac{3}{8}$ **c.** $\dfrac{5}{8}$ **d.** $\dfrac{7}{8}$

Getting Ready for the Next Section

Multiply.

81. $2 \cdot 2 \cdot 3 \cdot 3 \cdot 3$ **82.** $2^2 \cdot 3^3$ **83.** $2^2 \cdot 3 \cdot 5$ **84.** $2 \cdot 3^2 \cdot 5$

Divide.

85. $12 \div 3$ **86.** $15 \div 3$ **87.** $20 \div 4$ **88.** $24 \div 4$

89. $42 \div 6$ **90.** $72 \div 8$ **91.** $102 \div 2$ **92.** $105 \div 7$

Prime Numbers, Factors, and Reducing to Lowest Terms

Suppose you and a friend decide to split a medium-sized pizza for lunch. When the pizza is delivered you find that it has been cut into eight equal pieces. If you eat four pieces, you have eaten $\frac{4}{8}$ of the pizza, but you also know that you have eaten $\frac{1}{2}$ of the pizza. The fraction $\frac{4}{8}$ is equivalent to the fraction $\frac{1}{2}$; that is, they both have the same value. The mathematical process we use to rewrite $\frac{4}{8}$ as $\frac{1}{2}$ is called *reducing to lowest terms*. Before we look at that process, we need to define some new terms. Here is our first one.

DEFINITION

A ***prime number*** is any whole number greater than 1 that has exactly two divisors—itself and 1. (A number is a divisor of another number if it divides it without a remainder.)

$$\text{Prime numbers} = \{2, 3, 5, 7, 11, 13, 17, 19, 23, 29, 31, 37, \ldots\}$$

The list goes on indefinitely. Each number in the list has exactly two distinct divisors—itself and 1.

DEFINITION

Any whole number greater than 1 that is not a prime number is called a ***composite number.*** A composite number always has at least one divisor other than itself and 1.

EXAMPLE 1

Identify each of the numbers below as either a prime number or a composite number. For those that are composite, give two divisors other than the number itself or 1.

    a. 43   **b.** 12

SOLUTION   **a.** 43 is a prime number, because the only numbers that divide it without a remainder are 43 and 1.

      b. 12 is a composite number, because it can be written as $12 = 4 \cdot 3$, which means that 4 and 3 are divisors of 12. (These are not the only divisors of 12; other divisors are 1, 2, 6, and 12.)

Note

You may have already noticed that the word *divisor* as we are using it here means the same as the word *factor*. A divisor and a factor of a number are the same thing. A number can't be a divisor of another number without also being a factor of it.

Every composite number can be written as the product of prime factors. Let's look at the composite number 108. We know we can write 108 as $2 \cdot 54$. The number 2 is a prime number, but 54 is not prime. Because 54 can be written as $2 \cdot 27$, we have

$$108 = 2 \cdot 54$$
$$= 2 \cdot 2 \cdot 27$$

Note

This process works by writing the original composite number as the product of any two of its factors and then writing any factor that is not prime as the product of any two of its factors. The process is continued until all factors are prime numbers.

Now the number 27 can be written as $3 \cdot 9$ or $3 \cdot 3 \cdot 3$ (because $9 = 3 \cdot 3$), so

$$108 = 2 \cdot 54$$
$$108 = 2 \cdot 2 \cdot 27$$
$$108 = 2 \cdot 2 \cdot 3 \cdot 9$$
$$108 = 2 \cdot 2 \cdot 3 \cdot 3 \cdot 3$$

This last line is the number 108 written as the product of prime factors. We can use exponents to rewrite the last line:

$$108 = 2^2 \cdot 3^3$$

EXAMPLE 2   Factor 60 into a product of prime factors.

SOLUTION   We begin by writing 60 as $6 \cdot 10$ and continue factoring until all factors are prime numbers:

$$60 = 6 \cdot 10$$
$$= 2 \cdot 3 \cdot 2 \cdot 5$$
$$= 2^2 \cdot 3 \cdot 5$$

Notice that if we had started by writing 60 as $3 \cdot 20$, we would have achieved the same result:

$$60 = 3 \cdot 20$$
$$= 3 \cdot 2 \cdot 10$$
$$= 3 \cdot 2 \cdot 2 \cdot 5$$
$$= 2^2 \cdot 3 \cdot 5$$

We can use the method of factoring numbers into prime factors to help reduce fractions to lowest terms. Here is the definition for lowest terms.

Note on Divisibility

There are some "shortcuts" to finding the divisors of a number. For instance, if a number ends in 0 or 5, then it is divisible by 5. If a number ends in an even number (0, 2, 4, 6, or 8), then it is divisible by 2. A number is divisible by 3 if the sum of its digits is divisible by 3. For example, 921 is divisible by 3 because the sum of its digits is $9 + 2 + 1 = 12$, which is divisible by 3.

DEFINITION

A fraction is said to be in *lowest terms* if the numerator and the denominator have no factors in common other than the number 1.

CLARIFICATION 1:   The fractions $\frac{1}{2}, \frac{1}{3}, \frac{2}{3}, \frac{1}{4}, \frac{3}{4}, \frac{1}{5}, \frac{2}{5}, \frac{3}{5}$, and $\frac{4}{5}$ are all in lowest terms, because in each case the numerator and the denominator have no factors other than 1 in common. That is, in each fraction, no number other than 1 divides both the numerator and the denominator exactly (without a remainder).

CLARIFICATION 2:   The fraction $\frac{6}{8}$ is not written in lowest terms, because the numerator and the denominator are both divisible by 2. To write $\frac{6}{8}$ in lowest terms, we apply Property 2 from Section 2.1 and divide both the numerator and the denominator by 2:

$$\frac{6}{8} = \frac{6 \div 2}{8 \div 2} = \frac{3}{4}$$

The fraction $\frac{3}{4}$ is in lowest terms, because 3 and 4 have no factors in common except the number 1.

Reducing a fraction to lowest terms is simply a matter of dividing the numerator and the denominator by all the factors they have in common. We know from Property 2 of Section 2.1 that this will produce an equivalent fraction.

EXAMPLE 3   Reduce the fraction $\frac{12}{15}$ to lowest terms by first factoring the numerator and the denominator into prime factors and then dividing both the numerator and the denominator by the factor they have in common.

SOLUTION   The numerator and the denominator factor as follows:

$$12 = 2 \cdot 2 \cdot 3 \quad \text{and} \quad 15 = 3 \cdot 5$$

The factor they have in common is 3. Property 2 tells us that we can divide both terms of a fraction by 3 to produce an equivalent fraction. So

$$\frac{12}{15} = \frac{2 \cdot 2 \cdot 3}{3 \cdot 5} \qquad \text{\textit{Factor the numerator and the denominator completely}}$$

$$= \frac{2 \cdot 2 \cdot 3 \div 3}{3 \cdot 5 \div 3} \qquad \text{\textit{Divide by 3}}$$

$$= \frac{2 \cdot 2}{5} = \frac{4}{5}$$

The fraction $\frac{4}{5}$ is equivalent to $\frac{12}{15}$ and is in lowest terms, because the numerator and the denominator have no factors other than 1 in common.

We can shorten the work involved in reducing fractions to lowest terms by using a slash to indicate division. For example, we can write the above problem this way:

$$\frac{12}{15} = \frac{2 \cdot 2 \cdot \cancel{3}}{\cancel{3} \cdot 5} = \frac{4}{5}$$

So long as we understand that the slashes through the 3's indicate that we have divided both the numerator and the denominator by 3, we can use this notation.

EXAMPLE 4   Laura is having a party. She puts 4 six-packs of diet soda in a cooler for her guests. At the end of the party she finds that only 4 sodas have been consumed. What fraction of the sodas are left? Write your answer in lowest terms.

SOLUTION   She had 4 six-packs of soda, which is 4(6) = 24 sodas. Only 4 were consumed at the party, so 20 are left. The fraction of sodas left is

$$\frac{20}{24}$$

Factoring 20 and 24 completely and then dividing out both the factors they have in common gives us

$$\frac{20}{24} = \frac{\cancel{2} \cdot \cancel{2} \cdot 5}{\cancel{2} \cdot \cancel{2} \cdot 2 \cdot 3} = \frac{5}{6}$$

Note The slashes in Example 4 indicate that we have divided both the numerator and the denominator by 2 · 2, which is equal to 4. With some fractions it is apparent at the start what number divides the numerator and the denominator. For instance, you may have recognized that both 20 and 24 in Example 4 are divisible by 4. We can divide both terms by 4 without factoring first, just as we did in Section 2.1. Property 2 guarantees that dividing both terms of a fraction by 4 will produce an equivalent fraction:

$$\frac{20}{24} = \frac{20 \div 4}{24 \div 4} = \frac{5}{6}$$

EXAMPLE 5 Reduce $\frac{6}{42}$ to lowest terms.

SOLUTION We begin by factoring both terms. We then divide through by any factors common to both terms:

$$\frac{6}{42} = \frac{\cancel{2} \cdot \cancel{3}}{\cancel{2} \cdot \cancel{3} \cdot 7} = \frac{1}{7}$$

We must be careful in a problem like this to remember that the slashes indicate division. They are used to indicate that we have divided both the numerator and the denominator by $2 \cdot 3 = 6$. The result of dividing the numerator 6 by $2 \cdot 3$ is 1. It is a very common mistake to call the numerator 0 instead of 1 or to leave the numerator out of the answer.

EXAMPLE 6 Reduce $\frac{4}{40}$ to lowest terms.

$$\frac{4}{40} = \frac{\cancel{2} \cdot \cancel{2} \cdot 1}{\cancel{2} \cdot \cancel{2} \cdot 2 \cdot 5}$$

$$= \frac{1}{10}$$

EXAMPLE 7 Reduce $\frac{105}{30}$ to lowest terms.

$$\frac{105}{30} = \frac{\cancel{3} \cdot \cancel{5} \cdot 7}{2 \cdot \cancel{3} \cdot \cancel{5}}$$

$$= \frac{7}{2}$$

GETTING READY FOR CLASS

After reading through the preceding section, respond in your own words and in complete sentences.

1. What is a prime number?
2. Why is the number 22 a composite number?
3. Factor 120 into a product of prime factors.
4. What is meant by the phrase "a fraction in lowest possible terms"?

Problem Set 2.2

Identify each of the numbers below as either a prime number or a composite number. For those that are composite, give at least one divisor (factor) other than the number itself or the number 1.

1. 11   **2.** 23   **3.** 105   **4.** 41

5. 81   **6.** 50   **7.** 13   **8.** 219

Factor each of the following into a product of prime factors.

9. 12   **10.** 8   **11.** 81   **12.** 210

13. 215   **14.** 75   **15.** 15   **16.** 42

Reduce each fraction to lowest terms.

17. $\dfrac{5}{10}$ **18.** $\dfrac{3}{6}$ **19.** $\dfrac{4}{6}$ **20.** $\dfrac{4}{10}$

21. $\dfrac{8}{10}$ **22.** $\dfrac{6}{10}$ **23.** $\dfrac{36}{20}$ **24.** $\dfrac{32}{12}$

25. $\dfrac{42}{66}$ **26.** $\dfrac{36}{60}$ **27.** $\dfrac{24}{40}$ **28.** $\dfrac{50}{75}$

29. $\dfrac{14}{98}$ **30.** $\dfrac{12}{84}$ **31.** $\dfrac{70}{90}$ **32.** $\dfrac{80}{90}$

33. $\dfrac{42}{30}$ **34.** $\dfrac{60}{36}$ **35.** $\dfrac{18}{90}$ **36.** $\dfrac{150}{210}$

37. $\dfrac{110}{70}$ **38.** $\dfrac{45}{75}$ **39.** $\dfrac{180}{108}$ **40.** $\dfrac{105}{30}$

41. $\dfrac{96}{108}$ **42.** $\dfrac{66}{84}$ **43.** $\dfrac{126}{165}$ **44.** $\dfrac{210}{462}$

45. $\dfrac{102}{114}$ **46.** $\dfrac{255}{285}$ **47.** $\dfrac{294}{693}$ **48.** $\dfrac{273}{385}$

49. Reduce each fraction to lowest terms.

     a. $\dfrac{6}{51}$ **b.** $\dfrac{6}{52}$ **c.** $\dfrac{6}{54}$ **d.** $\dfrac{6}{56}$ **e.** $\dfrac{6}{57}$

50. Reduce each fraction to lowest terms.

     a. $\dfrac{6}{42}$ **b.** $\dfrac{6}{44}$ **c.** $\dfrac{6}{45}$ **d.** $\dfrac{6}{46}$ **e.** $\dfrac{6}{48}$

51. Reduce each fraction to lowest terms.

     a. $\dfrac{2}{90}$ **b.** $\dfrac{3}{90}$ **c.** $\dfrac{5}{90}$ **d.** $\dfrac{6}{90}$ **e.** $\dfrac{9}{90}$

52. Reduce each fraction to lowest terms.

     a. $\dfrac{3}{105}$ **b.** $\dfrac{5}{105}$ **c.** $\dfrac{7}{105}$ **d.** $\dfrac{15}{105}$ **e.** $\dfrac{21}{105}$

53. The answer to each problem below is wrong. Give the correct answer.

a. $\dfrac{5}{15} = \dfrac{\cancel{5}}{3 \cdot \cancel{5}} = \dfrac{0}{3}$ **b.** $\dfrac{5}{6} = \dfrac{3 + \cancel{2}}{4 + \cancel{2}} = \dfrac{3}{4}$ **c.** $\dfrac{6}{30} = \dfrac{\cancel{2} \cdot \cancel{3}}{\cancel{2} \cdot \cancel{3} \cdot 5} = 5$

54. The answer to each problem below is wrong. Give the correct answer.

a. $\dfrac{10}{20} = \dfrac{7 + \cancel{3}}{17 + \cancel{3}} = \dfrac{7}{17}$ **b.** $\dfrac{9}{36} = \dfrac{\cancel{3} \cdot \cancel{3}}{2 \cdot 2 \cdot \cancel{3} \cdot \cancel{3}} = \dfrac{0}{4}$ **c.** $\dfrac{4}{12} = \dfrac{\cancel{2} \cdot \cancel{2}}{\cancel{2} \cdot \cancel{2} \cdot 3} = 3$

55. Which of the fractions $\frac{6}{8}, \frac{15}{20}, \frac{9}{16},$ and $\frac{21}{28}$ does not reduce to $\frac{3}{4}$?

56. Which of the fractions $\frac{4}{9}, \frac{10}{15}, \frac{8}{12},$ and $\frac{6}{12}$ do not reduce to $\frac{2}{3}$?

The number line below extends from 0 to 2, with the segment from 0 to 1 and the segment from 1 to 2 each divided into 8 equal parts. Locate each of the following numbers on this number line.

57. $\dfrac{1}{2}, \dfrac{2}{4}, \dfrac{4}{8},$ and $\dfrac{8}{16}$ **58.** $\dfrac{3}{2}, \dfrac{6}{4}, \dfrac{12}{8},$ and $\dfrac{24}{16}$

59. $\dfrac{5}{4}, \dfrac{10}{8},$ and $\dfrac{20}{16}$ **60.** $\dfrac{1}{4}, \dfrac{2}{8},$ and $\dfrac{4}{16}$

Applying the Concepts

61. Income A family's monthly income is $2,400, and they spend $600 each month on food. Write the amount they spend on food as a fraction of their monthly income in lowest terms.

62. Hours and Minutes There are 60 minutes in 1 hour. What fraction of an hour is 20 minutes? Write your answer in lowest terms.

63. Final Exam Suppose 33 people took the final exam in a math class. If 11 people got an A on the final exam, what fraction of the students did not get an A on the exam? Write your answer in lowest terms.

64. Income Tax A person making $21,000 a year pays $3,000 in income tax. What fraction of the person's income is paid as income tax? Write your answer in lowest terms.

Nutrition The nutrition labels below are from two different granola bars. Use them to work Problems 65–70.

GRANOLA BAR 1

Nutrition Facts

Serving Size 2 bars (47g)
Servings Per Container: 6

Amount Per Serving

Calories 210	Calories from fat 70

	% Daily Value*
Total Fat 8g	**12%**
Saturated Fat 1g	**5%**
Cholesterol 0mg	**0%**
Sodium 150mg	**6%**
Total Carbohydrate 32g	**11%**
Fiber 2g	**10%**
Sugars 12g	
Protein 4g	

*Percent Daily Values are based on a 2,000 calorie diet. Your daily values may be higher or lower depending on your calorie needs.

GRANOLA BAR 2

Nutrition Facts

Serving Size 1 bar (21g)
Servings Per Container: 8

Amount Per Serving

Calories 80	Calories from fat 15

	% Daily Value*
Total Fat 1.5g	**2%**
Saturated Fat 0g	**0%**
Cholesterol 0mg	**0%**
Sodium 60mg	**3%**
Total Carbohydrate 16g	**5%**
Fiber 1g	**4%**
Sugars 5g	
Protein 2g	

*Percent Daily Values are based on a 2,000 calorie diet. Your daily values may be higher or lower depending on your calorie needs.

65. What fraction of the calories in Bar 1 comes from fat?

66. What fraction of the calories in Bar 2 comes from fat?

67. For Bar 1, what fraction of the total fat is from saturated fat?

68. For Bar 2, what fraction of the total fat is from saturated fat?

69. What fraction of the total carbohydrates in Bar 1 is from sugar?

70. What fraction of the total carbohydrates in Bar 2 is from sugar?

Getting Ready for the Next Section

Multiply.

71. $1 \cdot 3 \cdot 1$

72. $2 \cdot 4 \cdot 5$

73. $3 \cdot 5 \cdot 3$

74. $1 \cdot 4 \cdot 1$

75. $5 \cdot 5 \cdot 1$

76. $6 \cdot 6 \cdot 2$

Factor into prime factors.

77. 60

78. 72

79. $15 \cdot 4$

80. $8 \cdot 9$

Expand and multiply.

81. 3^2

82. 4^2

83. 5^2

84. 6^2

Multiplication with Fractions, and the Area of a Triangle

A recipe calls for $\frac{3}{4}$ cup of flour. If you are making only $\frac{1}{2}$ the recipe, how much flour do you use? This question can be answered by multiplying $\frac{1}{2}$ and $\frac{3}{4}$. Here is the problem written in symbols:

$$\frac{1}{2} \cdot \frac{3}{4} = \frac{3}{8}$$

As you can see from this example, to multiply two fractions, we multiply the numerators and then multiply the denominators. We begin this section with the rule for multiplication of fractions.

> **RULE**
>
> The product of two fractions is a fraction whose numerator is the product of the two numerators and whose denominator is the product of the two denominators. We can write this rule in symbols as follows:
>
> If a, b, c, and d represent any numbers and b and d are not zero, then
>
> $$\frac{a}{b} \cdot \frac{c}{d} = \frac{a \cdot c}{b \cdot d}$$

EXAMPLE 1   Multiply $\frac{3}{5} \cdot \frac{2}{7}$.

SOLUTION   Using our rule for multiplication, we multiply the numerators and multiply the denominators:

$$\frac{3}{5} \cdot \frac{2}{7} = \frac{3 \cdot 2}{5 \cdot 7} = \frac{6}{35}$$

The product of $\frac{3}{5}$ and $\frac{2}{7}$ is the fraction $\frac{6}{35}$. The numerator 6 is the product of 3 and 2, and the denominator 35 is the product of 5 and 7.

EXAMPLE 2   Multiply $\frac{3}{8} \cdot 5$.

SOLUTION   The number 5 can be written as $\frac{5}{1}$. That is, 5 can be considered a fraction with numerator 5 and denominator 1. Writing 5 this way enables us to apply the rule for multiplying fractions.

$$\frac{3}{8} \cdot 5 = \frac{3}{8} \cdot \frac{5}{1}$$
$$= \frac{3 \cdot 5}{8 \cdot 1}$$
$$= \frac{15}{8}$$

EXAMPLE 3   Multiply $\frac{1}{2}\left(\frac{3}{4}\cdot\frac{1}{5}\right)$.

SOLUTION   We find the product inside the parentheses first and then multiply the result by $\frac{1}{2}$:

$$\frac{1}{2}\left(\frac{3}{4}\cdot\frac{1}{5}\right)=\frac{1}{2}\left(\frac{3}{20}\right)$$

$$=\frac{1\cdot3}{2\cdot20}=\frac{3}{40}$$

The properties of multiplication that we developed in Chapter 1 for whole numbers apply to fractions as well. That is, if a, b, and c are fractions, then

$$a\cdot b=b\cdot a \qquad \text{Multiplication with fractions is commutative}$$

$$a\cdot(b\cdot c)=(a\cdot b)\cdot c \qquad \text{Multiplication with fractions is associative}$$

To demonstrate the associative property for fractions, let's do Example 3 again, but this time we will apply the associative property first:

$$\frac{1}{2}\left(\frac{3}{4}\cdot\frac{1}{5}\right)=\left(\frac{1}{2}\cdot\frac{3}{4}\right)\cdot\frac{1}{5} \qquad \text{Associative property}$$

$$=\left(\frac{1\cdot3}{2\cdot4}\right)\cdot\frac{1}{5}$$

$$=\left(\frac{3}{8}\right)\cdot\frac{1}{5}$$

$$=\frac{3\cdot1}{8\cdot5}=\frac{3}{40}$$

The result is identical to that of Example 3.

Here is another example that involves the associative property. Problems like this will be useful when we solve equations.

The answers to all the examples so far in this section have been in lowest terms. Let's see what happens when we multiply two fractions to get a product that is not in lowest terms.

EXAMPLE 4   Multiply $\frac{15}{8}\cdot\frac{4}{9}$.

SOLUTION   Multiplying the numerators and multiplying the denominators, we have

$$\frac{15}{8}\cdot\frac{4}{9}=\frac{15\cdot4}{8\cdot9}$$

$$=\frac{60}{72}$$

The product is $\frac{60}{72}$, which can be reduced to lowest terms by factoring 60 and 72 and then dividing out any factors they have in common:

$$\frac{60}{72}=\frac{\cancel{2}\cdot\cancel{2}\cdot\cancel{3}\cdot5}{\cancel{2}\cdot\cancel{2}\cdot2\cdot\cancel{3}\cdot3}$$

$$=\frac{5}{6}$$

We can actually save ourselves some time by factoring before we multiply. Here's how it is done:

$$\frac{15}{8} \cdot \frac{4}{9} = \frac{15 \cdot 4}{8 \cdot 9}$$

$$= \frac{(3 \cdot 5) \cdot (2 \cdot 2)}{(2 \cdot 2 \cdot 2) \cdot (3 \cdot 3)}$$

$$= \frac{\cancel{3} \cdot 5 \cdot \cancel{2} \cdot \cancel{2}}{\cancel{2} \cdot \cancel{2} \cdot 2 \cdot \cancel{3} \cdot 3}$$

$$= \frac{5}{6}$$

The result is the same in both cases. Reducing to lowest terms before we actually multiply takes less time. Here are some additional examples.

EXAMPLE 5

$$\frac{9}{2} \cdot \frac{8}{18} = \frac{9 \cdot 8}{2 \cdot 18}$$

$$= \frac{(3 \cdot 3) \cdot (2 \cdot 2 \cdot 2)}{2 \cdot (2 \cdot 3 \cdot 3)}$$

$$= \frac{\cancel{3} \cdot \cancel{3} \cdot \cancel{2} \cdot \cancel{2} \cdot 2}{\cancel{2} \cdot \cancel{2} \cdot \cancel{3} \cdot \cancel{3}}$$

$$= \frac{2}{1}$$

$$= 2$$

> **Note**
>
> Although $\frac{2}{1}$ is in lowest terms, it is still simpler to write the answer as just 2. We will always do this when the denominator is the number 1.

EXAMPLE 6

$$\frac{2}{3} \cdot \frac{6}{5} \cdot \frac{5}{8} = \frac{2 \cdot 6 \cdot 5}{3 \cdot 5 \cdot 8}$$

$$= \frac{2 \cdot (2 \cdot 3) \cdot 5}{3 \cdot 5 \cdot (2 \cdot 2 \cdot 2)}$$

$$= \frac{\cancel{2} \cdot \cancel{2} \cdot \cancel{3} \cdot \cancel{5}}{\cancel{3} \cdot \cancel{5} \cdot \cancel{2} \cdot 2 \cdot 2}$$

$$= \frac{1}{2}$$

In Chapter 1 we did some work with exponents. We can extend our work with exponents to include fractions, as the following examples indicate.

EXAMPLE 7

$$\left(\frac{3}{4}\right)^2 = \frac{3}{4}\left(\frac{3}{4}\right)$$

$$= \frac{3 \cdot 3}{4 \cdot 4}$$

$$= \frac{9}{16}$$

EXAMPLE 8

$$\left(\frac{5}{6}\right)^2 \cdot \frac{1}{2} = \frac{5}{6} \cdot \frac{5}{6} \cdot \frac{1}{2}$$

$$= \frac{5 \cdot 5 \cdot 1}{6 \cdot 6 \cdot 2}$$

$$= \frac{25}{72}$$

The word *of* used in connection with fractions indicates multiplication. If we want to find $\frac{1}{2}$ of $\frac{2}{3}$, then what we do is multiply $\frac{1}{2}$ and $\frac{2}{3}$.

EXAMPLE 9   Find $\frac{1}{2}$ of $\frac{2}{3}$.

SOLUTION   Knowing the word *of,* as used here, indicates multiplication, we have

$$\frac{1}{2} \text{ of } \frac{2}{3} = \frac{1}{2} \cdot \frac{2}{3}$$

$$= \frac{1 \cdot 2}{2 \cdot 3} = \frac{1}{3}$$

This seems to make sense. Logically, $\frac{1}{2}$ of $\frac{2}{3}$ should be $\frac{1}{3}$, as Figure 1 shows.

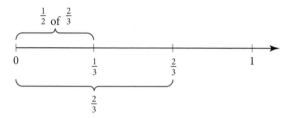

FIGURE 1

EXAMPLE 10   What is $\frac{3}{4}$ of 12?

SOLUTION   Again, *of* means multiply.

$$\frac{3}{4} \text{ of } 12 = \frac{3}{4}(12)$$

$$= \frac{3}{4}\left(\frac{12}{1}\right)$$

$$= \frac{3 \cdot 12}{4 \cdot 1}$$

$$= \frac{3 \cdot 2 \cdot 2 \cdot 3}{2 \cdot 2 \cdot 1}$$

$$= \frac{9}{1} = 9$$

Note on Shortcuts

As you become familiar with multiplying fractions, you may notice shortcuts that reduce the number of steps in the problems. It's okay to use these shortcuts if you understand why they work and are consistently getting correct answers. If you are using shortcuts and not consistently getting correct answers, then go back to showing all the work until you completely understand the process.

FACTS FROM GEOMETRY   *The Area of a Triangle*

The formula for the area of a triangle is one application of multiplication with fractions. Figure 2 shows a triangle with base b and height h. Below the triangle is the formula for its area. As you can see, it is a product containing the fraction $\frac{1}{2}$.

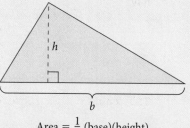

$$\text{Area} = \frac{1}{2}\,(\text{base})(\text{height})$$

$$A = \frac{1}{2}\,bh$$

FIGURE 2   The area of a triangle

The formula for the area of a triangle is one application of multiplication with fractions. Figure 2 shows a triangle with base b and height h. Below the triangle is the formula for its area. As you can see, it is a product containing the fraction $\frac{1}{2}$.

EXAMPLE 11   Find the area of the triangle in Figure 3.

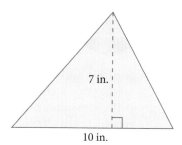

10 in.

FIGURE 3   A triangle with base 10 inches and height 7 inches

SOLUTION   Applying the formula for the area of a triangle, we have

$$A = \frac{1}{2}\,bh = \frac{1}{2} \cdot 10 \cdot 7 = 5 \cdot 7 = 35 \text{ in}^2$$

GETTING READY FOR CLASS

After reading through the preceding section, respond in your own words and in complete sentences.

1. When we multiply the fractions $\frac{3}{5}$ and $\frac{2}{7}$, the numerator in the answer will be what number?

2. When we ask for $\frac{1}{2}$ of $\frac{2}{3}$, are we asking for an addition problem or a multiplication problem?

3. True or false? Reducing to lowest terms before you multiply two fractions will give the same answer as if you were to reduce after you multiply.

4. Write the formula for the area of a triangle with base x and height y.

SPOTLIGHT ON SUCCESS   *University of North Alabama*

Pride is a personal commitment.
It is an attitude which separates excellence from mediocrity.
—William Blake

The University of Northern Alabama places its Pride Rock, a 60-pound granite stone engraved with a lion's paw print, behind the north end zone at all home football games. The rock reminds current Lion players of the proud athletic traditions that has been established at the school, and to take pride in their efforts on the field.

The same idea holds true for your work in your math class. Take pride in it. When you turn in an assignment, it should be accurate and easy for the instructor to read. It shows that you care about your progress in the course and that you take pride in your work. The work that you turn

Photo courtesy UNA

in to your instructor is a reflection of you. As the quote from William Blake indicates, pride is a personal commitment; a decision that you make, yourself. And once you make that commitment to take pride in the work you do in your math class, you have directed yourself toward excellence, and away from mediocrity.

Problem Set 2.3

Find each of the following products. (Multiply.)

1. $\dfrac{2}{3} \cdot \dfrac{4}{5}$

2. $\dfrac{5}{6} \cdot \dfrac{7}{4}$

3. $\dfrac{1}{2} \cdot \dfrac{7}{4}$

4. $\dfrac{3}{5} \cdot \dfrac{4}{7}$

5. $\dfrac{5}{3} \cdot \dfrac{3}{5}$

6. $\dfrac{4}{7} \cdot \dfrac{7}{4}$

7. $\dfrac{3}{4} \cdot 9$

8. $\dfrac{2}{3} \cdot 5$

9. $\dfrac{6}{7}\left(\dfrac{7}{6}\right)$

10. $\dfrac{2}{9}\left(\dfrac{9}{2}\right)$

11. $\dfrac{1}{2} \cdot \dfrac{1}{3} \cdot \dfrac{1}{4}$

12. $\dfrac{2}{3} \cdot \dfrac{4}{5} \cdot \dfrac{1}{3}$

13. $\dfrac{2}{5} \cdot \dfrac{3}{5} \cdot \dfrac{4}{5}$

14. $\dfrac{1}{4} \cdot \dfrac{3}{4} \cdot \dfrac{3}{4}$

15. $\dfrac{3}{2} \cdot \dfrac{5}{2} \cdot \dfrac{7}{2}$

16. $\dfrac{4}{3} \cdot \dfrac{5}{3} \cdot \dfrac{7}{3}$

Complete the following tables.

17.

First Number x	Second Number y	Their Product xy
$\dfrac{1}{2}$	$\dfrac{2}{3}$	
$\dfrac{2}{3}$	$\dfrac{3}{4}$	
$\dfrac{3}{4}$	$\dfrac{4}{5}$	
$\dfrac{5}{a}$	$\dfrac{a}{6}$	

18.

First Number x	Second Number y	Their Product xy
12	$\dfrac{1}{2}$	
12	$\dfrac{1}{3}$	
12	$\dfrac{1}{4}$	
12	$\dfrac{1}{6}$	

19.

First Number x	Second Number y	Their Product xy
$\dfrac{1}{2}$	30	
$\dfrac{1}{5}$	30	
$\dfrac{1}{6}$	30	
$\dfrac{1}{15}$	30	

20.

First Number x	Second Number y	Their Product xy
$\dfrac{1}{3}$	$\dfrac{3}{5}$	
$\dfrac{3}{5}$	$\dfrac{5}{7}$	
$\dfrac{5}{7}$	$\dfrac{7}{9}$	
$\dfrac{7}{b}$	$\dfrac{b}{11}$	

Multiply each of the following. Be sure all answers are written in lowest terms.

21. $\dfrac{9}{20} \cdot \dfrac{4}{3}$

22. $\dfrac{135}{16} \cdot \dfrac{2}{45}$

23. $\dfrac{3}{4} \cdot 12$

24. $\dfrac{3}{4} \cdot 20$

25. $\dfrac{1}{3}(3)$

26. $\dfrac{1}{5}(5)$

27. $\dfrac{2}{5} \cdot 20$

28. $\dfrac{3}{5} \cdot 15$

29. $\dfrac{72}{35} \cdot \dfrac{55}{108} \cdot \dfrac{7}{110}$

30. $\dfrac{32}{27} \cdot \dfrac{72}{49} \cdot \dfrac{1}{40}$

Expand and simplify each of the following.

31. $\left(\dfrac{2}{3}\right)^2$ **32.** $\left(\dfrac{3}{5}\right)^2$ **33.** $\left(\dfrac{3}{4}\right)^2$ **34.** $\left(\dfrac{2}{7}\right)^2$

35. $\left(\dfrac{1}{2}\right)^2$ **36.** $\left(\dfrac{1}{3}\right)^2$ **37.** $\left(\dfrac{2}{3}\right)^3$ **38.** $\left(\dfrac{3}{5}\right)^3$

39. $\left(\dfrac{3}{4}\right)^2 \cdot \dfrac{8}{9}$ **40.** $\left(\dfrac{5}{6}\right)^2 \cdot \dfrac{12}{15}$ **41.** $\left(\dfrac{1}{2}\right)^2\left(\dfrac{3}{5}\right)^2$ **42.** $\left(\dfrac{3}{8}\right)^2\left(\dfrac{4}{3}\right)^2$

43. $\left(\dfrac{1}{2}\right)^2 \cdot 8 + \left(\dfrac{1}{3}\right)^2 \cdot 9$ **44.** $\left(\dfrac{2}{3}\right)^2 \cdot 9 + \left(\dfrac{1}{2}\right)^2 \cdot 4$

45. Find $\dfrac{3}{8}$ of 64.   **46.** Find $\dfrac{2}{3}$ of 18.

47. What is $\dfrac{1}{3}$ of the sum of 8 and 4?   **48.** What is $\dfrac{3}{5}$ of the sum of 8 and 7?

49. Find $\dfrac{1}{2}$ of $\dfrac{3}{4}$ of 24.   **50.** Find $\dfrac{3}{5}$ of $\dfrac{1}{3}$ of 15.

Find the mistakes in Problems 51–52. Correct the right-hand side of each one.

51. $\dfrac{1}{2} \cdot \dfrac{3}{5} = \dfrac{4}{10}$ **52.** $\dfrac{2}{7} \cdot \dfrac{3}{5} = \dfrac{5}{35}$

53. a. Complete the following table.

   b. Using the results of part a, fill in the blank in the following statement:

   For numbers larger than 1, the square of the number is _____ than the number.

Number x	Square x^2
1	
2	
3	
4	
5	
6	
7	
8	

54. a. Complete the following table.

   b. Using the results of part a, fill in the blank in the following statement:

   For numbers between 0 and 1, the square of the number is _____ than the number.

Number x	Square x^2
$\dfrac{1}{2}$	
$\dfrac{1}{3}$	
$\dfrac{1}{4}$	
$\dfrac{1}{5}$	
$\dfrac{1}{6}$	
$\dfrac{1}{7}$	
$\dfrac{1}{8}$	

Apply the distributive property, then simplify.

55. $4\left(3 + \dfrac{1}{2}\right)$ **56.** $4\left(2 - \dfrac{3}{4}\right)$ **57.** $12\left(\dfrac{1}{2} + \dfrac{2}{3}\right)$ **58.** $12\left(\dfrac{3}{4} - \dfrac{1}{6}\right)$

59. Find the area of the triangle with base 19 inches and height 14 inches.

60. Find the area of the triangle with base 13 inches and height 8 inches.

61. The base of a triangle is $\dfrac{4}{3}$ feet and the height is $\dfrac{2}{3}$ feet. Find the area.

62. The base of a triangle is $\dfrac{8}{7}$ feet and the height is $\dfrac{14}{5}$ feet. Find the area.

Find the area of each figure.

63.

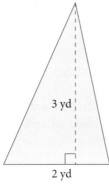

3 yd

2 yd

64.

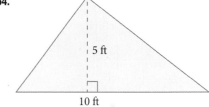

5 ft

10 ft

Applying the Concepts

Use the information in the pie chart to answer questions 65 and 66.

65. Reading a Pie Chart If there are approximately 15,800 students attending Cal Poly, approximately how many of them are studying agriculture?

66. Reading a Pie Chart If there are exactly 15,828 students attending Cal Poly, exactly how many of them are studying engineering?

67. Hot Air Balloon Aerostar International makes a hot air balloon called the Rally 105 that has a volume of 105,400 cubic feet. Another balloon, the Rally 126, was designed with a volume that is approximately $\dfrac{6}{5}$ the volume of the Rally 105. Find the volume of the Rally 126 to the nearest hundred cubic feet.

68. Health Care According to a study reported on MSNBC in 2004, almost one-third of the people diagnosed with diabetes don't seek proper medical care. If there are 12 million Americans with diabetes, about how many of them are seeking proper medical care?

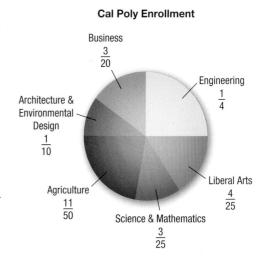

Cal Poly Enrollment

Business $\dfrac{3}{20}$

Engineering $\dfrac{1}{4}$

Architecture & Environmental Design $\dfrac{1}{10}$

Liberal Arts $\dfrac{4}{25}$

Agriculture $\dfrac{11}{50}$

Science & Mathematics $\dfrac{3}{25}$

iStockPhoto/©Elemental Imaging

iStockPhoto/©Roberto Caucino

69. Bicycle Safety The National Safe Kids Campaign and Bell Sports sponsored a study that surveyed 8,159 children ages 5 to 14 who were riding bicycles. Approximately $\frac{2}{5}$ of the children were wearing helmets, and of those, only $\frac{13}{20}$ were wearing the helmets correctly. About how many of the children were wearing helmets correctly?

70. Bicycle Safety From the information in Problem 69, how many of the children surveyed do not wear helmets?

Estimating For each problem below, mentally estimate if the answer will be closest to 0, 1, 2 or 3. Make your estimate without using pencil and paper or a calculator.

71. $\dfrac{11}{5} \cdot \dfrac{19}{20}$ **72.** $\dfrac{3}{5} \cdot \dfrac{1}{20}$ **73.** $\dfrac{16}{5} \cdot \dfrac{23}{24}$ **74.** $\dfrac{9}{8} \cdot \dfrac{31}{32}$

Getting Ready for the Next Section

In the next section we will do division with fractions. As you already know, division and multiplication are closely related. These review problems are intended to let you see more of the relationship between multiplication and division.

Perform the indicated operations.

75. $8 \div 4$ **76.** $8 \cdot \dfrac{1}{4}$ **77.** $15 \div 3$ **78.** $15 \cdot \dfrac{1}{3}$ **79.** $18 \div 6$ **80.** $18 \cdot \dfrac{1}{6}$

For each number below, find a number to multiply it by to obtain 1.

81. $\dfrac{3}{4}$ **82.** $\dfrac{9}{5}$ **83.** $\dfrac{1}{3}$ **84.** $\dfrac{1}{4}$ **85.** 7 **86.** 2

Division with Fractions

A few years ago our 4-H club was making blankets to keep their lambs clean at the county fair. Each blanket required $\frac{3}{4}$ yard of material. We had 9 yards of material left over from the year before. To see how many blankets we could make, we divided 9 by $\frac{3}{4}$. The result was 12, meaning that we could make 12 lamb blankets for the fair.

Before we define division with fractions, we must first introduce the idea of *reciprocals*. Look at the following multiplication problems:

$$\frac{3}{4} \cdot \frac{4}{3} = \frac{12}{12} = 1 \qquad \frac{7}{8} \cdot \frac{8}{7} = \frac{56}{56} = 1$$

In each case the product is 1. Whenever the product of two numbers is 1, we say the two numbers are *reciprocals*.

DEFINITION

Two numbers whose product is 1 are said to be **reciprocals.** In symbols, the reciprocal of $\frac{a}{b}$ is $\frac{b}{a}$, because

$$\frac{a}{b} \cdot \frac{b}{a} = \frac{a \cdot b}{b \cdot a} = \frac{a \cdot b}{a \cdot b} = 1 \qquad (a \neq 0, b \neq 0)$$

Every number has a reciprocal except 0. The reason 0 does not have a reciprocal is because the product of *any* number with 0 is 0. It can never be 1. Reciprocals of whole numbers are fractions with 1 as the numerator. For example, the reciprocal of 5 is $\frac{1}{5}$, because

$$5 \cdot \frac{1}{5} = \frac{5}{1} \cdot \frac{1}{5} = \frac{5}{5} = 1$$

Table 1 lists some numbers and their reciprocals.

TABLE 1

Number	Reciprocal	Reason
$\frac{3}{4}$	$\frac{4}{3}$	Because $\frac{3}{4} \cdot \frac{4}{3} = \frac{12}{12} = 1$
$\frac{9}{5}$	$\frac{5}{9}$	Because $\frac{9}{5} \cdot \frac{5}{9} = \frac{45}{45} = 1$
$\frac{1}{3}$	3	Because $\frac{1}{3} \cdot 3 = \frac{1}{3} \cdot \frac{3}{1} = \frac{3}{3} = 1$
7	$\frac{1}{7}$	Because $7 \cdot \frac{1}{7} = \frac{7}{1} \cdot \frac{1}{7} = \frac{7}{7} = 1$

Note

Defining division to be the same as multiplication by the reciprocal does make sense. If we divide 6 by 2, we get 3. On the other hand, if we multiply 6 by $\frac{1}{2}$ (the reciprocal of 2), we also get 3. Whether we divide by 2 or multiply by $\frac{1}{2}$, we get the same result.

Division with fractions is accomplished by using reciprocals. More specifically, we can define division by a fraction to be the same as multiplication by its reciprocal. Here is the precise definition:

DEFINITION

If a, b, c, and d are numbers and b, c, and d are all not equal to 0, then

$$\frac{a}{b} \div \frac{c}{d} = \frac{a}{b} \cdot \frac{d}{c}$$

This definition states that dividing by the fraction $\frac{c}{d}$ is exactly the same as multiplying by its reciprocal $\frac{d}{c}$. Because we developed the rule for multiplying fractions in Section 2.3, we do not need a new rule for division. We simply *replace the divisor by its reciprocal* and multiply. Here are some examples to illustrate the procedure.

EXAMPLE 1   Divide $\frac{1}{2} \div \frac{1}{4}$.

SOLUTION   The divisor is $\frac{1}{4}$, and its reciprocal is $\frac{4}{1}$. Applying the definition of division for fractions, we have

$$\frac{1}{2} \div \frac{1}{4} = \frac{1}{2} \cdot \frac{4}{1}$$

$$= \frac{1 \cdot 4}{2 \cdot 1}$$

$$= \frac{1 \cdot 2 \cdot 2}{2 \cdot 1}$$

$$= \frac{2}{1}$$

$$= 2$$

The quotient of $\frac{1}{2}$ and $\frac{1}{4}$ is 2. Or, $\frac{1}{4}$ "goes into" $\frac{1}{2}$ two times. Logically, our definition for division of fractions seems to be giving us answers that are consistent with what we know about fractions from previous experience. Because 2 times $\frac{1}{4}$ is $\frac{2}{4}$ or $\frac{1}{2}$, it seems logical that $\frac{1}{2}$ divided by $\frac{1}{4}$ should be 2.

EXAMPLE 2   Divide $\frac{3}{8} \div \frac{9}{4}$.

SOLUTION   Dividing by $\frac{9}{4}$ is the same as multiplying by its reciprocal, which is $\frac{4}{9}$:

$$\frac{3}{8} \div \frac{9}{4} = \frac{3}{8} \cdot \frac{4}{9}$$

$$= \frac{3 \cdot 2 \cdot 2}{2 \cdot 2 \cdot 2 \cdot 3 \cdot 3}$$

$$= \frac{1}{6}$$

The quotient of $\frac{3}{8}$ and $\frac{9}{4}$ is $\frac{1}{6}$.

EXAMPLE 3 Divide $\frac{2}{3} \div 2$.

SOLUTION The reciprocal of 2 is $\frac{1}{2}$. Applying the definition for division of fractions, we have

$$\frac{2}{3} \div 2 = \frac{2}{3} \cdot \frac{1}{2}$$

$$= \frac{\cancel{2} \cdot 1}{3 \cdot \cancel{2}}$$

$$= \frac{1}{3}$$

EXAMPLE 4 Divide $2 \div \frac{1}{3}$.

SOLUTION We replace $\frac{1}{3}$ by its reciprocal, which is 3, and multiply:

$$2 \div \frac{1}{3} = 2(3)$$

$$= 6$$

Here are some further examples of division with fractions. Notice in each case that the first step is the only new part of the process.

EXAMPLE 5

$$\frac{4}{27} \div \frac{16}{9} = \frac{4}{27} \cdot \frac{9}{16}$$

$$= \frac{\cancel{4} \cdot \cancel{9}}{3 \cdot \cancel{9} \cdot \cancel{4} \cdot 4}$$

$$= \frac{1}{12}$$

In this example we did not factor the numerator and the denominator completely in order to reduce to lowest terms because, as you have probably already noticed, it is not necessary to do so. We need to factor only enough to show what numbers are common to the numerator and the denominator. If we factored completely in the second step, it would look like this:

$$= \frac{\cancel{2} \cdot \cancel{2} \cdot \cancel{3} \cdot \cancel{3}}{\cancel{3} \cdot \cancel{3} \cdot 3 \cdot \cancel{2} \cdot \cancel{2} \cdot 2 \cdot 2}$$

$$= \frac{1}{12}$$

The result is the same in both cases. From now on we will factor numerators and denominators only enough to show the factors we are dividing out.

EXAMPLE 6 Divide.

a. $\frac{16}{35} \div 8$ **b.** $27 \div \frac{3}{2}$

SOLUTION   **a.** $\frac{16}{35} \div 8 = \frac{16}{35} \cdot \frac{1}{8}$ **b.** $27 \div \frac{3}{2} = 27 \cdot \frac{2}{3}$

$$= \frac{2 \cdot \cancel{8} \cdot 1}{35 \cdot \cancel{8}} \qquad\qquad = \frac{\cancel{3} \cdot 9 \cdot 2}{\cancel{3}}$$

$$= \frac{2}{35} \qquad\qquad\qquad\quad = 18$$

The next two examples combine what we have learned about division of fractions with the rule for order of operations.

EXAMPLE 7   The quotient of $\frac{8}{3}$ and $\frac{1}{6}$ is increased by 5. What number results?

SOLUTION   Translating to symbols, we have

$$\frac{8}{3} \div \frac{1}{6} + 5 = \left[\frac{8}{3} \cdot \frac{6}{1}\right] + 5$$
$$= 16 + 5$$
$$= 21$$

EXAMPLE 8   Simplify: $32 \div \left(\frac{4}{3}\right)^2 + 75 \div \left(\frac{5}{2}\right)^2$

SOLUTION   According to the rule for order of operations, we must first evaluate the numbers with exponents, then we divide, and finally we add.

$$\left[32 \div \left(\frac{4}{3}\right)^2\right] + \left[75 \div \left(\frac{5}{2}\right)^2\right] = 32 \div \frac{16}{9} + 75 \div \frac{25}{4}$$
$$= 32 \cdot \frac{9}{16} + 75 \cdot \frac{4}{25}$$
$$= 18 + 12$$
$$= 30$$

EXAMPLE 9   A 4-H Club is making blankets to keep their lambs clean at the county fair. If each blanket requires $\frac{3}{4}$ yard of material, how many blankets can they make from 9 yards of material?

SOLUTION   To answer this question we must divide 9 by $\frac{3}{4}$.

$$9 \div \frac{3}{4} = 9 \cdot \frac{4}{3}$$
$$= 3 \cdot 4$$
$$= 12$$

They can make 12 blankets from the 9 yards of material.

GETTING READY FOR CLASS

After reading through the preceding section, respond in your own words and in complete sentences.

1. What do we call two numbers whose product is 1?
2. True or false? The quotient of $\frac{3}{5}$ and $\frac{3}{8}$ is the same as the product of $\frac{3}{5}$ and $\frac{8}{3}$.
3. How are multiplication and division of fractions related?
4. Dividing by $\frac{19}{9}$ is the same as multiplying by what number?

Problem Set 2.4

Find the quotient in each case by replacing the divisor by its reciprocal and multiplying.

1. $\dfrac{3}{4} \div \dfrac{1}{5}$

2. $\dfrac{1}{3} \div \dfrac{1}{2}$

3. $\dfrac{2}{3} \div \dfrac{1}{2}$

4. $\dfrac{5}{8} \div \dfrac{1}{4}$

5. $6 \div \dfrac{2}{3}$

6. $8 \div \dfrac{3}{4}$

7. $20 \div \dfrac{1}{10}$

8. $16 \div \dfrac{1}{8}$

9. $\dfrac{3}{4} \div 2$

10. $\dfrac{3}{5} \div 2$

11. $\dfrac{7}{8} \div \dfrac{7}{8}$

12. $\dfrac{4}{3} \div \dfrac{4}{3}$

13. $\dfrac{7}{8} \div \dfrac{8}{7}$

14. $\dfrac{4}{3} \div \dfrac{3}{4}$

15. $\dfrac{9}{16} \div \dfrac{3}{4}$

16. $\dfrac{25}{36} \div \dfrac{5}{6}$

17. $\dfrac{25}{46} \div \dfrac{40}{69}$

18. $\dfrac{25}{24} \div \dfrac{15}{36}$

19. $\dfrac{13}{28} \div \dfrac{39}{14}$

20. $\dfrac{28}{125} \div \dfrac{5}{2}$

21. $\dfrac{27}{196} \div \dfrac{9}{392}$

22. $\dfrac{16}{135} \div \dfrac{2}{45}$

23. $\dfrac{25}{18} \div 5$

24. $\dfrac{30}{27} \div 6$

25. $6 \div \dfrac{4}{3}$

26. $12 \div \dfrac{4}{3}$

27. $\dfrac{4}{3} \div 6$

28. $\dfrac{4}{3} \div 12$

29. $\dfrac{3}{4} \div \dfrac{1}{2} \cdot 6$

30. $12 \div \dfrac{6}{7} \cdot 7$

31. $\dfrac{2}{3} \cdot \dfrac{3}{4} \div \dfrac{5}{8}$

32. $4 \cdot \dfrac{7}{6} \div 7$

33. $\dfrac{35}{110} \cdot \dfrac{80}{63} \div \dfrac{16}{27}$

34. $\dfrac{20}{72} \cdot \dfrac{42}{18} \div \dfrac{20}{16}$

Simplify each expression as much as possible.

35. $10 \div \left(\dfrac{1}{2}\right)^2$

36. $12 \div \left(\dfrac{1}{4}\right)^2$

37. $\dfrac{18}{35} \div \left(\dfrac{6}{7}\right)^2$

38. $\dfrac{48}{55} \div \left(\dfrac{8}{11}\right)^2$

39. $\dfrac{4}{5} \div \dfrac{1}{10} + 5$

40. $\dfrac{3}{8} \div \dfrac{1}{16} + 4$

41. $10 + \dfrac{11}{12} \div \dfrac{11}{24}$

42. $15 + \dfrac{13}{14} \div \dfrac{13}{42}$

43. $24 \div \left(\dfrac{2}{5}\right)^2 + 25 \div \left(\dfrac{5}{6}\right)^2$

44. $18 \div \left(\dfrac{3}{4}\right)^2 + 49 \div \left(\dfrac{7}{9}\right)^2$

45. $100 \div \left(\dfrac{5}{7}\right)^2 + 200 \div \left(\dfrac{2}{3}\right)^2$

46. $64 \div \left(\dfrac{8}{11}\right)^2 + 81 \div \left(\dfrac{9}{11}\right)^2$

47. What is the quotient of $\dfrac{3}{8}$ and $\dfrac{5}{8}$?

48. Find the quotient of $\dfrac{4}{5}$ and $\dfrac{16}{25}$.

49. If the quotient of 18 and $\dfrac{3}{5}$ is increased by 10, what number results?

50. If the quotient of 50 and $\dfrac{5}{3}$ is increased by 8, what number results?

51. Show that multiplying 3 by 5 is the same as dividing 3 by $\dfrac{1}{5}$.

52. Show that multiplying 8 by $\dfrac{1}{2}$ is the same as dividing 8 by 2.

Applying the Concepts

Although many of the application problems that follow involve division with fractions, some do not. Be sure to read the problems carefully.

53. Sewing If $\frac{6}{7}$ yard of material is needed to make a blanket, how many blankets can be made from 12 yards of material?

54. Manufacturing A clothing manufacturer is making scarves that require $\frac{3}{8}$ yard of material each. How many can be made from 27 yards of material?

55. Capacity Suppose a bag of candy holds exactly $\frac{1}{4}$ pound of candy. How many of these bags can be filled from 12 pounds of candy?

56. Capacity A certain size bottle holds exactly $\frac{4}{5}$ pint of liquid. How many of these bottles can be filled from a 20-pint container?

57. Cooking A man is making cookies from a recipe that calls for $\frac{3}{4}$ teaspoon of oil. If the only measuring spoon he can find is a $\frac{1}{8}$ teaspoon, how many of these will he have to fill with oil in order to have a total of $\frac{3}{4}$ teaspoon of oil?

58. Cooking A cake recipe calls for $\frac{1}{2}$ cup of sugar. If the only measuring cup available is a $\frac{1}{8}$ cup, how many of these will have to be filled with sugar to make a total of $\frac{1}{2}$ cup of sugar?

59. Student Population If 14 of every 32 students attending Cuesta College are female, what fraction of the students is female? (Simplify your answer.)

60. Population If 27 of every 48 residents of a small town are male, what fraction of the population is male? (Simplify your answer.)

Getting Ready for the Next Section

Write each fraction as an equivalent fraction with denominator 6.

61. $\frac{1}{2}$ **62.** $\frac{1}{3}$ **63.** $\frac{3}{2}$ **64.** $\frac{2}{3}$

Write each fraction as an equivalent fraction with denominator 12.

65. $\frac{1}{3}$ **66.** $\frac{1}{2}$ **67.** $\frac{2}{3}$ **68.** $\frac{3}{4}$

Write each fraction as an equivalent fraction with denominator 30.

69. $\frac{7}{15}$ **70.** $\frac{3}{10}$ **71.** $\frac{3}{5}$ **72.** $\frac{1}{6}$

Write each fraction as an equivalent fraction with denominator 24.

73. $\frac{1}{2}$ **74.** $\frac{1}{4}$ **75.** $\frac{1}{6}$ **76.** $\frac{1}{8}$

Write each fraction as an equivalent fraction with denominator 36.

77. $\frac{5}{12}$ **78.** $\frac{7}{18}$

Chapter 2 Summary

Definition of Fractions [2.1]

1. Each of the following is a fraction:

$$\frac{1}{2}, \frac{3}{4}, \frac{8}{1}, \frac{7}{3}$$

A fraction is any number that can be written in the form $\frac{a}{b}$, where a and b are numbers and b is not 0. The number a is called the *numerator*, and the number b is called the *denominator*.

Properties of Fractions [2.1]

2. Change $\frac{3}{4}$ to an equivalent fraction with denominator 12.

$$\frac{3}{4} = \frac{3 \cdot 3}{4 \cdot 3} = \frac{9}{12}$$

Multiplying the numerator and the denominator of a fraction by the same non-zero number will produce an equivalent fraction. The same is true for dividing the numerator and denominator by the same nonzero number. In symbols the properties look like this: If a, b, and c are numbers and b and c are not 0, then

$$\text{Property 1} \quad \frac{a}{b} = \frac{a \cdot c}{b \cdot c} \qquad \text{Property 2} \quad \frac{a}{b} = \frac{a \div c}{b \div c}$$

Fractions and the Number 1 [2.1]

3. $\frac{5}{1} = 5, \frac{5}{5} = 1$

If a represents any number, then

$$\frac{a}{1} = a \quad \text{and} \quad \frac{a}{a} = 1 \quad \text{(where } a \text{ is not 0)}$$

Reducing Fractions to Lowest Terms [2.2]

4.
$$\frac{90}{588} = \frac{2 \cdot 3 \cdot 3 \cdot 5}{2 \cdot 2 \cdot 3 \cdot 7 \cdot 7}$$
$$= \frac{3 \cdot 5}{2 \cdot 7 \cdot 7}$$
$$= \frac{15}{98}$$

To reduce a fraction to lowest terms, factor the numerator and the denominator, and then divide both the numerator and denominator by any factors they have in common.

Multiplying Fractions [2.3]

5. $\frac{3}{5} \cdot \frac{4}{7} = \frac{3 \cdot 4}{5 \cdot 7} = \frac{12}{35}$

To multiply fractions, multiply numerators and multiply denominators.

The Area of a Triangle [2.3]

6. If the base of a triangle is 10 inches and the height is 7 inches, then the area is

$$A = \frac{1}{2}bh$$
$$= \frac{1}{2} \cdot 10 \cdot 7$$
$$= 5 \cdot 7$$
$$= 35 \text{ square inches}$$

The formula for the area of a triangle with base b and height h is

$$A = \frac{1}{2}bh$$

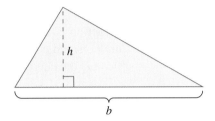

Reciprocals [2.4]

Any two numbers whose product is 1 are called *reciprocals*. The numbers $\frac{2}{3}$ and $\frac{3}{2}$ are reciprocals, because their product is 1.

Division with Fractions [2.4]

7. $\frac{3}{8} \div \frac{1}{3} = \frac{3}{8} \cdot \frac{3}{1} = \frac{9}{8}$

To divide by a fraction, multiply by its reciprocal. That is, the quotient of two fractions is defined to be the product of the first fraction with the reciprocal of the second fraction (the divisor).

> **COMMON MISTAKES**
>
> **1.** A common mistake made with division of fractions occurs when we multiply by the reciprocal of the first fraction instead of the reciprocal of the divisor. For example,
>
> $$\frac{2}{3} \div \frac{5}{6} \neq \frac{3}{2} \cdot \frac{5}{6}$$
>
> Remember, we perform division by multiplying by the reciprocal of the divisor (the fraction to the right of the division symbol).
>
> **2.** If the answer to a problem turns out to be a fraction, that fraction should always be written in lowest terms. It is a mistake not to reduce to lowest terms.

Chapter 2 Test Form A

These problems are all taken from examples in your text. Work each problem and check your answers. If you have made a mistake, work the problem again. If you cannot get the correct answer after two tries, look up the correct solution in your text.

1. Write $\dfrac{3}{4}$ as an equivalent fraction with denominator 20.

2. Write $\dfrac{3}{4}$ as an equivalent fraction with denominator $12x$.

3. Factor 60 into a product of prime factors.

4. Reduce $\dfrac{6}{42}$ to lowest terms.

5. Multiply $\dfrac{15}{8} \cdot \dfrac{4}{9}$

6. Multiply $\dfrac{1}{2}\left(\dfrac{3}{4} \cdot \dfrac{1}{5}\right)$

7. Find $\dfrac{1}{2}$ of $\dfrac{2}{3}$

8. Divide $\dfrac{3}{8} \div \dfrac{9}{4}$

9. Divide $2 \div \dfrac{1}{3}$

10. Divide $\dfrac{2}{3} \div 2$

Chapter Test, Form B

For an alternate, more comprehensive, chapter test, go to MathTV.com and select the test and summary for this chapter of the textbook. Click the worksheet labeled Chapter 2 Test, Form B to download it.

Fractions 2: Addition and Subtraction

iStockPhoto/©Tom Tomczyk

I f you have ever made pesto from scratch, you know that you can vary the kinds of herbs used in the mixture, depending on your taste or what you have available. You also know that it is important to keep the overall amount of herbs in the total mixture the same so that the resulting sauce has the same consistency from batch to batch. Suppose you want to use both basil and parsley to make your pesto, and the amount of herbs used must equal 2 cups to get the desired texture. The table below shows some possible combinations of the two herbs needed to make pesto if you need a total of 2 cups.

Possible Parsley + Basil Combinations		
Parsley	Basil	Total
$\frac{1}{2}$	$1\frac{1}{2}$	2
$\frac{3}{4}$	$1\frac{1}{4}$	2
1	1	2
$1\frac{1}{4}$	$\frac{3}{4}$	2
$1\frac{1}{2}$	$\frac{1}{2}$	2

If you bake or cook, the ability to combine different amounts of ingredients by adding or subtracting fractions is necessary. In this chapter we will continue our look at fractions and expand our work to include addition and subtraction of fractions and mixed numbers.

Study Skills

The study skills for this chapter are about attitude. They are points of view that point toward success.

1. **Be Focused, Not Distracted** I have students who begin their assignments by asking themselves, "Why am I taking this class?" If you are asking yourself similar questions, you are distracting yourself from doing the things that will produce the results you want in this course. Don't dwell on questions and evaluations of the class that can be used as excuses for not doing well. If you want to succeed in this course, focus your energy and efforts toward success, rather than distracting yourself from your goals.

2. **Be Resilient** Don't let setbacks keep you from your goals. You want to put yourself on the road to becoming a person who can succeed in this class, or any class in college. Failing a test or quiz, or having a difficult time on some topics, is normal. No one goes through college without some setbacks. Don't let a temporary disappointment keep you from succeeding in this course. A low grade on a test or quiz is simply a signal that you need to reevaluate your study habits.

3. **Intend to Succeed** I have a few students who simply go through the motions of studying without intending to master the material. It is more important to them to look like they are studying than to actually study. You need to study with the intention of being successful in the course. Intend to master the material, no matter what it takes.

Addition and Subtraction with Fractions

Adding and subtracting fractions is actually just another application of the distributive property. The distributive property looks like this:

$$a(b + c) = a(b) + a(c)$$

where a, b, and c may be whole numbers or fractions. We will want to apply this property to expressions like

$$\frac{2}{7} + \frac{3}{7}$$

But before we do, we must make one additional observation about fractions. The fraction $\frac{2}{7}$ can be written as $2 \cdot \frac{1}{7}$, because

$$2 \cdot \frac{1}{7} = \frac{2}{1} \cdot \frac{1}{7} = \frac{2}{7}$$

Likewise, the fraction $\frac{3}{7}$ can be written as $3 \cdot \frac{1}{7}$, because

$$3 \cdot \frac{1}{7} = \frac{3}{1} \cdot \frac{1}{7} = \frac{3}{7}$$

In general, we can say that the fraction $\frac{a}{b}$ can always be written as $a \cdot \frac{1}{b}$, because

$$a \cdot \frac{1}{b} = \frac{a}{1} \cdot \frac{1}{b} = \frac{a}{b}$$

To add the fractions $\frac{2}{7}$ and $\frac{3}{7}$, we simply rewrite each of them as we have done above and apply the distributive property. Here is how it works:

$$\frac{2}{7} + \frac{3}{7} = 2 \cdot \frac{1}{7} + 3 \cdot \frac{1}{7} \qquad \text{Rewrite each fraction}$$

$$= (2 + 3) \cdot \frac{1}{7} \qquad \text{Apply the distributive property}$$

$$= 5 \cdot \frac{1}{7} \qquad \text{Add 2 and 3 to get 5}$$

$$= \frac{5}{7} \qquad \text{Rewrite } 5 \cdot \frac{1}{7} \text{ as } \frac{5}{7}$$

> **Note**
>
> Most people who have done any work with adding fractions know that you add fractions that have the same denominator by adding their numerators, but not their denominators. However, most people don't know why this works. The reason why we add numerators but not denominators is because of the distributive property. And that is what the discussion on the right is all about. If you really want to understand addition of fractions, pay close attention to this discussion.

We can visualize the process shown above by using circles that are divided into 7 equal parts:

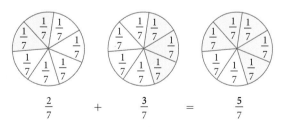

The fraction $\frac{5}{7}$ is the sum of $\frac{2}{7}$ and $\frac{3}{7}$. The steps and diagrams on the previous page show why we add numerators, *but do not add denominators.* Using this example as justification, we can write a rule for adding two fractions that have the same denominator.

RULE

To add two fractions that have the same denominator, we add their numerators to get the numerator of the answer. The denominator in the answer is the same denominator as in the original fractions.

What we have here is the sum of the numerators placed over the *common denominator.* In symbols we have the following:

Addition and Subtraction of Fractions

If a, b, and c are numbers, and c is not equal to 0, then

$$\frac{a}{c} + \frac{b}{c} = \frac{a+b}{c}$$

This rule holds for subtraction as well. That is,

$$\frac{a}{c} - \frac{b}{c} = \frac{a-b}{c}$$

EXAMPLE 1   Add or subtract.

a. $\dfrac{3}{8} + \dfrac{1}{8}$ **b.** $\dfrac{9}{5} - \dfrac{3}{5}$ **c.** $\dfrac{3}{7} + \dfrac{2}{7} + \dfrac{9}{7}$

SOLUTION

a. $\dfrac{3}{8} + \dfrac{1}{8} = \dfrac{3+1}{8}$ Add numerators; keep the same denominator

$= \dfrac{4}{8}$ The sum of 3 and 1 is 4

$= \dfrac{1}{2}$ Reduce to lowest terms

b. $\dfrac{9}{5} - \dfrac{3}{5} = \dfrac{9-3}{5}$ Subtract numerators; keep the same denominator

$= \dfrac{6}{5}$ The difference of 9 and 3 is 6

c. $\dfrac{3}{7} + \dfrac{2}{7} + \dfrac{9}{7} = \dfrac{3+2+9}{7}$

$= \dfrac{14}{7}$

$= 2$

As Example 1 indicates, addition and subtraction are simple, straightforward processes when all the fractions have the same denominator. We will now turn our attention to the process of adding fractions that have different denominators. In order to get started, we need the following definition:

> **DEFINITION**
>
> The *least common denominator* (LCD) for a set of denominators is the smallest number that is exactly divisible by each denominator. (Note that, in some books, the least common denominator is also called the *least common multiple*.)

In other words, all the denominators of the fractions involved in a problem must divide into the least common denominator exactly. That is, they divide it without leaving a remainder.

EXAMPLE 2   Find the LCD for the fractions $\dfrac{5}{12}$ and $\dfrac{7}{18}$.

SOLUTION   The least common denominator for the denominators 12 and 18 must be the smallest number divisible by both 12 and 18. We can factor 12 and 18 completely and then build the LCD from these factors. Factoring 12 and 18 completely gives us

$$12 = 2 \cdot 2 \cdot 3 \qquad 18 = 2 \cdot 3 \cdot 3$$

Now, if 12 is going to divide the LCD exactly, then the LCD must have factors of $2 \cdot 2 \cdot 3$. If 18 is to divide it exactly, it must have factors of $2 \cdot 3 \cdot 3$. We don't need to repeat the factors that 12 and 18 have in common:

$$\left.\begin{array}{l} 12 = 2 \cdot 2 \cdot 3 \\ 18 = 2 \cdot 3 \cdot 3 \end{array}\right\} \qquad \overset{\text{12 divides the LCD}}{\text{LCD} = 2 \cdot 2 \cdot 3 \cdot 3 = 36}_{\text{18 divides the LCD}}$$

The LCD for 12 and 18 is 36. It is the smallest number that is divisible by both 12 and 18; 12 divides it exactly three times, and 18 divides it exactly two times.

Note

The ability to find least common denominators is very important in mathematics. The discussion here is a detailed explanation of how to find an LCD.

We can visualize the results in Example 2 with the diagram below. It shows that 36 is the smallest number that both 12 and 18 divide evenly. As you can see, 12 divides 36 exactly 3 times, and 18 divides 36 exactly 2 times.

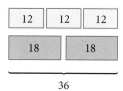

EXAMPLE 3   Add $\dfrac{5}{12} + \dfrac{7}{18}$.

SOLUTION   We can add fractions only when they have the same denominators. In Example 2, we found the LCD for $\frac{5}{12}$ and $\frac{7}{18}$ to be 36. We change $\frac{5}{12}$ and $\frac{7}{18}$ to equivalent fractions that have 36 for a denominator by applying Property 1 for fractions:

$$\frac{5}{12} = \frac{5 \cdot 3}{12 \cdot 3} = \frac{15}{36}$$

$$\frac{7}{18} = \frac{7 \cdot 2}{18 \cdot 2} = \frac{14}{36}$$

The fraction $\frac{15}{36}$ is equivalent to $\frac{5}{12}$, because it was obtained by multiplying both the numerator and the denominator by 3. Likewise, $\frac{14}{36}$ is equivalent to $\frac{7}{18}$, because it was obtained by multiplying the numerator and the denominator by 2. All we have left to do is to add numerators.

$$\frac{15}{36} + \frac{14}{36} = \frac{29}{36}$$

The sum of $\frac{5}{12}$ and $\frac{7}{18}$ is the fraction $\frac{29}{36}$. Let's write the complete problem again step by step.

$$\frac{5}{12} + \frac{7}{18} = \frac{5 \cdot 3}{12 \cdot 3} + \frac{7 \cdot 2}{18 \cdot 2} \qquad \text{\small Rewrite each fraction as an equivalent fraction with denominator 36}$$

$$= \frac{15}{36} + \frac{14}{36}$$

$$= \frac{29}{36} \qquad \text{\small Add numerators; keep the common denominator}$$

EXAMPLE 4   Find the LCD for $\dfrac{3}{4}$ and $\dfrac{1}{6}$.

SOLUTION   We factor 4 and 6 into products of prime factors and build the LCD from these factors.

$$\left.\begin{array}{l} 4 = 2 \cdot 2 \\ 6 = 2 \cdot 3 \end{array}\right\} \quad \text{LCD} = 2 \cdot 2 \cdot 3 = 12$$

The LCD is 12. Both denominators divide it exactly; 4 divides 12 exactly 3 times, and 6 divides 12 exactly 2 times.

Note

We can visualize the work in Example 5 using circles and shading:

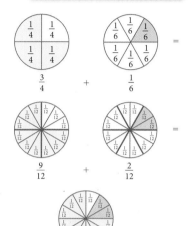

EXAMPLE 5   Add $\frac{3}{4} + \frac{1}{6}$.

SOLUTION   In Example 4, we found that the LCD for these two fractions is 12. We begin by changing $\frac{3}{4}$ and $\frac{1}{6}$ to equivalent fractions with denominator 12:

$$\frac{3}{4} = \frac{3 \cdot 3}{4 \cdot 3} = \frac{9}{12}$$

$$\frac{1}{6} = \frac{1 \cdot 2}{6 \cdot 2} = \frac{2}{12}$$

The fraction $\frac{9}{12}$ is equal to the fraction $\frac{3}{4}$, because it was obtained by multiplying the numerator and the denominator of $\frac{3}{4}$ by 3. Likewise, $\frac{2}{12}$ is equivalent to $\frac{1}{6}$, because it was obtained by multiplying the numerator and the denominator of $\frac{1}{6}$ by 2. To complete the problem we add numerators:

$$\frac{9}{12} + \frac{2}{12} = \frac{11}{12}$$

The sum of $\frac{3}{4}$ and $\frac{1}{6}$ is $\frac{11}{12}$. Here is how the complete problem looks:

$$\frac{3}{4} + \frac{1}{6} = \frac{3 \cdot 3}{4 \cdot 3} + \frac{1 \cdot 2}{6 \cdot 2}$$ Rewrite each fraction as an equivalent fraction with denominator 12

$$= \frac{9}{12} + \frac{2}{12}$$

$$= \frac{11}{12}$$ Add numerators; keep the same denominator

EXAMPLE 6   Subtract $\frac{7}{15} - \frac{3}{10}$.

SOLUTION   Let's factor 15 and 10 completely and use these factors to build the LCD:

15 divides the LCD

$$\left.\begin{array}{l} 15 = 3 \cdot 5 \\ 10 = 2 \cdot 5 \end{array}\right\} \quad \text{LCD} = 2 \cdot 3 \cdot 5 = 30$$

10 divides the LCD

Changing to equivalent fractions and subtracting, we have

$$\frac{7}{15} - \frac{3}{10} = \frac{7 \cdot 2}{15 \cdot 2} - \frac{3 \cdot 3}{10 \cdot 3}$$ Rewrite as equivalent fractions with the LCD for the denominator

$$= \frac{14}{30} - \frac{9}{30}$$

$$= \frac{5}{30}$$ Subtract numerators; keep the LCD

$$= \frac{1}{6}$$ Reduce to lowest terms

As a summary of what we have done so far, and as a guide to working other problems, we now list the steps involved in adding and subtracting fractions with different denominators.

To Add or Subtract Any Two Fractions

Step 1   Factor each denominator completely, and use the factors to build the LCD. (Remember, the LCD is the smallest number divisible by each of the denominators in the problem.)

Step 2   Rewrite each fraction as an equivalent fraction with the LCD. This is done by multiplying both the numerator and the denominator of the fraction in question by the appropriate whole number.

Step 3   Add or subtract the numerators of the fractions produced in Step 2. This is the numerator of the sum or difference. The denominator of the sum or difference is the LCD.

Step 4   Reduce the fraction produced in Step 3 to lowest terms if it is not already in lowest terms.

The idea behind adding or subtracting fractions is really very straight-forward. We can only add or subtract fractions that have the same denominators. If the fractions we are trying to add or subtract do not have the same denominators, we rewrite each of them as an equivalent fraction with the LCD for a denominator.

Here are some additional examples of sums and differences of fractions.

EXAMPLE 7   Subtract $\dfrac{3}{5} - \dfrac{1}{6}$.

SOLUTION   The LCD for 5 and 6 is their product, 30. We begin by rewriting each fraction with this common denominator:

$$\frac{3}{5} - \frac{1}{6} = \frac{3 \cdot 6}{5 \cdot 6} - \frac{1 \cdot 5}{6 \cdot 5}$$

$$= \frac{18}{30} - \frac{5}{30}$$

$$= \frac{13}{30}$$

EXAMPLE 8   Add $\dfrac{1}{6} + \dfrac{1}{8} + \dfrac{1}{4}$.

SOLUTION   We begin by factoring the denominators completely and building the LCD from the factors that result:

$$\left. \begin{aligned} 6 &= 2 \cdot 3 \\ 8 &= 2 \cdot 2 \cdot 2 \\ 4 &= 2 \cdot 2 \end{aligned} \right\} \qquad \text{LCD} = 2 \cdot 2 \cdot 2 \cdot 3 = 24$$

8 divides the LCD
4 divides the LCD   *6 divides the LCD*

We then change to equivalent fractions and add as usual:

$$\frac{1}{6} + \frac{1}{8} + \frac{1}{4} = \frac{1 \cdot 4}{6 \cdot 4} + \frac{1 \cdot 3}{8 \cdot 3} + \frac{1 \cdot 6}{4 \cdot 6} = \frac{4}{24} + \frac{3}{24} + \frac{6}{24} = \frac{13}{24}$$

EXAMPLE 9   Subtract $3 - \dfrac{5}{6}$.

SOLUTION   The denominators are 1 $\left(\text{because } 3 = \frac{3}{1}\right)$ and 6. The smallest number divisible by both 1 and 6 is 6.

$$3 - \frac{5}{6} = \frac{3}{1} - \frac{5}{6} = \frac{3 \cdot 6}{1 \cdot 6} - \frac{5}{6} = \frac{18}{6} - \frac{5}{6} = \frac{13}{6}$$ ∎

Comparing Fractions

As we have shown previously, we can compare fractions to see which is larger or smaller when they have the same denominator. Now that we know how to find the LCD for a set of fractions, we can use the LCD to write equivalent fractions with the intention of comparing them.

EXAMPLE 10   Find the LCD for the fractions below, then write each fraction as an equivalent fraction with the LCD for a denominator. Then write them in order from smallest to largest.

$$\frac{5}{8} \qquad\qquad \frac{5}{16} \qquad\qquad \frac{3}{4} \qquad\qquad \frac{1}{2}$$

SOLUTION   The LCD for the four fractions is 16. We begin by writing each fraction as an equivalent fraction with denominator 16.

$$\frac{5}{8} = \frac{10}{16} \qquad \frac{5}{16} = \frac{5}{16} \qquad \frac{3}{4} = \frac{12}{16} \qquad \frac{1}{2} = \frac{8}{16}$$

Now that they all have the same denominator, the smallest fraction is the one with the smallest numerator, and the largest fraction is the one with the largest numerator. Writing them in order from smallest to largest we have:

$$\frac{5}{16} \quad < \quad \frac{8}{16} \quad < \quad \frac{10}{16} \quad < \quad \frac{12}{16}$$

$$\frac{5}{16} \quad < \quad \frac{1}{2} \quad < \quad \frac{5}{8} \quad < \quad \frac{3}{4}$$ ∎

GETTING READY FOR CLASS

After reading through the preceding section, respond in your own words and in complete sentences.

1. When adding two fractions with the same denominators, we always add their _____, but we never add their _____.

2. What does the abbreviation LCD stand for?

3. What is the first step when finding the LCD for the fractions $\dfrac{5}{12}$ and $\dfrac{7}{18}$?

4. When adding fractions, what is the last step?

Problem Set 3.1

Find the following sums and differences, and reduce to lowest terms. (Add or subtract as indicated.)

1. $\dfrac{3}{6} + \dfrac{1}{6}$

2. $\dfrac{2}{5} + \dfrac{3}{5}$

3. $\dfrac{5}{8} - \dfrac{3}{8}$

4. $\dfrac{6}{7} - \dfrac{1}{7}$

5. $\dfrac{3}{4} - \dfrac{1}{4}$

6. $\dfrac{7}{9} - \dfrac{4}{9}$

7. $\dfrac{2}{3} - \dfrac{1}{3}$

8. $\dfrac{9}{8} - \dfrac{1}{8}$

9. $\dfrac{1}{4} + \dfrac{2}{4} + \dfrac{3}{4}$

10. $\dfrac{2}{5} + \dfrac{3}{5} + \dfrac{4}{5}$

11. $\dfrac{x+7}{2} - \dfrac{1}{2}$

12. $\dfrac{x+5}{4} - \dfrac{3}{4}$

13. $\dfrac{1}{10} + \dfrac{3}{10} + \dfrac{4}{10}$

14. $\dfrac{3}{20} + \dfrac{1}{20} + \dfrac{4}{20}$

15. $\dfrac{1}{3} + \dfrac{4}{3} + \dfrac{5}{3}$

16. $\dfrac{5}{4} + \dfrac{4}{4} + \dfrac{3}{4}$

Complete the following tables.

17.

First Number a	Second Number b	The Sum of a and b $a + b$
$\dfrac{1}{2}$	$\dfrac{1}{3}$	
$\dfrac{1}{3}$	$\dfrac{1}{4}$	
$\dfrac{1}{4}$	$\dfrac{1}{5}$	
$\dfrac{1}{5}$	$\dfrac{1}{6}$	

18.

First Number a	Second Number b	The Sum of a and b $a + b$
1	$\dfrac{1}{2}$	
1	$\dfrac{1}{3}$	
1	$\dfrac{1}{4}$	
1	$\dfrac{1}{5}$	

19.

First Number a	Second Number b	The Sum of a and b $a + b$
$\dfrac{1}{12}$	$\dfrac{1}{2}$	
$\dfrac{1}{12}$	$\dfrac{1}{3}$	
$\dfrac{1}{12}$	$\dfrac{1}{4}$	
$\dfrac{1}{12}$	$\dfrac{1}{6}$	

20.

First Number a	Second Number b	The Sum of a and b $a + b$
$\dfrac{1}{8}$	$\dfrac{1}{2}$	
$\dfrac{1}{8}$	$\dfrac{1}{4}$	
$\dfrac{1}{8}$	$\dfrac{1}{16}$	
$\dfrac{1}{8}$	$\dfrac{1}{24}$	

Find the LCD for each of the following; then use the methods developed in this section to add or subtract as indicated.

21. $\dfrac{4}{9} + \dfrac{1}{3}$

22. $\dfrac{1}{2} + \dfrac{1}{4}$

23. $2 + \dfrac{1}{3}$

24. $3 + \dfrac{1}{2}$

25. $\dfrac{3}{4} + 1$

26. $\dfrac{3}{4} + 2$

27. $\dfrac{1}{2} + \dfrac{2}{3}$

28. $\dfrac{1}{8} + \dfrac{3}{4}$

29. $\dfrac{1}{4} + \dfrac{1}{5}$

30. $\dfrac{1}{3} + \dfrac{1}{5}$

31. $\dfrac{1}{2} + \dfrac{1}{5}$

32. $\dfrac{1}{2} - \dfrac{1}{5}$

33. $\dfrac{5}{12} + \dfrac{3}{8}$

34. $\dfrac{9}{16} + \dfrac{7}{12}$

35. $\dfrac{8}{30} - \dfrac{1}{20}$

36. $\dfrac{9}{40} - \dfrac{1}{30}$

37. $\dfrac{3}{10} + \dfrac{1}{100}$

38. $\dfrac{9}{100} + \dfrac{7}{10}$

39. $\dfrac{10}{36} + \dfrac{9}{48}$

40. $\dfrac{12}{28} + \dfrac{9}{20}$

41. $\dfrac{17}{30} + \dfrac{11}{42}$

42. $\dfrac{19}{42} + \dfrac{13}{70}$

43. $\dfrac{25}{84} + \dfrac{41}{90}$

44. $\dfrac{23}{70} + \dfrac{29}{84}$

45. $\dfrac{13}{126} - \dfrac{13}{180}$

46. $\dfrac{17}{84} - \dfrac{17}{90}$

47. $\dfrac{3}{4} + \dfrac{1}{8} + \dfrac{5}{6}$

48. $\dfrac{3}{8} + \dfrac{2}{5} + \dfrac{1}{4}$

49. $\dfrac{3}{10} + \dfrac{5}{12} + \dfrac{1}{6}$

50. $\dfrac{5}{21} + \dfrac{1}{7} + \dfrac{3}{14}$

51. $\dfrac{1}{2} + \dfrac{1}{3} + \dfrac{1}{4} + \dfrac{1}{6}$

52. $\dfrac{1}{8} + \dfrac{1}{4} + \dfrac{1}{5} + \dfrac{1}{10}$

53. $10 - \dfrac{2}{9}$

54. $9 - \dfrac{3}{5}$

55. $\dfrac{1}{10} + \dfrac{4}{5} - \dfrac{3}{20}$

56. $\dfrac{1}{2} + \dfrac{3}{4} - \dfrac{5}{8}$

57. $\dfrac{1}{4} - \dfrac{1}{8} + \dfrac{1}{2} - \dfrac{3}{8}$

58. $\dfrac{7}{8} - \dfrac{3}{4} + \dfrac{5}{8} - \dfrac{1}{2}$

There are two ways to work the problems below. You can combine the fractions inside the parentheses first and then multiply, or you can apply the distributive property first, then add.

59. $15\left(\dfrac{2}{3} + \dfrac{3}{5}\right)$

60. $15\left(\dfrac{4}{5} - \dfrac{1}{3}\right)$

61. $4\left(\dfrac{1}{2} + \dfrac{1}{4}\right)$

62. $6\left(\dfrac{1}{3} + \dfrac{1}{2}\right)$

63. Write the fractions in order from smallest to largest.

$\dfrac{3}{4} \qquad \dfrac{3}{8} \qquad \dfrac{1}{2} \qquad \dfrac{1}{4}$

64. Write the fractions in order from smallest to largest.

$\dfrac{1}{2} \qquad \dfrac{1}{6} \qquad \dfrac{1}{4} \qquad \dfrac{1}{3}$

65. Find the sum of $\dfrac{3}{7}$, 2, and $\dfrac{1}{9}$.

66. Find the sum of 6, $\dfrac{6}{11}$, and 11.

67. Give the difference of $\dfrac{7}{8}$ and $\dfrac{1}{4}$.

68. Give the difference of $\dfrac{9}{10}$ and $\dfrac{1}{100}$.

Applying the Concepts

Some of the application problems below involve multiplication or division, while others involve addition or subtraction.

69. Capacity One carton of milk contains $\frac{1}{2}$ pint while another contains 4 pints. How much milk is contained in both cartons?

70. Baking A recipe calls for $\frac{2}{3}$ cup of flour and $\frac{3}{4}$ cup of sugar. What is the total amount of flour and sugar called for in the recipe?

71. Budget A family decides that they can spend $\frac{5}{8}$ of their monthly income on house payments. If their monthly income is $2,120, how much can they spend for house payments?

72. Savings A family saves $\frac{3}{16}$ of their income each month. If their monthly income is $1,264, how much do they save each month?

Reading a Pie Chart The pie chart below shows how the students at one of the universities in California are distributed among the different schools at the university. Use the information in the pie chart to answer questions 73 and 74.

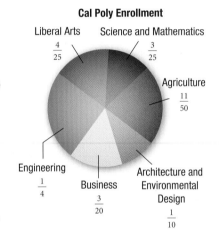

Cal Poly Enrollment

Liberal Arts $\frac{4}{25}$

Science and Mathematics $\frac{3}{25}$

Agriculture $\frac{11}{50}$

Engineering $\frac{1}{4}$

Business $\frac{3}{20}$

Architecture and Environmental Design $\frac{1}{10}$

73. If the students in the Schools of Engineering and Business are combined, what fraction results?

74. What fraction of the university's students are enrolled in the Schools of Agriculture, Engineering, and Business combined?

75. Final Exam Grades The table gives the fraction of students in a class of 40 that received grades of A, B, or C on the final exam. Fill in all the missing parts of the table.

Grade	Number of Students	Fraction of Students
A		$\frac{1}{8}$
B		$\frac{1}{5}$
C		$\frac{1}{2}$
below C		
Total	40	1

76. Flu During a flu epidemic a company with 200 employees has $\frac{1}{10}$ of their employees call in sick on Monday and another $\frac{3}{10}$ call in sick on Tuesday. What is the total number of employees calling in sick during this 2-day period?

77. Subdivision A 6-acre piece of land is subdivided into $\frac{3}{5}$-acre lots. How many lots are there?

78. Cutting Wood A 12-foot piece of wood is cut into shelves. If each is $\frac{3}{4}$ foot in length, how many shelves are there?

Find the perimeter of each figure.

79.

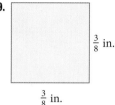

$\frac{3}{8}$ in.

$\frac{3}{8}$ in.

80.

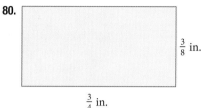

$\frac{3}{8}$ in.

$\frac{3}{4}$ in.

81.

$\frac{3}{10}$ ft

$\frac{3}{5}$ ft

82.

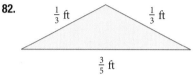

$\frac{1}{3}$ ft $\frac{1}{3}$ ft

$\frac{3}{5}$ ft

Arithmetic Sequences Recall that an arithmetic sequence is a sequence in which each term comes from the previous term by adding the same number each time. For example, the sequence $1, \frac{3}{2}, 2, \frac{5}{2}, \ldots$ is an arithmetic sequence that starts with the number 1. Then each term after that is found by adding $\frac{1}{2}$ to the previous term. By observing this fact, we know that the next term in the sequence will be $\frac{5}{2} + \frac{1}{2} = \frac{6}{2} = 3$.

Find the next number in each arithmetic sequence below.

83. $1, \frac{4}{3}, \frac{5}{3}, 2, \ldots$

84. $1, \frac{5}{4}, \frac{3}{2}, \frac{7}{4}, \ldots$

85. $\frac{3}{2}, 2, \frac{5}{2}, \ldots$

86. $\frac{2}{3}, 1, \frac{4}{3}, \ldots$

Getting Ready for the Next Section

Simplify.

87. $9 \cdot 6 + 5$

88. $4 \cdot 6 + 3$

89. Write 2 as a fraction with denominator 8.

90. Write 2 as a fraction with denominator 4.

91. Write 1 as a fraction with denominator 8.

92. Write 5 as a fraction with denominator 4.

Add.

93. $\frac{8}{4} + \frac{3}{4}$

94. $\frac{16}{8} + \frac{1}{8}$

95. $2 + \frac{1}{8}$

96. $2 + \frac{3}{4}$

97. $1 + \frac{1}{8}$

98. $5 + \frac{3}{4}$

Divide.

99. $11 \div 4$

100. $10 \div 3$

101. $208 \div 24$

102. $207 \div 26$

Mixed-Number Notation

If you are interested in the stock market, you know that, prior to the year 2000, stock prices were given in eighths. For example, on the day I wrote this introduction, one share of Intel Corporation was selling at \$73 $\frac{5}{8}$, or seventy-three and five-eighths dollars. The number $73\frac{5}{8}$ is called a ***mixed number***. It is the sum of a whole number and a proper fraction. With mixed-number notation, we leave out the addition sign.

Notation

A number such as $5\frac{3}{4}$ is called a *mixed number* and is equal to $5 + \frac{3}{4}$. It is simply the sum of the whole number 5 and the proper fraction $\frac{3}{4}$, written without an addition sign. Here are some further examples:

$$2\frac{1}{8} = 2 + \frac{1}{8}, \quad 6\frac{5}{9} = 6 + \frac{5}{9}, \quad 11\frac{2}{3} = 11 + \frac{2}{3}$$

The notation used in writing mixed numbers (writing the whole number and the proper fraction next to each other) must always be interpreted as addition. It is a mistake to read $5\frac{3}{4}$ as meaning 5 times $\frac{3}{4}$. If we want to indicate multiplication, we must use parentheses or a multiplication symbol. That is:

$$5\frac{3}{4} \text{ is not the same as } 5\left(\frac{3}{4}\right)$$

This implies addition   These imply multiplication

$$5\frac{3}{4} \text{ is not the same as } 5 \cdot \frac{3}{4}$$

Changing Mixed Numbers to Improper Fractions

To change a mixed number to an improper fraction, we write the mixed number with the + sign showing and then add the two numbers, as we did earlier.

EXAMPLE 1   Change $2\frac{3}{4}$ to an improper fraction.

SOLUTION

$$2\frac{3}{4} = 2 + \frac{3}{4}$$ Write the mixed number as a sum

$$= \frac{2}{1} + \frac{3}{4}$$ Show that the denominator of 2 is 1

$$= \frac{4 \cdot 2}{4 \cdot 1} + \frac{3}{4}$$ Multiply the numerator and the denominator of $\frac{2}{1}$ by 4 so both fractions will have the same denominator

$$= \frac{8}{4} + \frac{3}{4}$$

$$= \frac{11}{4}$$ Add the numerators; keep the common denominator

The mixed number $2\frac{3}{4}$ is equal to the improper fraction $\frac{11}{4}$. The diagram that follows further illustrates the equivalence of $2\frac{3}{4}$ and $\frac{11}{4}$.

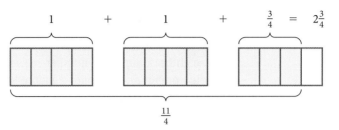

EXAMPLE 2 Change $2\frac{1}{8}$ to an improper fraction.

SOLUTION

$$2\frac{1}{8} = 2 + \frac{1}{8}$$ *Write as addition*

$$= \frac{2}{1} + \frac{1}{8}$$ *Write the whole number 2 as a fraction*

$$= \frac{8 \cdot 2}{8 \cdot 1} + \frac{1}{8}$$ *Change $\frac{2}{1}$ to a fraction with denominator 8*

$$= \frac{16}{8} + \frac{1}{8}$$

$$= \frac{17}{8}$$ *Add the numerators*

If we look closely at Examples 1 and 2, we can see a shortcut that will let us change a mixed number to an improper fraction without so many steps.

Shortcut: To change a mixed number to an improper fraction, simply multiply the whole number by the denominator of the fraction, and add the result to the numerator of the fraction. The result is the numerator of the improper fraction we are looking for. The denominator is the same as the original denominator.

EXAMPLE 3 Use the shortcut to change $5\frac{3}{4}$ to an improper fraction.

SOLUTION **1.** First, we multiply 4×5 to get 20.

2. Next, we add 20 to 3 to get 23.

3. The improper fraction equal to $5\frac{3}{4}$ is $\frac{23}{4}$.

Here is a diagram showing what we have done:

Step 1 Multiply $4 \times 5 = 20$.

$$5\,\frac{3}{4}$$

Step 2 Add $20 + 3 = 23$.

Mathematically, our shortcut is written like this:

$$5\frac{3}{4} = \frac{(4 \cdot 5) + 3}{4} = \frac{20 + 3}{4} = \frac{23}{4}$$ *The result will always have the same denominator as the original mixed number*

The shortcut shown in Example 3 works because the whole-number part of a mixed number can always be written with a denominator of 1. Therefore, the LCD for a whole number and fraction will always be the denominator of the fraction. That is why we multiply the whole number by the denominator of the fraction:

$$5\frac{3}{4} = 5 + \frac{3}{4} = \frac{5}{1} + \frac{3}{4} = \frac{4 \cdot 5}{4 \cdot 1} + \frac{3}{4} = \frac{4 \cdot 5 + 3}{4} = \frac{23}{4}$$

EXAMPLE 4   Change $6\frac{5}{9}$ to an improper fraction.

SOLUTION   Using the first method, we have

$$6\frac{5}{9} = 6 + \frac{5}{9} = \frac{6}{1} + \frac{5}{9} = \frac{9 \cdot 6}{9 \cdot 1} + \frac{5}{9} = \frac{54}{9} + \frac{5}{9} = \frac{59}{9}$$

Using the shortcut method, we have

$$6\frac{5}{9} = \frac{(9 \cdot 6) + 5}{9} = \frac{54 + 5}{9} = \frac{59}{9}$$

Calculator Note

The sequence of keys to press on a calculator to obtain the numerator in Example 4 looks like this:

$9\ \boxed{\times}\ 6\ \boxed{+}\ 5\ \boxed{=}$

Changing Improper Fractions to Mixed Numbers

To change an improper fraction to a mixed number, we divide the numerator by the denominator. The result is used to write the mixed number.

EXAMPLE 5   Change $\frac{11}{4}$ to a mixed number.

SOLUTION   Dividing 11 by 4 gives us

$$\begin{array}{r} 2 \\ 4\overline{)11} \\ \underline{8} \\ 3 \end{array}$$

Note

This division process shows us how many ones are in $\frac{11}{4}$ and, when the ones are taken out, how many fourths are left.

We see that 4 goes into 11 two times with 3 for a remainder. We write this result as

$$\frac{11}{4} = 2 + \frac{3}{4} = 2\frac{3}{4}$$

The improper fraction $\frac{11}{4}$ is equivalent to the mixed number $2\frac{3}{4}$.

An easy way to visualize the results in Example 5 is to imagine having 11 quarters. Your 11 quarters are equivalent to $\frac{11}{4}$ dollars. In dollars, your quarters are worth 2 dollars plus 3 quarters, or $2\frac{3}{4}$ dollars.

EXAMPLE 6   Write as a mixed number.

a. $\dfrac{10}{3}$ **b.** $\dfrac{208}{24}$

SOLUTION

a.
$$\begin{array}{r} 3 \\ 3\overline{)10} \\ \underline{9} \\ 1 \end{array} \quad \text{so} \quad \frac{10}{3} = 3 + \frac{1}{3} = 3\frac{1}{3}$$

b.
$$\begin{array}{r} 8 \\ 24\overline{)208} \\ \underline{192} \\ 16 \end{array} \quad \text{so} \quad \frac{208}{24} = 8 + \frac{16}{24} = 8 + \frac{2}{3} = 8\frac{2}{3}$$

Reduce to lowest terms

Long Division, Remainders, and Mixed Numbers

Mixed numbers give us another way of writing the answers to long division problems that contain remainders. Here is how we divided 1,690 by 67 in Chapter 1:

$$\begin{array}{r} 25\ \text{R } 15 \\ 67\overline{)1{,}690} \\ \underline{1\ 34} \\ 350 \\ \underline{335} \\ 15 \end{array}$$

The answer is 25 with a remainder of 15. Using mixed numbers, we can now write the answer as $25\frac{15}{67}$. That is,

$$\frac{1{,}690}{67} = 25\frac{15}{67}$$

The quotient of 1,690 and 67 is $25\frac{15}{67}$.

GETTING READY FOR CLASS

After reading through the preceding section, respond in your own words and in complete sentences.

1. What is a mixed number?

2. The expression $5\frac{3}{4}$ is equivalent to what addition problem?

3. The improper fraction $\frac{11}{4}$ is equivalent to what mixed number?

4. Why is $\frac{13}{5}$ an improper fraction, but $\frac{3}{5}$ is not an improper fraction?

Problem Set 3.2

Change each mixed number to an improper fraction.

1. $4\dfrac{2}{3}$ **2.** $3\dfrac{5}{8}$ **3.** $5\dfrac{1}{4}$ **4.** $7\dfrac{1}{2}$ **5.** $1\dfrac{5}{8}$ **6.** $1\dfrac{6}{7}$

7. $15\dfrac{2}{3}$ **8.** $17\dfrac{3}{4}$ **9.** $4\dfrac{20}{21}$ **10.** $5\dfrac{18}{19}$ **11.** $12\dfrac{31}{33}$ **12.** $14\dfrac{29}{31}$

Change each improper fraction to a mixed number.

13. $\dfrac{9}{8}$ **14.** $\dfrac{10}{9}$ **15.** $\dfrac{19}{4}$ **16.** $\dfrac{23}{5}$ **17.** $\dfrac{29}{6}$ **18.** $\dfrac{7}{2}$

19. $\dfrac{13}{4}$ **20.** $\dfrac{41}{15}$ **21.** $\dfrac{109}{27}$ **22.** $\dfrac{319}{23}$ **23.** $\dfrac{428}{15}$ **24.** $\dfrac{769}{27}$

Applying the Concepts

25. Sounds Around Us Use the chart shown here to answer the following.

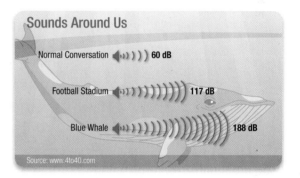

a. The decibel level of a blue whale is what fraction of the decibel level of normal conversation?

b. The decibel level of a football stadium is what fraction of the decibel level of normal conversation?

26. Annual Earnings Use the chart shown here to answer the following questions. Write improper fractions as mixed numbers.

a. The number of The Beatles hits is what fraction of The Beach Boys hits?

b. The number of The Rolling Stones hits is what fraction of The Beach Boys hits?

27. Stocks Suppose a stock is selling on a stock exchange for $5\frac{1}{4}$ dollars per share. If the price increases $\frac{3}{4}$ dollar per share, what is the new price of the stock?

28. Stocks Suppose a stock is selling on a stock exchange for $5\frac{1}{4}$ dollars per share. If the price increases 2 dollars per share, what is the new price of the stock?

29. Height If a man is 71 inches tall, then in feet his height is $5\frac{11}{12}$ feet. Change $5\frac{11}{12}$ to an improper fraction.

30. Height If a woman is 63 inches tall, then her height in feet is $\frac{63}{12}$. Write $\frac{63}{12}$ as a mixed number.

31. Sleeping Infants The table below shows the average number of hours an infant sleeps each day. Construct a line graph from the information in the table.

Age (Months)	Daily Sleep (Hours)
1	$18\frac{1}{2}$
2	17
3	16
4	15
5	$14\frac{1}{2}$
6	14

32. Shoe Sales The table below shows the number of pairs of shoes sold, in various sizes, during a one-day sale.

Size	Number of Pairs
8	5
$8\frac{1}{2}$	7
9	8
$9\frac{1}{2}$	10
10	15
$10\frac{1}{2}$	12
11	6

One Day Only!

SALE!

Available in sizes 8-11, including half sizes

a. Find the mode of the shoe size for all pairs of shoes sold.

b. Find the median shoe size for all pairs of shoes sold. (Do this by looking at the numbers in the second column to decide which shoe size is in the middle of all the sizes of the pairs of shoes sold.)

33. Gasoline Prices The price of unleaded gasoline is $135\frac{9}{10}$¢ per gallon. Write this number as an improper fraction.

34. Gasoline Prices Suppose the price of gasoline is $127\frac{9}{10}$¢ if purchased with a credit card, but 5¢ less if purchased with cash. What is the cash price of the gasoline?

Getting Ready for the Next Section

Change to improper fractions.

35. $2\frac{3}{4}$ **36.** $3\frac{1}{5}$ **37.** $4\frac{5}{8}$ **38.** $1\frac{3}{5}$ **39.** $2\frac{4}{5}$ **40.** $5\frac{9}{10}$

Find the following products. (Multiply.)

41. $\frac{3}{8} \cdot \frac{3}{5}$ **42.** $\frac{11}{4} \cdot \frac{16}{5}$ **43.** $\frac{2}{3}\left(\frac{9}{16}\right)$ **44.** $\frac{7}{10}\left(\frac{5}{21}\right)$

Find the quotients. (Divide.)

45. $\frac{4}{5} \div \frac{7}{8}$ **46.** $\frac{3}{4} \div \frac{1}{2}$ **47.** $\frac{8}{5} \div \frac{14}{5}$ **48.** $\frac{59}{10} \div 2$

Multiplication and Division with Mixed Numbers

The figure here shows one of the nutrition labels we worked with in Chapter 1. It is from a can of Italian tomatoes. Notice toward the top of the label, the number of servings in the can is $3\frac{1}{2}$. The number $3\frac{1}{2}$ is called a *mixed number*. If we want to know how many calories are in the whole can of tomatoes, we must be able to multiply $3\frac{1}{2}$ by 25 (the number of calories per serving). Multiplication with mixed numbers is one of the topics we will cover in this section.

The procedures for multiplying and dividing mixed numbers are the same as those we used in Sections 2.3 and 2.4 to multiply and divide fractions. The only additional work involved is in changing the mixed numbers to improper fractions before we actually multiply or divide.

CANNED ITALIAN TOMATOES

Nutrition Facts	
Serving Size 1/2 cup (121g)	
Servings Per Container: about 3 1/2	

Amount Per Serving	
Calories 25	Calories from fat 0

	% Daily Value*
Total Fat 0g	0%
Saturated Fat 0g	0%
Cholesterol 0mg	0%
Sodium 300mg	12%
Potassium 145mg	4%
Total Carbohydrate 4g	2%
Dietary Fiber 1g	4%
Sugars 4g	
Protein 1g	

Vitamin A 20%	•	Vitamin C 15%
Calcium 4%	•	Iron 15%

*Percent Daily Values are based on a 2,000 calorie diet. Your daily values may be higher or lower depending on your calorie needs.

EXAMPLE 1   Multiply: $2\frac{3}{4} \cdot 3\frac{1}{5}$

SOLUTION   We begin by changing each mixed number to an improper fraction:

$$2\frac{3}{4} = \frac{11}{4} \quad \text{and} \quad 3\frac{1}{5} = \frac{16}{5}$$

Using the resulting improper fractions, we multiply as usual. (That is, we multiply numerators and multiply denominators.)

$$\frac{11}{4} \cdot \frac{16}{5} = \frac{11 \cdot 16}{4 \cdot 5}$$

$$= \frac{11 \cdot \cancel{4} \cdot 4}{\cancel{4} \cdot 5}$$

$$= \frac{44}{5} \quad \text{or} \quad 8\frac{4}{5}$$

EXAMPLE 2   Multiply: $3 \cdot 4\frac{5}{8}$

SOLUTION   Writing each number as an improper fraction, we have

$$3 = \frac{3}{1} \quad \text{and} \quad 4\frac{5}{8} = \frac{37}{8}$$

The complete problem looks like this:

$$3 \cdot 4\frac{5}{8} = \frac{3}{1} \cdot \frac{37}{8} \qquad \textit{Change to improper fractions}$$

$$= \frac{111}{8} \qquad \textit{Multiply numerators and multiply denominators}$$

$$= 13\frac{7}{8} \qquad \textit{Write the answer as a mixed number}$$

Note

As you can see, once you have changed each mixed number to an improper fraction, you multiply the resulting fractions the same way you did in Section 2.3.

Dividing mixed numbers also requires that we change all mixed numbers to improper fractions before we actually do the division.

EXAMPLE 3   Divide: $1\frac{3}{5} \div 2\frac{4}{5}$

SOLUTION   We begin by rewriting each mixed number as an improper fraction:

$$1\frac{3}{5} = \frac{8}{5} \quad \text{and} \quad 2\frac{4}{5} = \frac{14}{5}$$

We then divide using the same method we used in Section 2.4. Remember? We multiply by the reciprocal of the divisor. Here is the complete problem:

$$1\frac{3}{5} \div 2\frac{4}{5} = \frac{8}{5} \div \frac{14}{5} \qquad \text{Change to improper fractions}$$

$$= \frac{8}{5} \cdot \frac{5}{14} \qquad \text{To divide by } \frac{14}{5}, \text{ multiply by } \frac{5}{14}$$

$$= \frac{8 \cdot 5}{5 \cdot 14} \qquad \text{Multiply numerators and multiply denominators}$$

$$= \frac{4 \cdot 2 \cdot 5}{5 \cdot 2 \cdot 7} \qquad \text{Divide out factors common to the numerator and denominator}$$

$$= \frac{4}{7} \qquad \text{Answer in lowest terms}$$

EXAMPLE 4   Divide: $5\frac{9}{10} \div 2$

SOLUTION   We change to improper fractions and proceed as usual:

$$5\frac{9}{10} \div 2 = \frac{59}{10} \div \frac{2}{1} \qquad \text{Write each number as an improper fraction}$$

$$= \frac{59}{10} \cdot \frac{1}{2} \qquad \text{Write division as multiplication by the reciprocal}$$

$$= \frac{59}{20} \qquad \text{Multiply numerators and multiply denominators}$$

$$= 2\frac{19}{20} \qquad \text{Change to a mixed number}$$

GETTING READY FOR CLASS

After reading through the preceding section, respond in your own words and in complete sentences.

1. What is the first step when multiplying or dividing mixed numbers?
2. What is the reciprocal of $2\frac{4}{5}$?
3. Dividing $5\frac{9}{10}$ by 2 is equivalent to multiplying $5\frac{9}{10}$ by what number?

4. Find $4\frac{5}{8}$ of 3.

Problem Set 3.3

Write your answers as proper fractions or mixed numbers, not as improper fractions.

Find the following products. (Multiply.)

1. $3\dfrac{2}{5} \cdot 1\dfrac{1}{2}$ **2.** $2\dfrac{1}{3} \cdot 6\dfrac{3}{4}$ **3.** $5\dfrac{1}{8} \cdot 2\dfrac{2}{3}$ **4.** $1\dfrac{5}{6} \cdot 1\dfrac{4}{5}$

5. $2\dfrac{1}{10} \cdot 3\dfrac{3}{10}$ **6.** $4\dfrac{7}{10} \cdot 3\dfrac{1}{10}$ **7.** $1\dfrac{1}{4} \cdot 4\dfrac{2}{3}$ **8.** $3\dfrac{1}{2} \cdot 2\dfrac{1}{6}$

9. $2 \cdot 4\dfrac{7}{8}$ **10.** $10 \cdot 1\dfrac{1}{4}$ **11.** $\dfrac{3}{5} \cdot 5\dfrac{1}{3}$ **12.** $\dfrac{2}{3} \cdot 4\dfrac{9}{10}$

13. $2\dfrac{1}{2} \cdot 3\dfrac{1}{3} \cdot 1\dfrac{1}{2}$ **14.** $3\dfrac{1}{5} \cdot 5\dfrac{1}{6} \cdot 1\dfrac{1}{8}$ **15.** $\dfrac{3}{4} \cdot 7 \cdot 1\dfrac{4}{5}$ **16.** $\dfrac{7}{8} \cdot 6 \cdot 1\dfrac{5}{6}$

Find the following quotients. (Divide.)

17. $3\dfrac{1}{5} \div 4\dfrac{1}{2}$ **18.** $1\dfrac{4}{5} \div 2\dfrac{5}{6}$ **19.** $6\dfrac{1}{4} \div 3\dfrac{3}{4}$ **20.** $8\dfrac{2}{3} \div 4\dfrac{1}{3}$

21. $10 \div 2\dfrac{1}{2}$ **22.** $12 \div 3\dfrac{1}{6}$ **23.** $8\dfrac{3}{5} \div 2$ **24.** $12\dfrac{6}{7} \div 3$

25. $\left(\dfrac{3}{4} \div 2\dfrac{1}{2}\right) \div 3$ **26.** $\dfrac{7}{8} \div \left(1\dfrac{1}{4} \div 4\right)$

27. $\left(8 \div 1\dfrac{1}{4}\right) \div 2$ **28.** $8 \div \left(1\dfrac{1}{4} \div 2\right)$

29. $2\dfrac{1}{2} \cdot \left(3\dfrac{2}{5} \div 4\right)$ **30.** $4\dfrac{3}{5} \cdot \left(2\dfrac{1}{4} \div 5\right)$

31. Find the product of $2\dfrac{1}{2}$ and 3. **32.** Find the product of $\dfrac{1}{5}$ and $3\dfrac{2}{3}$.

33. What is the quotient of $2\dfrac{3}{4}$ and $3\dfrac{1}{4}$? **34.** What is the quotient of $1\dfrac{1}{5}$ and $2\dfrac{2}{5}$?

Applying the Concepts

35. Cooking A certain recipe calls for $2\dfrac{3}{4}$ cups of sugar. If the recipe is to be doubled, how much sugar should be used?

36. Cooking If a recipe calls for $3\dfrac{1}{2}$ cups of flour, how much flour will be needed if the recipe is tripled?

37. Cooking If a recipe calls for $2\dfrac{1}{2}$ cups of sugar, how much sugar is needed to make $\dfrac{1}{3}$ of the recipe?

38. Cooking A recipe calls for $3\dfrac{1}{4}$ cups of flour. If Diane is using only half the recipe, how much flour should she use?

39. Number Problem Find $\dfrac{3}{4}$ of $1\dfrac{7}{9}$. (Remember that *of* means multiply.)

40. Number Problem Find $\dfrac{5}{6}$ of $2\dfrac{4}{15}$.

41. Cost of Gasoline If a gallon of gas costs $135\dfrac{9}{10}$¢, how much does 8 gallons cost?

42. **Cost of Gasoline** If a gallon of gas costs $159\frac{9}{10}$¢, how much does $\frac{1}{2}$ gallon cost?

43. **Distance Traveled** If a car can travel $32\frac{3}{4}$ miles on a gallon of gas, how far will it travel on 5 gallons of gas?

44. **Distance Traveled** If a new car can travel $20\frac{3}{10}$ miles on 1 gallon of gas, how far can it travel on $\frac{1}{2}$ gallon of gas?

45. **Sewing** If it takes $1\frac{1}{2}$ yards of material to make a pillow cover, how much material will it take to make 3 pillow covers?

46. **Sewing** If the material for the pillow covers in Problem 45 costs $2 a yard, how much will it cost for the material for the 3 pillow covers?

Find the area of each figure.

47.

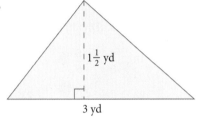

$1\frac{1}{2}$ yd

3 yd

48.

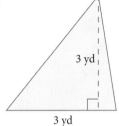

3 yd

3 yd

49. Write the numbers in order from smallest to largest.

$$2\frac{1}{8} \qquad \frac{5}{4} \qquad \frac{3}{4} \qquad 1\frac{1}{2}$$

50. Write the numbers in order from smallest to largest.

$$1\frac{3}{8} \qquad \frac{7}{8} \qquad \frac{7}{4} \qquad 1\frac{11}{16}$$

Nutrition The figure below shows nutrition labels for two different cans of Italian tomatoes.

CANNED TOMATOES 1

Nutrition Facts

Serving Size 1/2 cup (121g)
Servings Per Container: about 3 1/2

Amount Per Serving	
Calories 45	Calories from fat 0

	% Daily Value*
Total Fat 0g	0%
Saturated Fat 0g	0%
Cholesterol 0mg	0%
Sodium 560mg	23%
Total Carbohydrate 11g	4%
Dietary Fiber 2g	8%
Sugars 9g	
Protein 1g	

Vitamin A 10%	•	Vitamin C 25%
Calcium 2%	•	Iron 2%

*Percent Daily Values are based on a 2,000 calorie diet.

CANNED TOMATOES 2

Nutrition Facts

Serving Size 1/2 cup (121g)
Servings Per Container: about 3 1/2

Amount Per Serving	
Calories 25	Calories from fat 0

	% Daily Value*
Total Fat 0g	0%
Saturated Fat 0g	0%
Cholesterol 0mg	0%
Sodium 300mg	12%
Potassium 145mg	4%
Total Carbohydrate 4g	2%
Dietary Fiber 1g	4%
Sugars 4g	
Protein 1g	

Vitamin A 20%	•	Vitamin C 15%
Calcium 4%	•	Iron 15%

*Percent Daily Values are based on a 2,000 calorie diet. Your daily values may be higher or lower depending on your calorie needs.

51. Compare the total number of calories in the two cans of tomatoes.

52. Compare the total amount of sugar in the two cans of tomatoes.

53. Compare the total amount of sodium in the two cans of tomatoes.

54. Compare the total amount of protein in the two cans of tomatoes.

Getting Ready for the Next Section

55. Write as equivalent fractions with denominator 15.

    a. $\dfrac{2}{3}$ **b.** $\dfrac{1}{5}$ **c.** $\dfrac{3}{5}$ **d.** $\dfrac{1}{3}$

56. Write as equivalent fractions with denominator 12.

    a. $\dfrac{3}{4}$ **b.** $\dfrac{1}{3}$ **c.** $\dfrac{5}{6}$ **d.** $\dfrac{1}{4}$

57. Write as equivalent fractions with denominator 20.

    a. $\dfrac{1}{4}$ **b.** $\dfrac{3}{5}$ **c.** $\dfrac{9}{10}$ **d.** $\dfrac{1}{10}$

58. Write as equivalent fractions with denominator 24.

    a. $\dfrac{3}{4}$ **b.** $\dfrac{7}{8}$ **c.** $\dfrac{5}{8}$ **d.** $\dfrac{3}{8}$

Add or subtract the following fractions, as indicated.

59. $\dfrac{2}{3} + \dfrac{1}{5}$ **60.** $\dfrac{3}{4} + \dfrac{5}{6}$ **61.** $\dfrac{2}{3} + \dfrac{8}{9}$ **62.** $\dfrac{1}{4} + \dfrac{3}{5} + \dfrac{9}{10}$

63. $\dfrac{9}{10} - \dfrac{3}{10}$ **64.** $\dfrac{7}{10} - \dfrac{3}{5}$

Addition and Subtraction with Mixed Numbers

3.4

iStockPhoto/©Arthur Kwiatkowski

In March 1995, rumors that Michael Jordan would return to basketball sent stock prices for the companies whose products he endorsed higher. The price of one share of General Mills, the maker of Wheaties, which Michael Jordan endorses, went from $60\frac{1}{2}$ to $63\frac{3}{8}$. To find the increase in the price of this stock, we must be able to subtract mixed numbers.

The notation we use for mixed numbers is especially useful for addition and subtraction. When adding and subtracting mixed numbers, we will assume you recall how to go about finding a least common denominator (LCD). (If you don't remember, then review Section 3.1.)

EXAMPLE 1 Add: $3\frac{2}{3} + 4\frac{1}{5}$

SOLUTION We begin by writing each mixed number showing the + sign. We then apply the commutative and associative properties to rearrange the order and grouping:

$$3\frac{2}{3} + 4\frac{1}{5} = 3 + \frac{2}{3} + 4 + \frac{1}{5}$$ Expand each number to show the + sign

$$= 3 + 4 + \frac{2}{3} + \frac{1}{5}$$ Commutative property

$$= (3 + 4) + \left(\frac{2}{3} + \frac{1}{5}\right)$$ Associative property

$$= 7 + \left(\frac{5 \cdot 2}{5 \cdot 3} + \frac{3 \cdot 1}{3 \cdot 5}\right)$$ Add $3 + 4 = 7$; then multiply to get the LCD

$$= 7 + \left(\frac{10}{15} + \frac{3}{15}\right)$$ Write each fraction with the LCD

$$= 7 + \frac{13}{15}$$ Add the numerators

$$= 7\frac{13}{15}$$ Write the answer in mixed-number notation

As you can see, we obtain our result by adding the whole-number parts $(3 + 4 = 7)$ and the fraction parts $\left(\frac{2}{3} + \frac{1}{5} = \frac{13}{15}\right)$ of each mixed number. Knowing this, we can save ourselves some writing by doing the same problem in columns:

$$3\frac{2}{3} = 3\frac{2 \cdot 5}{3 \cdot 5} = 3\frac{10}{15}$$ Add whole numbers

$$+ 4\frac{1}{5} = 4\frac{1 \cdot 3}{5 \cdot 3} = 4\frac{3}{15}$$ Then add fractions

$$7\frac{13}{15}$$

Write each fraction with LCD 15

The second method shown above requires less writing and lends itself to mixed-number notation. We will use this method for the rest of this section.

EXAMPLE 2   Add: $5\dfrac{3}{4} + 9\dfrac{5}{6}$

SOLUTION   The LCD for 4 and 6 is 12. Writing the mixed numbers in a column and then adding looks like this:

$$5\dfrac{3}{4} = 5\dfrac{3 \cdot 3}{4 \cdot 3} = 5\dfrac{9}{12}$$

$$+\, 9\dfrac{5}{6} = 9\dfrac{5 \cdot 2}{6 \cdot 2} = 9\dfrac{10}{12}$$

$$14\dfrac{19}{12}$$

The fraction part of the answer is an improper fraction. We rewrite it as a whole number and a proper fraction:

$$14\dfrac{19}{12} = 14 + \dfrac{19}{12} \qquad \text{Write the mixed number with a + sign}$$

$$= 14 + 1\dfrac{7}{12} \qquad \text{Write } \tfrac{19}{12} \text{ as a mixed number}$$

$$= 15\dfrac{7}{12} \qquad \text{Add 14 and 1}$$

EXAMPLE 3   Add: $5\dfrac{2}{3} + 6\dfrac{8}{9}$

SOLUTION

$$5\dfrac{2}{3} = 5\dfrac{2 \cdot 3}{3 \cdot 3} = 5\dfrac{6}{9}$$

$$+\, 6\dfrac{8}{9} = 6\dfrac{8}{9} = 6\dfrac{8}{9}$$

$$11\dfrac{14}{9} = 12\dfrac{5}{9}$$

The last step involves writing $\dfrac{14}{9}$ as $1\dfrac{5}{9}$ and then adding 11 and 1 to get 12.

EXAMPLE 4   Add: $3\dfrac{1}{4} + 2\dfrac{3}{5} + 1\dfrac{9}{10}$

SOLUTION   The LCD is 20. We rewrite each fraction as an equivalent fraction with denominator 20 and add:

$$3\frac{1}{4} = 3\frac{1\cdot 5}{4\cdot 5} = 3\frac{5}{20}$$

$$2\frac{3}{5} = 2\frac{3\cdot 4}{5\cdot 4} = 2\frac{12}{20}$$

$$+\,1\frac{9}{10} = 1\frac{9\cdot 2}{10\cdot 2} = 1\frac{18}{20}$$

$$6\frac{35}{20} = 7\frac{15}{20} = 7\frac{3}{4} \qquad \textit{Reduce to lowest terms}$$

$$\frac{35}{20} = 1\frac{15}{20}$$

Change to a mixed number

We should note here that we could have worked each of the first four examples in this section by first changing each mixed number to an improper fraction and then adding as we did in Section 3.1. To illustrate, if we were to work Example 4 this way, it would look like this:

$$3\frac{1}{4} + 2\frac{3}{5} + 1\frac{9}{10} = \frac{13}{4} + \frac{13}{5} + \frac{19}{10} \qquad \textit{Change to improper fractions}$$

$$= \frac{13\cdot 5}{4\cdot 5} + \frac{13\cdot 4}{5\cdot 4} + \frac{19\cdot 2}{10\cdot 2} \qquad \textit{LCD is 20}$$

$$= \frac{65}{20} + \frac{52}{20} + \frac{38}{20} \qquad \textit{Equivalent fractions}$$

$$= \frac{155}{20} \qquad \textit{Add numerators}$$

$$= 7\frac{15}{20} = 7\frac{3}{4} \qquad \textit{Change to a mixed number, and reduce}$$

As you can see, the result is the same as the result we obtained in Example 4.

There are advantages to both methods. The method just shown works well when the whole-number parts of the mixed numbers are small. The vertical method shown in Examples 1–4 works well when the whole-number parts of the mixed numbers are large.

Subtraction with mixed numbers is very similar to addition with mixed numbers.

EXAMPLE 5   Subtract: $3\frac{9}{10} - 1\frac{3}{10}$

SOLUTION   Because the denominators are the same, we simply subtract the whole numbers and subtract the fractions:

$$
\begin{array}{r}
3\dfrac{9}{10} \\[2ex]
-\,1\dfrac{3}{10} \\[1ex]
\hline
2\dfrac{6}{10} = 2\dfrac{3}{5}
\end{array}
\qquad \text{Reduce to lowest terms}
$$

An easy way to visualize the results in Example 5 is to imagine 3 dollar bills and 9 dimes in your pocket. If you spend 1 dollar and 3 dimes, you will have 2 dollars and 6 dimes left.

EXAMPLE 6   Subtract: $12\frac{7}{10} - 8\frac{3}{5}$

SOLUTION   The common denominator is 10. We must rewrite $\frac{3}{5}$ as an equivalent fraction with denominator 10:

$$
\begin{array}{rcccl}
12\dfrac{7}{10} & = & 12\dfrac{7}{10} & = & 12\dfrac{7}{10} \\[2ex]
-\,8\dfrac{3}{5} & = & -\,8\dfrac{3\cdot 2}{5\cdot 2} & = & -\,8\dfrac{6}{10} \\[1ex]
\hline
& & & & 4\dfrac{1}{10}
\end{array}
$$

EXAMPLE 7   Subtract: $10 - 5\frac{2}{7}$

SOLUTION   In order to have a fraction from which to subtract $\frac{2}{7}$, we borrow 1 from 10 and rewrite the 1 we borrow as $\frac{7}{7}$. The process looks like this:

$$
\begin{array}{rcl}
10 = & 9\dfrac{7}{7} & \longleftarrow \text{We rewrite 10 as } 9+1, \text{which is } 9+\tfrac{7}{7}=9\tfrac{7}{7} \\[2ex]
-\,5\dfrac{2}{7} = & -\,5\dfrac{2}{7} & \text{Then we can subtract as usual} \\[1ex]
\hline
& 4\dfrac{5}{7} &
\end{array}
$$

Note

Convince yourself that 10 is the same as $9\frac{7}{7}$. The reason we choose to write the 1 we borrowed as $\frac{7}{7}$ is that the fraction we eventually subtracted from $\frac{7}{7}$ was $\frac{2}{7}$. Both fractions must have the same denominator, 7, so that we can subtract.

EXAMPLE 8   Subtract: $8\frac{1}{4} - 3\frac{3}{4}$

SOLUTION   Because $\frac{3}{4}$ is larger than $\frac{1}{4}$, we again need to borrow 1 from the whole number. The 1 that we borrow from the 8 is rewritten as $\frac{4}{4}$, because 4 is the denominator of both fractions:

$$8\frac{1}{4} = 7\frac{5}{4} \longleftarrow \quad \text{Borrow 1 in the form } \frac{4}{4};$$
$$\text{then } \frac{4}{4} + \frac{1}{4} = \frac{5}{4}$$
$$-3\frac{3}{4} = -3\frac{3}{4}$$
$$\overline{\qquad\qquad 4\frac{2}{4} = 4\frac{1}{2}} \quad \text{Reduce to lowest terms}$$

EXAMPLE 9   Subtract: $4\frac{3}{4} - 1\frac{5}{6}$

SOLUTION   This is about as complicated as it gets with subtraction of mixed numbers. We begin by rewriting each fraction with the common denominator 12:

$$4\frac{3}{4} = 4\frac{3\cdot3}{4\cdot3} = 4\frac{9}{12}$$
$$-1\frac{5}{6} = -1\frac{5\cdot2}{6\cdot2} = -1\frac{10}{12}$$

Because $\frac{10}{12}$ is larger than $\frac{9}{12}$, we must borrow 1 from 4 in the form $\frac{12}{12}$ before we subtract:

$$4\frac{9}{12} = 3\frac{21}{12} \longleftarrow 4 = 3+1 = 3+\frac{12}{12}, \text{ so } 4\frac{9}{12} = \left(3+\frac{12}{12}\right)+\frac{9}{12}$$
$$-1\frac{10}{12} = -1\frac{10}{12} \qquad\qquad\qquad\qquad = 3+\left(\frac{12}{12}+\frac{9}{12}\right)$$
$$\overline{\qquad\qquad 2\frac{11}{12}} \qquad\qquad\qquad\qquad\qquad = 3+\frac{21}{12}$$
$$\qquad\qquad\qquad\qquad\qquad\qquad\qquad = 3\frac{21}{12}$$

GETTING READY FOR CLASS

After reading through the preceding section, respond in your own words and in complete sentences.

1. Is it necessary to "borrow" when subtracting $1\frac{3}{10}$ from $3\frac{9}{10}$?

2. To subtract $1\frac{2}{7}$ from 10 it is necessary to rewrite 10 as what mixed number?

3. To subtract $11\frac{20}{30}$ from $15\frac{3}{30}$ it is necessary to rewrite $15\frac{3}{30}$ as what mixed number?

4. Rewrite $14\frac{19}{12}$ so that the fraction part is a proper fraction instead of an improper fraction.

Problem Set 3.4

Add and subtract the following mixed numbers as indicated.

1. $2\frac{1}{5} + 3\frac{3}{5}$ 2. $8\frac{2}{9} + 1\frac{5}{9}$ 3. $4\frac{3}{10} + 8\frac{1}{10}$ 4. $5\frac{2}{7} + 3\frac{3}{7}$

5. $6\frac{8}{9} - 3\frac{4}{9}$ 6. $12\frac{5}{12} - 7\frac{1}{12}$ 7. $9\frac{1}{6} + 2\frac{5}{6}$ 8. $9\frac{1}{4} + 5\frac{3}{4}$

9. $3\frac{5}{8} - 2\frac{1}{4}$ 10. $7\frac{9}{10} - 6\frac{3}{5}$ 11. $11\frac{1}{3} + 2\frac{5}{6}$ 12. $1\frac{5}{8} + 2\frac{1}{2}$

13. $7\frac{5}{12} - 3\frac{1}{3}$ 14. $7\frac{3}{4} - 3\frac{5}{12}$ 15. $6\frac{1}{3} - 4\frac{1}{4}$ 16. $5\frac{4}{5} - 3\frac{1}{3}$

17. $10\frac{5}{6} + 15\frac{3}{4}$ 18. $11\frac{7}{8} + 9\frac{1}{6}$

19. $5\frac{2}{3}$ 20. $8\frac{5}{6}$ 21. $10\frac{13}{16}$ 22. $17\frac{7}{12}$

$+ 6\frac{1}{3}$ $+ 9\frac{5}{6}$ $- 8\frac{5}{16}$ $- 9\frac{5}{12}$

23. $6\frac{1}{2}$ 24. $9\frac{11}{12}$ 25. $1\frac{5}{8}$ 26. $7\frac{6}{7}$

$+ 2\frac{5}{14}$ $+ 4\frac{1}{6}$ $+ 1\frac{3}{4}$ $+ 2\frac{3}{14}$

27. $4\frac{2}{3}$ 28. $9\frac{4}{9}$ 29. $5\frac{4}{10}$ 30. $12\frac{7}{8}$

$+ 5\frac{3}{5}$ $+ 1\frac{1}{6}$ $- 3\frac{1}{3}$ $- 3\frac{5}{6}$

Find the following sums. (Add.)

31. $1\frac{1}{4} + 2\frac{3}{4} + 5$ 32. $6 + 5\frac{3}{5} + 8\frac{2}{5}$

33. $7\frac{1}{10} + 8\frac{3}{10} + 2\frac{7}{10}$ 34. $5\frac{2}{7} + 8\frac{1}{7} + 3\frac{5}{7}$

35. $\frac{3}{4} + 8\frac{1}{4} + 5$ 36. $\frac{5}{8} + 1\frac{1}{8} + 7$

37. $3\frac{1}{2} + 8\frac{1}{3} + 5\frac{1}{6}$ 38. $4\frac{1}{5} + 7\frac{1}{3} + 8\frac{1}{15}$

39. $8\frac{2}{3}$ 40. $7\frac{3}{5}$ 41. $6\frac{1}{7}$ 42. $1\frac{5}{6}$

$9\frac{1}{8}$ $8\frac{2}{3}$ $9\frac{3}{14}$ $2\frac{3}{4}$

$+ 6\frac{1}{4}$ $+ 1\frac{1}{5}$ $+ 12\frac{1}{2}$ $+ 5\frac{1}{2}$

43. $10\frac{1}{20}$ 44. $18\frac{7}{12}$

$11\frac{4}{5}$ $19\frac{3}{16}$

$+ 15\frac{3}{10}$ $+ 10\frac{2}{3}$

The following problems all involve the concept of borrowing. Subtract in each case.

45. $8 - 1\frac{3}{4}$ **46.** $5 - 3\frac{1}{3}$ **47.** $15 - 5\frac{3}{10}$ **48.** $24 - 10\frac{5}{12}$

49. $8\frac{1}{4} - 2\frac{3}{4}$ **50.** $12\frac{3}{10} - 5\frac{7}{10}$ **51.** $9\frac{1}{3} - 8\frac{2}{3}$ **52.** $7\frac{1}{6} - 6\frac{5}{6}$

53. $4\frac{1}{4} - 2\frac{1}{3}$ **54.** $6\frac{1}{5} - 1\frac{2}{3}$ **55.** $9\frac{2}{3} - 5\frac{3}{4}$ **56.** $12\frac{5}{6} - 8\frac{7}{8}$

57. $16\frac{3}{4} - 10\frac{4}{5}$ **58.** $18\frac{5}{12} - 9\frac{3}{4}$ **59.** $10\frac{3}{10} - 4\frac{4}{5}$ **60.** $9\frac{4}{7} - 7\frac{2}{3}$

61. $13\frac{1}{6} - 12\frac{5}{8}$ **62.** $21\frac{2}{5} - 20\frac{5}{6}$

63. Find the difference between $6\frac{1}{5}$ and $2\frac{7}{10}$.

64. Give the difference between $5\frac{1}{3}$ and $1\frac{5}{6}$.

65. Find the sum of $3\frac{1}{8}$ and $2\frac{3}{5}$.

66. Find the sum of $1\frac{5}{6}$ and $3\frac{4}{9}$.

Applying the Concepts

67. Building Two pieces of molding $5\frac{7}{8}$ inches and $6\frac{3}{8}$ inches long are placed end to end. What is the total length of the two pieces of molding together?

68. Jogging A jogger runs $2\frac{1}{2}$ miles on Monday, $3\frac{1}{4}$ miles on Tuesday, and $2\frac{2}{5}$ miles on Wednesday. What is the jogger's total mileage for this 3-day period?

69. Horse Racing According to the Daily Racing Form, in 2004 the horse New Dreams ran $1\frac{3}{8}$ miles at Churchill Downs, and $1\frac{1}{2}$ miles at Keeneland. How much further did the horse run at Keeneland?

70. Triple Crown The three races that constitute the Triple Crown in horse racing are shown in the table. The information comes from the ESPN website.
   a. Write the distances in order from smallest to largest.
   b. How much longer is the Belmont Stakes race than the Preakness Stakes?

Race	Distance (miles)
Kentucky Derby	$1\frac{1}{4}$
Preakness Stakes	$1\frac{3}{16}$
Belmont Stakes	$1\frac{1}{2}$

71. Length of Jeans A pair of jeans is $32\frac{1}{2}$ inches long. How long are the jeans after they have been washed if they shrink $1\frac{1}{3}$ inches?

72. Manufacturing A clothing manufacturer has two rolls of cloth. One roll is $35\frac{1}{2}$ yards, and the other is $62\frac{5}{8}$ yards. What is the total number of yards in the two rolls?

Area and Perimeter The diagrams below show the dimensions of playing fields for the National Football League (NFL), the Canadian Football League, and arena football.

Football Fields

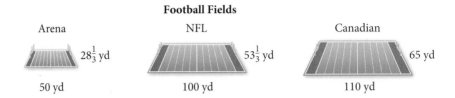

Arena   NFL   Canadian

$28\frac{1}{3}$ yd $53\frac{1}{3}$ yd   65 yd

50 yd   100 yd   110 yd

73. Find the perimeter of each football field.

74. Find the area of each football field.

Stock Prices As we mentioned in the introduction to this section, in March 1995, rumors that Michael Jordan would return to basketball sent stock prices for the companies whose products he endorses higher. The table at the right gives some of the details of those increases. Use the table to work Problems 75–78.

STOCK PRICES FOR COMPANIES WITH MICHAEL JORDAN ENDORSEMENTS			
Company	Product Endorsed	Stock Price (Dollars) 3/8/95	3/13/95
Nike	Air Jordans	$74\frac{7}{8}$	$77\frac{3}{8}$
Quaker Oats	Gatorade	$32\frac{1}{4}$	$32\frac{5}{8}$
General Mills	Wheaties	$60\frac{1}{2}$	$63\frac{3}{8}$
McDonald's		$32\frac{7}{8}$	$34\frac{3}{8}$

75. a. Find the difference in the price of Nike stock between March 13 and March 8.
    b. If you owned 100 shares of Nike stock, how much more are the 100 shares worth on March 13 than on March 8?

76. a. Find the difference in price of General Mills stock between March 13 and March 8.
    b. If you owned 1,000 shares of General Mills stock on March 8, how much more would they be worth on March 13?

77. If you owned 200 shares of McDonald's stock on March 8, how much more would they be worth on March 13?

78. If you owned 100 shares of McDonald's stock on March 8, how much more would they be worth on March 13?

Getting Ready for the Next Section

Multiply or divide as indicated.

79. $\dfrac{11}{8} \cdot \dfrac{29}{8}$

80. $\dfrac{3}{4} \div \dfrac{5}{6}$

81. $\dfrac{7}{6} \cdot \dfrac{12}{7}$

82. $10\dfrac{1}{3} \div 8\dfrac{2}{3}$

Combine.

83. $\dfrac{3}{4} + \dfrac{5}{8}$

84. $\dfrac{1}{2} + \dfrac{2}{3}$

85. $2\dfrac{3}{8} + 1\dfrac{1}{4}$

86. $3\dfrac{2}{3} + 4\dfrac{1}{3}$

Combinations of Operations and Complex Fractions

Now that we have developed skills with both fractions and mixed numbers, we can simplify expressions that contain both types of numbers.

EXAMPLE 1

Simplify the expression: $5 + \left(2\frac{1}{2}\right)\left(3\frac{2}{3}\right)$

SOLUTION The rule for order of operations indicates that we should multiply $2\frac{1}{2}$ times $3\frac{2}{3}$ and then add 5 to the result:

$$5 + \left(2\frac{1}{2}\right)\left(3\frac{2}{3}\right) = 5 + \left(\frac{5}{2}\right)\left(\frac{11}{3}\right) \qquad \text{Change the mixed numbers to improper fractions}$$

$$= 5 + \frac{55}{6} \qquad \text{Multiply the improper fractions}$$

$$= \frac{30}{6} + \frac{55}{6} \qquad \text{Write 5 as } \tfrac{30}{6} \text{ so both numbers have the same denominator}$$

$$= \frac{85}{6} \qquad \text{Add fractions by adding their numerators}$$

$$= 14\frac{1}{6} \qquad \text{Write the answer as a mixed number}$$

EXAMPLE 2

Simplify: $\left(\frac{3}{4} + \frac{5}{8}\right)\left(2\frac{3}{8} + 1\frac{1}{4}\right)$

SOLUTION We begin by combining the numbers inside the parentheses:

$$\frac{3}{4} + \frac{5}{8} = \frac{3 \cdot 2}{4 \cdot 2} + \frac{5}{8} \qquad \text{and} \qquad 2\frac{3}{8} = \ 2\frac{3}{8} \ = \ 2\frac{3}{8}$$

$$= \frac{6}{8} + \frac{5}{8} \qquad\qquad\qquad\qquad + 1\frac{1}{4} = \ + 1\frac{1 \cdot 2}{4 \cdot 2} = \ + 1\frac{2}{8}$$

$$= \frac{11}{8} \qquad\qquad\qquad\qquad\qquad\qquad\qquad\qquad\qquad 3\frac{5}{8}$$

Now that we have combined the expressions inside the parentheses, we can complete the problem by multiplying the results:

$$\left(\frac{3}{4} + \frac{5}{8}\right)\left(2\frac{3}{8} + 1\frac{1}{4}\right) = \left(\frac{11}{8}\right)\left(3\frac{5}{8}\right)$$

$$= \frac{11}{8} \cdot \frac{29}{8} \qquad \text{Change } 3\tfrac{5}{8} \text{ to an improper fraction}$$

$$= \frac{319}{64} \qquad \text{Multiply fractions}$$

$$= 4\frac{63}{64} \qquad \text{Write the answer as a mixed number}$$

EXAMPLE 3   Simplify: $\dfrac{3}{5} + \dfrac{1}{2}\left(3\dfrac{2}{3} + 4\dfrac{1}{3}\right)^2$

SOLUTION   We begin by combining the expressions inside the parentheses:

$$\dfrac{3}{5} + \dfrac{1}{2}\left(3\dfrac{2}{3} + 4\dfrac{1}{3}\right)^2 = \dfrac{3}{5} + \dfrac{1}{2}(8)^2 \qquad \text{The sum inside the parentheses is 8}$$

$$= \dfrac{3}{5} + \dfrac{1}{2}(64) \qquad \text{The square of 8 is 64}$$

$$= \dfrac{3}{5} + 32 \qquad \tfrac{1}{2}\text{ of 64 is 32}$$

$$= 32\dfrac{3}{5} \qquad \text{The result is a mixed number}$$

Complex Fractions

> **DEFINITION**
>
> A *complex fraction* is a fraction in which the numerator and/or the denominator are themselves fractions or combinations of fractions.

Each of the following is a complex fraction:

$$\dfrac{\frac{3}{4}}{\frac{5}{6}}, \quad \dfrac{3+\frac{1}{2}}{2-\frac{3}{4}}, \quad \dfrac{\frac{1}{2}+\frac{2}{3}}{\frac{3}{4}-\frac{1}{6}}$$

EXAMPLE 4   Simplify: $\dfrac{\frac{3}{4}}{\frac{5}{6}}$

SOLUTION   This is actually the same as the problem $\frac{3}{4} \div \frac{5}{6}$, because the bar between $\frac{3}{4}$ and $\frac{5}{6}$ indicates division. Therefore, it must be true that

$$\dfrac{\frac{3}{4}}{\frac{5}{6}} = \dfrac{3}{4} \div \dfrac{5}{6}$$

$$= \dfrac{3}{4} \cdot \dfrac{6}{5}$$

$$= \dfrac{18}{20}$$

$$= \dfrac{9}{10}$$

As you can see, we continue to use properties we have developed previously when we encounter new situations. In Example 4 we use the fact that division by a number and multiplication by its reciprocal produce the same result. We are taking a new problem, simplifying a complex fraction, and thinking of it in terms of a problem we have done previously, division by a fraction.

EXAMPLE 5   Simplify: $\dfrac{\frac{1}{2} + \frac{2}{3}}{\frac{3}{4} - \frac{1}{6}}$

Note

We are going to simplify this complex fraction by two different methods. This is the first method.

SOLUTION   Let's decide to call the numerator of this complex fraction the *top* of the fraction and its denominator the *bottom* of the complex fraction. It will be less confusing if we name them this way. The LCD for all the denominators on the top and bottom is 12, so we can multiply the top and bottom of this complex fraction by 12 and be sure all the denominators will divide it exactly. This will leave us with only whole numbers on the top and bottom:

$$\dfrac{\frac{1}{2} + \frac{2}{3}}{\frac{3}{4} - \frac{1}{6}} = \dfrac{12\left(\frac{1}{2} + \frac{2}{3}\right)}{12\left(\frac{3}{4} - \frac{1}{6}\right)} \qquad \text{Multiply the top and bottom by the LCD}$$

$$= \dfrac{12 \cdot \frac{1}{2} + 12 \cdot \frac{2}{3}}{12 \cdot \frac{3}{4} - 12 \cdot \frac{1}{6}} \qquad \text{Distributive property}$$

$$= \dfrac{6 + 8}{9 - 2} \qquad \text{Multiply each fraction by 12}$$

$$= \dfrac{14}{7} \qquad \text{Add on top and subtract on bottom}$$

$$= 2 \qquad \text{Reduce to lowest terms}$$

Note

The fraction bar that separates the numerator of the complex fraction from its denominator works like parentheses. If we were to rewrite this problem without it, we would write it like this:

$$\left(\frac{1}{2} + \frac{2}{3}\right) \div \left(\frac{3}{4} - \frac{1}{6}\right)$$

That is why we simplify the top and bottom of the complex fraction separately and then divide.

The problem can be worked in another way also. We can simplify the top and bottom of the complex fraction separately. Simplifying the top, we have

$$\frac{1}{2} + \frac{2}{3} = \frac{1 \cdot 3}{2 \cdot 3} + \frac{2 \cdot 2}{3 \cdot 2} = \frac{3}{6} + \frac{4}{6} = \frac{7}{6}$$

Simplifying the bottom, we have

$$\frac{3}{4} - \frac{1}{6} = \frac{3 \cdot 3}{4 \cdot 3} - \frac{1 \cdot 2}{6 \cdot 2} = \frac{9}{12} - \frac{2}{12} = \frac{7}{12}$$

We now write the original complex fraction again using the simplified expressions for the top and bottom. Then we proceed as we did in Example 4.

$$\dfrac{\frac{1}{2} + \frac{2}{3}}{\frac{3}{4} - \frac{1}{6}} = \dfrac{\frac{7}{6}}{\frac{7}{12}}$$

$$= \frac{7}{6} \div \frac{7}{12} \qquad \text{The divisor is } \tfrac{7}{12}$$

$$= \frac{7}{6} \cdot \frac{12}{7} \qquad \text{Replace } \tfrac{7}{12} \text{ by its reciprocal and multiply}$$

$$= \frac{7 \cdot 2 \cdot 6}{6 \cdot 7} \qquad \text{Divide out common factors}$$

$$= 2$$

EXAMPLE 6   Simplify: $\dfrac{3 + \dfrac{1}{2}}{2 - \dfrac{3}{4}}$

SOLUTION   The simplest approach here is to multiply both the top and bottom by the LCD for all fractions, which is 4:

$$\frac{3 + \dfrac{1}{2}}{2 - \dfrac{3}{4}} = \frac{4\left(3 + \dfrac{1}{2}\right)}{4\left(2 - \dfrac{3}{4}\right)} \qquad \text{Multiply the top and bottom by 4}$$

$$= \frac{4 \cdot 3 + 4 \cdot \dfrac{1}{2}}{4 \cdot 2 - 4 \cdot \dfrac{3}{4}} \qquad \text{Distributive property}$$

$$= \frac{12 + 2}{8 - 3} \qquad \text{Multiply each number by 4}$$

$$= \frac{14}{5} \qquad \text{Add on top and subtract on bottom}$$

$$= 2\frac{4}{5}$$

EXAMPLE 7   Simplify: $\dfrac{10\dfrac{1}{3}}{8\dfrac{2}{3}}$

SOLUTION   The simplest way to simplify this complex fraction is to think of it as a division problem.

$$\frac{10\dfrac{1}{3}}{8\dfrac{2}{3}} = 10\frac{1}{3} \div 8\frac{2}{3} \qquad \text{Write with a} \div \text{symbol}$$

$$= \frac{31}{3} \div \frac{26}{3} \qquad \text{Change to improper fractions}$$

$$= \frac{31}{3} \cdot \frac{3}{26} \qquad \text{Write in terms of multiplication}$$

$$= \frac{31 \cdot \cancel{3}}{\cancel{3} \cdot 26} \qquad \text{Divide out the common factor 3}$$

$$= \frac{31}{26} = 1\frac{5}{26} \qquad \text{Answer as a mixed number}$$

GETTING READY FOR CLASS

After reading through the preceding section, respond in your own words and in complete sentences.

1. What is a complex fraction?

2. Rewrite $\dfrac{\frac{5}{6}}{\frac{1}{3}}$ as a multiplication problem.

3. True or false? The rules for order of operations tell us to work inside parentheses first.

4. True or false? We find the LCD when we add or subtract fractions, but not when we multiply them.

Problem Set 3.5

Use the rule for order of operations to simplify each of the following.

1. $3 + \left(1\frac{1}{2}\right)\left(2\frac{2}{3}\right)$

2. $7 - \left(1\frac{3}{5}\right)\left(2\frac{1}{2}\right)$

3. $8 - \left(\frac{6}{11}\right)\left(1\frac{5}{6}\right)$

4. $10 + \left(2\frac{4}{5}\right)\left(\frac{5}{7}\right)$

5. $\frac{2}{3}\left(1\frac{1}{2}\right) + \frac{3}{4}\left(1\frac{1}{3}\right)$

6. $\frac{2}{5}\left(2\frac{1}{2}\right) + \frac{5}{8}\left(3\frac{1}{5}\right)$

7. $2\left(1\frac{1}{2}\right) + 5\left(6\frac{2}{5}\right)$

8. $4\left(5\frac{3}{4}\right) + 6\left(3\frac{5}{6}\right)$

9. $\left(\frac{3}{5} + \frac{1}{10}\right)\left(\frac{1}{2} + \frac{3}{4}\right)$

10. $\left(\frac{2}{9} + \frac{1}{3}\right)\left(\frac{1}{5} + \frac{1}{10}\right)$

11. $\left(2 + \frac{2}{3}\right)\left(3 + \frac{1}{8}\right)$

12. $\left(3 - \frac{3}{4}\right)\left(3 + \frac{1}{3}\right)$

13. $\left(1 + \frac{5}{6}\right)\left(1 - \frac{5}{6}\right)$

14. $\left(2 - \frac{1}{4}\right)\left(2 + \frac{1}{4}\right)$

15. $\frac{2}{3} + \frac{1}{3}\left(2\frac{1}{2} + \frac{1}{2}\right)^2$

16. $\frac{3}{5} + \frac{1}{4}\left(2\frac{1}{2} - \frac{1}{2}\right)^3$

17. $2\frac{3}{8} + \frac{1}{2}\left(\frac{1}{3} + \frac{5}{3}\right)^3$

18. $8\frac{2}{3} + \frac{1}{3}\left(\frac{8}{5} + \frac{7}{5}\right)^2$

19. $2\left(\frac{1}{2} + \frac{1}{3}\right) + 3\left(\frac{2}{3} + \frac{1}{4}\right)$

20. $5\left(\frac{1}{5} + \frac{3}{10}\right) + 2\left(\frac{1}{10} + \frac{1}{2}\right)$

Simplify each complex fraction as much as possible.

21. $\dfrac{\frac{2}{3}}{\frac{3}{4}}$

22. $\dfrac{\frac{5}{6}}{\frac{3}{12}}$

23. $\dfrac{\frac{2}{3}}{\frac{4}{3}}$

24. $\dfrac{\frac{7}{9}}{\frac{5}{9}}$

25. $\dfrac{\frac{11}{20}}{\frac{5}{10}}$

26. $\dfrac{\frac{9}{16}}{\frac{3}{4}}$

27. $\dfrac{\frac{1}{2} + \frac{1}{3}}{\frac{1}{2} - \frac{1}{3}}$

28. $\dfrac{\frac{1}{4} + \frac{1}{5}}{\frac{1}{4} - \frac{1}{5}}$

29. $\dfrac{\frac{5}{8} - \frac{1}{4}}{\frac{1}{8} + \frac{1}{2}}$

30. $\dfrac{\frac{3}{4} + \frac{1}{3}}{\frac{2}{3} + \frac{1}{6}}$

31. $\dfrac{\frac{9}{20} - \frac{1}{10}}{\frac{1}{10} + \frac{9}{20}}$

32. $\dfrac{\frac{1}{2} + \frac{2}{3}}{\frac{3}{4} + \frac{5}{6}}$

33. $\dfrac{1 + \frac{2}{3}}{1 - \frac{2}{3}}$

34. $\dfrac{5 - \frac{3}{4}}{2 + \frac{3}{4}}$

35. $\dfrac{2 + \frac{5}{6}}{5 - \frac{1}{3}}$

36. $\dfrac{9 - \frac{11}{5}}{3 + \frac{13}{10}}$

37. $\dfrac{3 + \frac{5}{6}}{1 + \frac{5}{3}}$

38. $\dfrac{10 + \frac{9}{10}}{5 + \frac{4}{5}}$

39. $\dfrac{\frac{1}{3} + \frac{3}{4}}{2 - \frac{1}{6}}$

40. $\dfrac{3 + \frac{5}{2}}{\frac{5}{6} + \frac{1}{4}}$

41. $\dfrac{\frac{5}{6}}{3 + \frac{2}{3}}$

42. $\dfrac{9 - \frac{3}{2}}{\frac{7}{4}}$

Simplify each of the following complex fractions.

43. $\dfrac{2\frac{1}{2} + \frac{1}{2}}{3\frac{3}{5} - \frac{2}{5}}$

44. $\dfrac{5\frac{3}{8} + \frac{5}{8}}{4\frac{1}{4} + 1\frac{3}{4}}$

45. $\dfrac{2 + 1\frac{2}{3}}{3\frac{5}{6} - 1}$

46. $\dfrac{5 + 8\frac{3}{5}}{2\frac{3}{10} + 4}$

47. $\dfrac{3\frac{1}{4} - 2\frac{1}{2}}{5\frac{3}{4} + 1\frac{1}{2}}$

48. $\dfrac{9\frac{3}{8} + 2\frac{5}{8}}{6\frac{1}{2} + 7\frac{1}{2}}$

49. $\dfrac{3\frac{1}{4} + 5\frac{1}{6}}{2\frac{1}{3} + 3\frac{1}{4}}$

50. $\dfrac{8\frac{5}{6} + 1\frac{2}{3}}{7\frac{1}{3} + 2\frac{1}{4}}$

51. $\dfrac{6\frac{2}{3} + 7\frac{3}{4}}{8\frac{1}{2} + 9\frac{7}{8}}$

52. $\dfrac{3\frac{4}{5} - 1\frac{9}{10}}{6\frac{5}{6} - 2\frac{3}{4}}$

53. What is twice the sum of $2\frac{1}{5}$ and $\frac{3}{6}$?

54. Find 3 times the difference of $1\frac{7}{9}$ and $\frac{2}{9}$.

55. Add $5\frac{1}{4}$ to the sum of $\frac{3}{4}$ and 2.

56. Subtract $\frac{7}{8}$ from the product of 2 and $3\frac{1}{2}$.

Applying the Concepts

57. Manufacturing A dress manufacturer usually buys two rolls of cloth, one of $32\frac{1}{2}$ yards and the other of $25\frac{1}{3}$ yards, to fill his weekly orders. If his orders double one week, how much of the cloth should he order? (Give the total yardage.)

58. Body Temperature Suppose your normal body temperature is $98\frac{3}{5}$ degrees Fahrenheit. If your temperature goes up $3\frac{1}{5}$ degrees on Monday and then down $1\frac{4}{5}$ degrees on Tuesday, what is your temperature on Tuesday?

Chapter 3 Summary

Least Common Denominator (LCD) [3.1]

The *least common denominator* (LCD) for a set of denominators is the smallest number that is exactly divisible by each denominator.

EXAMPLES

1. $\dfrac{1}{8} + \dfrac{3}{8} = \dfrac{1+3}{8}$

$= \dfrac{4}{8}$

$= \dfrac{1}{2}$

Addition and Subtraction of Fractions [3.1]

To add (or subtract) two fractions with a common denominator, add (or subtract) numerators and use the common denominator. In symbols: If a, b, and c are numbers with c not equal to 0, then

$$\frac{a}{c} + \frac{b}{c} = \frac{a+b}{c} \quad \text{and} \quad \frac{a}{c} - \frac{b}{c} = \frac{a-b}{c}$$

Mixed-Number Notation [3.2]

A mixed number is the sum of a whole number and a fraction. The + sign is not shown when we write mixed numbers; it is implied. The mixed number $4\frac{2}{3}$ is actually the sum $4 + \frac{2}{3}$.

Changing Mixed Numbers to Improper Fractions [3.2]

2. $4\dfrac{2}{3} = \dfrac{3 \cdot 4 + 2}{3} = \dfrac{14}{3}$

Mixed number   Improper fraction

To change a mixed number to an improper fraction, we write the mixed number showing the + sign and add as usual. The result is the same if we multiply the denominator of the fraction by the whole number and add what we get to the numerator of the fraction, putting this result over the denominator of the fraction.

Changing an Improper Fraction to a Mixed Number [3.2]

3. Change $\dfrac{14}{3}$ to a mixed number.

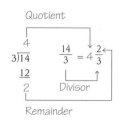

To change an improper fraction to a mixed number, divide the denominator into the numerator. The quotient is the whole-number part of the mixed number. The fraction part is the remainder over the divisor.

Multiplication and Division with Mixed Numbers [3.3]

4. $2\frac{1}{3} \cdot 1\frac{3}{4} = \frac{7}{3} \cdot \frac{7}{4} = \frac{49}{12} = 4\frac{1}{12}$

To multiply or divide two mixed numbers, change each to an improper fraction and multiply or divide as usual.

Addition and Subtraction with Mixed Numbers [3.4]

5.
$$3\frac{4}{9} = 3\frac{4}{9} = 3\frac{4}{9}$$
$$+2\frac{2}{3} = 2\frac{2 \cdot 3}{3 \cdot 3} = 2\frac{6}{9}$$
$$5\frac{10}{9} = 6\frac{1}{9}$$

Common denominator

Add fractions

Add whole numbers

To add or subtract two mixed numbers, add or subtract the whole-number parts and the fraction parts separately. This is best done with the numbers written in columns.

Borrowing in Subtraction with Mixed Numbers [3.4]

6.
$$4\frac{1}{3} = 4\frac{2}{6} = 3\frac{8}{6}$$
$$-1\frac{5}{6} = -1\frac{5}{6} = -1\frac{5}{6}$$
$$2\frac{3}{6} = 2\frac{1}{2}$$

It is sometimes necessary to borrow when doing subtraction with mixed numbers. We always change to a common denominator before we actually borrow.

Complex Fractions [3.5]

7.
$$\frac{4+\frac{1}{3}}{2-\frac{5}{6}} = \frac{6\left(4+\frac{1}{3}\right)}{6\left(2-\frac{5}{6}\right)}$$
$$= \frac{6 \cdot 4 + 6 \cdot \frac{1}{3}}{6 \cdot 2 - 6 \cdot \frac{5}{6}}$$
$$= \frac{24+2}{12-5}$$
$$= \frac{26}{7} = 3\frac{5}{7}$$

A fraction that contains a fraction in its numerator or denominator is called a *complex fraction*.

Additional Facts about Fractions

1. In some books fractions are called *rational numbers*.
2. Every whole number can be written as a fraction with a denominator of 1.
3. The commutative, associative, and distributive properties are true for fractions.
4. The word *of* as used in the expression "$\frac{2}{3}$ *of* 12" indicates that we are to multiply $\frac{2}{3}$ and 12.
5. Two fractions with the same value are called *equivalent fractions*.

COMMON MISTAKES

1. The most common mistake when working with fractions occurs when we try to add two fractions without using a common denominator. For example,

    $$\frac{2}{3} + \frac{4}{5} \neq \frac{2 + 4}{3 + 5}$$

    If the two fractions we are trying to add don't have the same denominators, then we *must* rewrite each one as an equivalent fraction with a common denominator. *We never add denominators when adding fractions.*

    Note We do *not* need a common denominator when multiplying fractions.

2. A common mistake when working with mixed numbers is to confuse mixed-number notation for multiplication of fractions. The notation $3\frac{2}{5}$ does *not* mean 3 *times* $\frac{2}{5}$. It means 3 *plus* $\frac{2}{5}$.

3. Another mistake occurs when multiplying mixed numbers. The mistake occurs when we don't change the mixed number to an improper fraction before multiplying and instead try to multiply the whole numbers and fractions separately. Like this:

    $$2\frac{1}{2} \cdot 3\frac{1}{3} = (2 \cdot 3) + \left(\frac{1}{2} \cdot \frac{1}{3}\right) \quad \text{Mistake}$$

    $$= 6 + \frac{1}{6}$$

    $$= 6\frac{1}{6}$$

    Remember, the correct way to multiply mixed numbers is to first change to improper fractions and then multiply numerators and multiply denominators. This is correct:

    $$2\frac{1}{2} \cdot 3\frac{1}{3} = \frac{5}{2} \cdot \frac{10}{3} = \frac{50}{6} = 8\frac{2}{6} = 8\frac{1}{3} \quad \text{Correct}$$

Chapter 3 Test Form A

These problems are all taken from examples in your text. Work each problem and check your answers. If you have made a mistake, work the problem again. If you cannot get the correct answer after two tries, look up the correct solution in your text.

1. Add: $\dfrac{3}{4} + \dfrac{1}{6}$

2. Subtract: $\dfrac{3}{5} - \dfrac{1}{6}$

3. Change $6\dfrac{5}{9}$ to an improper fraction

4. Change $\dfrac{11}{4}$ to a mixed number.

5. Multiply: $2\dfrac{3}{4} \cdot 3\dfrac{1}{5}$

6. Divide: $1\dfrac{3}{5} \div 2\dfrac{4}{5}$

7. Add: $5\dfrac{2}{3} + 6\dfrac{8}{9}$

8. Subtract: $10 - 5\dfrac{2}{7}$

9. Simplify: $5 + \left(2\dfrac{1}{2}\right)\left(3\dfrac{2}{3}\right)$

10. Simplify: $\dfrac{3 + \dfrac{1}{2}}{2 - \dfrac{3}{4}}$

Chapter Test, Form B

For an alternate, more comprehensive, chapter test, go to MathTV.com and select the test and summary for this chapter of the textbook. Click the worksheet labeled Chapter 3 Test, Form B to download it.

Decimals

4

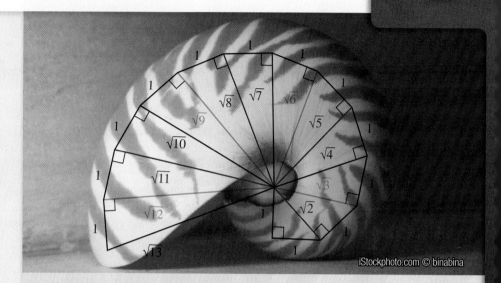

iStockphoto.com © binabina

The diagram shown here is called the *spiral of roots*. It is constructed using the Pythagorean theorem, which is one of the topics we will work with in this chapter. The spiral of roots gives us a way to visualize positive square roots, another of the topics we will cover in this chapter. The table below gives us the decimal equivalents (some of which are approximations) of the first 10 square roots in the spiral. The line graph can be constructed from the table or from the spiral.

As you can see from the preceding diagrams, there is an attractive visual component to square roots. If you think about it, spirals like the one above can be found in a number of places in our everyday world. With square roots, some of these spirals are easy to model with mathematics.

Approximate length of diagonals

Number	Positive Square Root
1	1
2	1.41
3	1.73
4	2
5	2.24
6	2.45
7	2.65
8	2.83
9	3
10	3.16

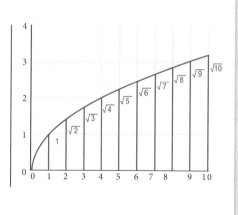

Study Skills

The study skills for this chapter are concerned with getting ready to take an exam.

1. **Getting Ready to Take an Exam** Try to arrange your daily study habits so you have little studying to do the night before your next exam. The next two goals will help you achieve goal number 1.

2. **Review With the Exam in Mind** You should review material that will be covered on the next exam every day. Your review should consist of working problems. Preferably, the problems you work should be problems from your list of difficult problems.

3. **Continue to List Difficult Problems** You should continue to list and rework the problems that give you the most trouble. It is this list that you will use to study for the next exam. Your goal is to go into that test knowing you can successfully work any problem from your list of hard problems.

4. **Pay Attention to Instructions** Taking a test is not like doing homework. On an exam, the problems will be varied. When you do your homework, you usually work a number of similar problems. I have some students who do very well on their homework, but become confused when they see the same problems on a test. The reason for their confusion is that they have not paid attention to the instructions on their homework. If an exam problem asks for the *mean* of some numbers, then you must know the definition of the word *mean*. Likewise, if an exam problem asks you to find a *sum* and then to *round* your answer to the nearest hundred, then you must know that the word *sum* indicates addition, and after you have added, you must *round* your answer as indicated.

In this chapter we will focus our attention on *decimals*. Anyone who has used money in the United States has worked with decimals already. For example, if you have been paid an hourly wage, such as

$6.25 per hour

Decimal point

you have had experience with decimals. What is interesting and useful about decimals is their relationship to fractions and to powers of ten. The work we have done up to now—especially our work with fractions—can be used to develop the properties of decimal numbers.

In Chapter 1 we developed the idea of place value for the digits in a whole number. At that time we gave the name and the place value of each of the first seven columns in our number system, as follows:

Millions Column	Hundred Thousands Column	Ten Thousands Column	Thousands Column	Hundreds Column	Tens Column	Ones Column
1,000,000	100,000	10,000	1,000	100	10	1

As we move from right to left, we multiply by 10 each time. The value of each column is 10 times the value of the column on its right, with the rightmost column being 1. Up until now we have always looked at place value as increasing by a factor of 10 each time we move one column to the left:

Ten Thousands		Thousands		Hundreds		Tens		Ones
10,000 ←		1,000 ←		100 ←		10 ←		1
	Multiply by 10		Multiply by 10		Multiply by 10		Multiply by 10	

To understand the idea behind decimal numbers, we notice that moving in the opposite direction, from left to right, we *divide* by 10 each time:

Ten Thousands		Thousands		Hundreds		Tens		Ones
10,000 →		1,000 →		100 →		10 →		1
	Divide by 10		Divide by 10		Divide by 10		Divide by 10	

If we keep going to the right, the next column will have to be

$$1 \div 10 = \frac{1}{10} \qquad \text{Tenths}$$

The next one after that will be

$$\frac{1}{10} \div 10 = \frac{1}{10} \cdot \frac{1}{10} = \frac{1}{100} \qquad \textit{Hundredths}$$

After that, we have

$$\frac{1}{100} \div 10 = \frac{1}{100} \cdot \frac{1}{10} = \frac{1}{1,000} \qquad \textit{Thousandths}$$

We could continue this pattern as long as we wanted. We simply divide by 10 to move one column to the right. (And remember, dividing by 10 gives the same result as multiplying by $\frac{1}{10}$.)

To show where the ones column is, we use a *decimal point* between the ones column and the tenths column.

Note

Because the digits to the right of the decimal point have fractional place values, numbers with digits to the right of the decimal point are called *decimal fractions*. In this book we will also call them *decimal numbers*, or simply *decimals* for short.

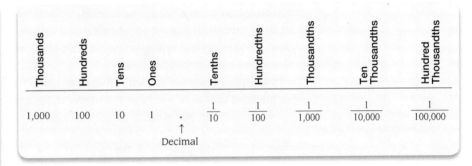

The ones column can be thought of as the middle column, with columns larger than 1 to the left and columns smaller than 1 to the right. The first column to the right of the ones column is the tenths column, the next column to the right is the hundredths column, the next is the thousandths column, and so on. The decimal point is always written between the ones column and the tenths column.

We can use the place value of decimal fractions to write them in expanded form.

EXAMPLE 1   Write 423.576 in expanded form.

SOLUTION $423.576 = 400 + 20 + 3 + \dfrac{5}{10} + \dfrac{7}{100} + \dfrac{6}{1,000}$

EXAMPLE 2   Write each number in words.
  a. 0.4   **b.** 0.04   **c.** 0.004

SOLUTION   **a.** 0.4 is "four tenths."
  b. 0.04 is "four hundredths."
  c. 0.004 is "four thousandths."

When a decimal fraction contains digits to the left of the decimal point, we use the word "and" to indicate where the decimal point is when writing the number in words.

Note

Sometimes we name decimal fractions by simply reading the digits from left to right and using the word "point" to indicate where the decimal point is. For example, using this method the number 5.04 is read "five point zero four."

EXAMPLE 3   Write each number in words.
  a. 5.4   **b.** 5.04   **c.** 5.004

SOLUTION   **a.** 5.4 is "five and four tenths."
  b. 5.04 is "five and four hundredths."
  c. 5.004 is "five and four thousandths."

EXAMPLE 4   Write 3.64 in words.

SOLUTION   The number 3.64 is read "three and sixty-four hundredths." The place values of the digits are as follows:

$$
\begin{array}{ccc}
3 & . & 6 \quad\quad 4 \\
\uparrow & & \uparrow \quad\quad \nwarrow \\
3\ \text{ones} & & 6\ \text{tenths} \quad 4\ \text{hundredths}
\end{array}
$$

We read the decimal part as "sixty-four hundredths" because

$$
6\ \text{tenths} + 4\ \text{hundredths} = \frac{6}{10} + \frac{4}{100} = \frac{60}{100} + \frac{4}{100} = \frac{64}{100}
$$

EXAMPLE 5   Write 25.4936 in words.

SOLUTION   Using the idea given in Example 4, we write 25.4936 in words as "twenty-five and four thousand, nine hundred thirty-six ten thousandths."

In order to understand addition and subtraction of decimals in the next section, we need to be able to convert decimal numbers to fractions or mixed numbers.

EXAMPLE 6   Write each number as a fraction or a mixed number. Do not reduce to lowest terms.

      a. 0.004   **b.** 3.64   **c.** 25.4936

SOLUTION   **a.** Because 0.004 is 4 thousandths, we write

$$
0.004 = \frac{4}{1{,}000}
$$

Three digits after the decimal point   Three zeros

b. Looking over the work in Example 4, we can write

$$
3.64 = 3\frac{64}{100}
$$

Two digits after the decimal point   Two zeros

c. From the way in which we wrote 25.4936 in words in Example 5, we have

$$
25.4936 = 25\frac{4936}{10{,}000}
$$

Four digits after the decimal point   Four zeros

Rounding Decimal Numbers

The rule for rounding decimal numbers is similar to the rule for rounding whole numbers. If the digit in the column to the right of the one we are rounding to is 5 or more, we add 1 to the digit in the column we are rounding to; otherwise, we leave it alone. We then replace all digits to the right of the column we are rounding to with zeros if they are to the left of the decimal point; otherwise, we simply delete them. Table 1 illustrates the procedure.

TABLE 1

| Number | Rounded to the Nearest | | |
	Whole Number	Tenth	Hundredth
24.785	25	24.8	24.79
2.3914	2	2.4	2.39
0.98243	1	1.0	0.98
14.0942	14	14.1	14.09
0.545	1	0.5	0.55

EXAMPLE 7   Round 9,235.492 to the nearest hundred.

SOLUTION   The number next to the hundreds column is 3, which is less than 5. We change all digits to the right to 0, and we can drop all digits to the right of the decimal point, so we write

9,200

EXAMPLE 8   Round 0.0034675 to the nearest ten thousandth.

SOLUTION   Because the number to the right of the ten thousandths column is more than 5, we add 1 to the 4 and get

0.0035

GETTING READY FOR CLASS

After reading through the preceding section, respond in your own words and in complete sentences.

1. Write 754.326 in expanded form.

2. Write $400 + 70 + 5 + \dfrac{1}{10} + \dfrac{3}{100} + \dfrac{7}{1,000}$ in decimal form.

3. Write seventy-two and three tenths in decimal form.

4. How many places to the right of the decimal point is the hundredths column?

Problem Set 4.1

Write out the name of each number in words.

1. 0.3 **2.** 0.03 **3.** 0.015 **4.** 0.0015

5. 3.4 **6.** 2.04 **7.** 52.7 **8.** 46.8

Write each number as a fraction or a mixed number. Do not reduce your answers.

9. 405.36 **10.** 362.78 **11.** 9.009 **12.** 60.06

13. 1.234 **14.** 12.045 **15.** 0.00305 **16.** 2.00106

Give the place value of the 5 in each of the following numbers.

17. 458.327 **18.** 327.458 **19.** 29.52 **20.** 25.92 **21.** 0.00375

22. 0.00532 **23.** 275.01 **24.** 0.356 **25.** 539.76 **26.** 0.123456

Write each of the following as a decimal number.

27. Fifty-five hundredths

28. Two hundred thirty-five ten thousandths

29. Six and nine tenths

30. Forty-five thousand and six hundred twenty-one thousandths

31. Eleven and eleven hundredths

32. Twenty-six thousand, two hundred forty-five and sixteen hundredths

33. One hundred and two hundredths

34. Seventy-five and seventy-five hundred thousandths

35. Three thousand and three thousandths

36. One thousand, one hundred eleven and one hundred eleven thousandths

Complete the following table.

	Rounded to the Nearest			
Number	Whole Number	Tenth	Hundredth	Thousandth
37. 47.5479				
38. 100.9256				
39. 0.8175				
40. 29.9876				
41. 0.1562				
42. 128.9115				
43. 2,789.3241				
44. 0.8743				
45. 99.9999				
46. 71.7634				

Applying the Concepts

1959-1982   1983 - present

47. Penny Weight If you have a penny dated anytime from 1959 through 1982, its original weight was 3.11 grams. If the penny has a date of 1983 or later, the original weight was 2.5 grams. Write the two weights in words.

48. 100 Meters At the 1928 Olympic Games in Amsterdam, the winning time for the women's 100 meters was 12.2 seconds. Since then, the time has continued to get faster. The chart shows the fastest times for the women's 100 meters in the Olympics.

Faster Than...

Florence Griffith Joyner, 1988	**10.49 sec**
Marion Jones, 1998	**10.65 sec**
Christine Arron, 1998	**10.73 sec**
Merlene Ottey, 1996	**10.74 sec**

Source: www.tenmojo.com

   a. What is the place value of the 3 in Christine Arron's time in 1998?

   b. Write Christine Arron's time using words.

NASA

49. Speed of Light The speed of light is 186,282.3976 miles per second. Round this number to the nearest hundredth.

50. Halley's Comet Halley's comet was seen from the earth during 1986. It will be another 76.1 years before it returns. Write 76.1 in words.

51. Nutrition A 50-gram egg contains 0.15 milligram of riboflavin. Write 0.15 in words.

52. Nutrition One medium banana contains 0.64 milligram of B_6. Write 0.64 in words.

53. Gasoline Prices The bar chart below was created from a survey by the U.S. Department of Energy's Energy Information Administration during the month of April 2004. It gives the average price of regular gasoline for the state of California on each Monday of the month. Use the information in the chart to fill in the table.

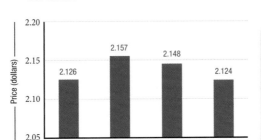

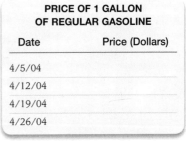

PRICE OF 1 GALLON OF REGULAR GASOLINE

Date	Price (Dollars)
4/5/04	
4/12/04	
4/19/04	
4/26/04	

54. Speed and Time The bar chart below was created from data given by *Car and Driver* magazine. It gives the minimum time in seconds for a Toyota Echo to reach various speeds from a complete stop. Use the information in the chart to fill in the table.

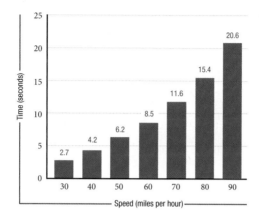

Speed (Miles per Hour)	Time (Seconds)
30	
40	
50	
60	
70	
80	
90	

For each pair of numbers, place the correct symbol, < or >, between the numbers.

55. a. 0.02   0.2

b. 0.3   0.032

56. a. 0.45   0.5

b. 0.5   0.56

57. Write the following numbers in order from smallest to largest.
0.02   0.05   0.025   0.052   0.005   0.002

58. Write the following numbers in order from smallest to largest.
0.2   0.02   0.4   0.04   0.42   0.24

59. Which of the following numbers will round to 7.5?
7.451   7.449   7.54   7.56

60. Which of the following numbers will round to 3.2?
3.14999   3.24999   3.279   3.16111

Change each decimal to a fraction, and then reduce to lowest terms.

61. 0.25   **62.** 0.75   **63.** 0.125   **64.** 0.375

65. 0.625   **66.** 0.0625   **67.** 0.875   **68.** 0.1875

Estimating   For each pair of numbers, choose the number that is closest to 10.

69. 9.9 and 9.99   **70.** 8.5 and 8.05   **71.** 10.5 and 10.05   **72.** 10.9 and 10.99

Estimating   For each pair of numbers, choose the number that is closest to 0.

73. 0.5 and 0.05   **74.** 0.10 and 0.05   **75.** 0.01 and 0.02   **76.** 0.1 and 0.01

Getting Ready for the Next Section

In the next section we will do addition and subtraction with decimals. To understand the process of addition and subtraction, we need to understand the process of addition and subtraction with mixed numbers.

Find each of the following sums and differences. (Add or subtract.)

77. $4\frac{3}{10} + 2\frac{1}{100}$

78. $5\frac{35}{100} + 2\frac{3}{10}$

79. $8\frac{5}{10} - 2\frac{4}{100}$

80. $6\frac{3}{100} - 2\frac{125}{1,000}$

81. $5\frac{1}{10} + 6\frac{2}{100} + 7\frac{3}{1,000}$

82. $4\frac{27}{100} + 6\frac{3}{10} + 7\frac{123}{1,000}$

Addition and Subtraction with Decimals

The chart shows the top recorded times for the women's 100-meter race. In order to analyze the different finishing times, it is important that you are able to add and subtract decimals, and that is what we will cover in this section.

Faster Than...

Florence Griffith Joyner, 1988	10.49 sec
Marion Jones, 1998	10.65 sec
Christine Arron, 1998	10.73 sec
Merlene Ottey, 1996	10.74 sec

Source: www.tenmojo.com

Suppose you are earning $8.50 an hour and you receive a raise of $1.25 an hour. Your new hourly rate of pay is

$$\begin{array}{r} \$8.50 \\ + \ \$1.25 \\ \hline \$9.75 \end{array}$$

To add the two rates of pay, we align the decimal points, and then add in columns.

To see why this is true in general, we can use mixed-number notation:

$$8.50 = 8\frac{50}{100}$$
$$+ \ 1.25 = 1\frac{25}{100}$$
$$\rule{2cm}{0.4pt}$$
$$9\frac{75}{100} = 9.75$$

We can visualize the mathematics above by thinking in terms of money:

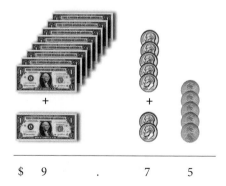

$$\begin{array}{ccccc} \$ & 9 & . & 7 & 5 \end{array}$$

EXAMPLE 1   Add by first changing to fractions: 25.43 + 2.897 + 379.6

SOLUTION   We first change each decimal to a mixed number. We then write each fraction using the least common denominator and add as usual:

$$25.43 = 25\frac{43}{100} = 25\frac{430}{1{,}000}$$

$$2.897 = 2\frac{897}{1{,}000} = 2\frac{897}{1{,}000}$$

$$+\ 379.6 = 379\frac{6}{10} = 379\frac{600}{1{,}000}$$

$$406\frac{1{,}927}{1{,}000} = 407\frac{927}{1{,}000} = 407.927$$

Again, the result is the same if we just line up the decimal points and add as if we were adding whole numbers:

$$\begin{array}{r} 25.430 \\ 2.897 \\ +\ 379.600 \\ \hline 407.927 \end{array}$$

Notice that we can fill in zeros on the right to help keep the numbers in the correct columns. Doing this does not change the value of any of the numbers.

Note: *The decimal point in the answer is directly below the decimal points in the problem*

The same thing would happen if we were to subtract two decimal numbers. We can use these facts to write a rule for addition and subtraction of decimal numbers.

> **DEFINITION**
>
> To add (or subtract) decimal numbers, we line up the decimal points and add (or subtract) as usual. The decimal point in the result is written directly below the decimal points in the problem.

We will use this rule for the rest of the examples in this section.

EXAMPLE 2   Subtract: 39.812 − 14.236

SOLUTION   We write the numbers vertically, with the decimal points lined up, and subtract as usual.

$$\begin{array}{r} 39.812 \\ -\ 14.236 \\ \hline 25.576 \end{array}$$

EXAMPLE 3   Add: 8 + 0.002 + 3.1 + 0.04

SOLUTION   To make sure we keep the digits in the correct columns, we can write zeros to the right of the rightmost digits.

$$\begin{array}{l} 8 = 8.000 \\ 3.1 = 3.100 \\ 0.04 = 0.040 \end{array}$$

Writing the extra zeros here is really equivalent to finding a common denominator for the fractional parts of the original four numbers—now we have a thousandths column in all the numbers

This doesn't change the value of any of the numbers, and it makes our task easier. Now we have

$$
\begin{array}{r}
8.000 \\
0.002 \\
3.100 \\
+\ 0.040 \\
\hline
11.142
\end{array}
$$

EXAMPLE 4   Subtract: $5.9 - 3.0814$

SOLUTION   In this case it is very helpful to write 5.9 as 5.9000, since we will have to re-group in order to subtract.

$$
\begin{array}{r}
5.9000 \\
-\ 3.0814 \\
\hline
2.8186
\end{array}
$$

EXAMPLE 5   Subtract 3.09 from the sum of 9 and 5.472.

SOLUTION   Writing the problem in symbols, we have

$$
\begin{aligned}
(9 + 5.472) - 3.09 &= 14.472 - 3.09 \\
&= 11.382
\end{aligned}
$$

Applications

EXAMPLE 6   While I was writing this section of the book, I stopped to have lunch with a friend at a coffee shop near my office. The bill for lunch was $15.64. I gave the person at the cash register a $20 bill. For change, I received four $1 bills, a quarter, a nickel, and a penny. Was my change correct?

SOLUTION   To find the total amount of money I received in change, we add:

Four $1 bills	=	$4.00
One quarter	=	0.25
One nickel	=	0.05
One penny	=	0.01
Total		= $4.31

To find out if this is the correct amount, we subtract the amount of the bill from $20.00.

$$
\begin{array}{r}
\$20.00 \\
-\ 15.64 \\
\hline
\$\ 4.36
\end{array}
$$

The change was not correct. It is off by 5 cents. Instead of the nickel, I should have been given a dime.

GETTING READY FOR CLASS

After reading through the preceding section, respond in your own words and in complete sentences.

1. When adding numbers with decimals, why is it important to line up the decimal points?

2. Write 379.6 in mixed-number notation.

3. Look at Example 6 in this section of your book. If I had given the person at the cash register a $20 bill and four pennies, how much change should I then have received?

4. How many quarters does the decimal 0.75 represent?

Problem Set 4.2

Find each of the following sums. (Add.)

1. 2.91 + 3.28

2. 8.97 + 2.04

3. 0.04 + 0.31 + 0.78

4. 0.06 + 0.92 + 0.65

5. 3.89 + 2.4

6. 7.65 + 3.8

7. 4.532 + 1.81 + 2.7

8. 9.679 + 3.49 + 6.5

9. 0.081 + 5 + 2.94

10. 0.396 + 7 + 3.96

11. 5.0003 + 6.78 + 0.004

12. 27.0179 + 7.89 + 0.009

13.
7.123
8.12
+ 9.1

14.
5.432
4.32
+ 3.2

15.
9.001
8.01
+ 7.1

16.
6.003
5.02
+ 4.1

17.
89.7854
3.4
65.35
+ 100.006

18.
57.4698
9.89
32.032
+ 572.0079

19.
543.21
+ 123.45

20.
987.654
+ 456.789

Find each of the following differences. (Subtract.)

21. 99.34 − 88.23

22. 47.69 − 36.58

23. 5.97 − 2.4

24. 9.87 − 1.04

25. 6.3 − 2.08

26. 7.5 − 3.04

27. 149.37 − 28.96

28. 796.45 − 32.68

29. 45 − 0.067

30. 48 − 0.075

31. 8 − 0.327

32. 12 − 0.962

33. 765.432 − 234.567

34. 654.321 − 123.456

Subtract.

35.
34.07
− 6.18

36.
25.008
− 3.119

37.
40.04
− 4.4

38.
50.05
− 5.5

39.
768.436
−356.998

40.
495.237
− 247.668

Add and subtract as indicated.

41. (7.8 − 4.3) + 2.5

42. (8.3 − 1.2) + 3.4

43. 7.8 − (4.3 + 2.5)

44. 8.3 − (1.2 + 3.4)

45. (9.7 − 5.2) − 1.4

46. (7.8 − 3.2) − 1.5

47. 9.7 − (5.2 − 1.4)

48. 7.8 − (3.2 − 1.5)

49. Subtract 5 from the sum of 8.2 and 0.072.

50. Subtract 8 from the sum of 9.37 and 2.5.

51. What number is added to 0.035 to obtain 4.036?

52. What number is added to 0.043 to obtain 6.054?

Applying the Concepts

53. Shopping A family buying school clothes for their two children spends $25.37 at one store, $39.41 at another, and $52.04 at a third store. What is the total amount spent at the three stores?

54. Fund Raising A 4-H Club member is raising a lamb to take to the county fair. If she spent $75 for the lamb, $25.60 for feed, and $35.89 for shearing tools, what was the total cost of the project?

55. Take-Home Pay A college professor making $2,105.96 per month has deducted from her check $311.93 for federal income tax, $158.21 for retirement, and $64.72 for state income tax. How much does the professor take home after the deductions have been taken from her monthly income?

56. Take-Home Pay A cook making $1,504.75 a month has deductions of $157.32 for federal income tax, $58.52 for Social Security, and $45.12 for state income tax. How much does the cook take home after the deductions have been taken from his check?

57. Rectangle The logo on a business letter is rectangular. The rectangle has a width of 0.84 inches and a length of 1.41 inches. Find the perimeter.

58. Rectangle A small sticky note is a rectangle. It has a width of 21.4 millimeters and a length of 35.8 millimeters. Find the perimeter.

59. Change A person buys $4.57 worth of candy. If he pays for the candy with a $10 bill, how much change should he receive?

60. Checking Account A checking account contains $342.38. If checks are written for $25.04, $36.71, and $210, how much money is left in the account?

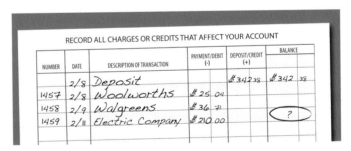

61. Sydney Olympics The chart show the top finishing times for the men's 400-meter freestyle swim during Sydney's Olympics. How much faster was Ian Thorpe than Emiliano Brembilla?

400-meter Freestyle

Ian Thorpe	3:40.59
Massimiliano Rosolino	3:43.40
Klete Keller	3:47.00
Emiliano Brembilla	3:47.01

Final times for the 400-meter freestyle swim.

62. Sydney Olympics The chart shows the top finishing times for the women's 400-meter race during the Sydney Olympics. How much faster was Lorraine Graham than Katharine Merry?

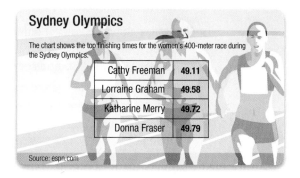

Sydney Olympics

The chart shows the top finishing times for the women's 400-meter race during the Sydney Olympics.

Cathy Freeman	**49.11**
Lorraine Graham	**49.58**
Katharine Merry	**49.72**
Donna Fraser	**49.79**

Source: espn.com

63. Wind Energy Use the chart to answer the following questions.

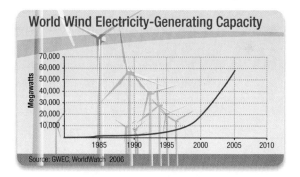

World Wind Electricity-Generating Capacity

Source: GWEC, WorldWatch 2006

   a. Was the capacity for 2005 more or less than 60,000 megawatts?
   b. Was the capacity for 2000 more or less than 15,000 megawatts?
   c. Estimate the increase in capacity from 1995 to 2000.

Minutes and Seconds The chart shows the times of the five fastest runners for the 2005 Fifth Avenue Mile run in New York City. The times are given in minutes and seconds, to the nearest tenth of a second.

Fastest on Fifth

Continental Airlines
Fifth Avenue Mile

Craig Mottram, AUS	**3:49.90**
Alan Webb, USA	**3:51.40**
Elkanah Angwenyi, KEN	**3:54.30**
Anthony Famiglietti, USA	**3:57.10**
Rui Silva, POR	**3:57.40**

Source: www.coolrunning.com, 2005

64. How much faster was Craig Mottram than Rui Silva?

65. How much faster Alan Webb than Elkanah Angwenyi?

66. **Geometry** A rectangle has a perimeter of 9.5 inches. If the length is 2.75 inches, find the width.

67. **Geometry** A rectangle has a perimeter of 11 inches. If the width is 2.5 inches, find the length.

68. **Change** Suppose you eat dinner in a restaurant and the bill comes to $16.76. If you give the cashier a $20 bill and a penny, how much change should you receive? List the bills and coins you should receive for change.

69. **Change** Suppose you buy some tools at the hardware store and the bill comes to $37.87. If you give the cashier two $20 bills and 2 pennies, how much change should you receive? List the bills and coins you should receive for change.

Sequences Find the next number in each sequence.

70. 2.5, 2.75, 3, . . .

71. 3.125, 3.375, 3.625, . . .

Getting Ready for the Next Section

To understand how to multiply decimals, we need to understand multiplication with whole numbers, fractions, and mixed numbers. The following problems review these concepts.

72. $\dfrac{1}{10} \cdot \dfrac{3}{10}$

73. $\dfrac{5}{10} \cdot \dfrac{6}{10}$

74. $\dfrac{3}{100} \cdot \dfrac{17}{100}$

75. $\dfrac{7}{100} \cdot \dfrac{31}{100}$

76. $5\left(\dfrac{3}{10}\right)$

77. $7 \cdot \dfrac{7}{10}$

78. $56 \cdot 25$

79. $39(48)$

80. $\dfrac{5}{10} \times \dfrac{3}{10}$

81. $\dfrac{5}{100} \times \dfrac{3}{1,000}$

82. $2\dfrac{1}{10} \times \dfrac{7}{100}$

83. $3\dfrac{5}{10} \times \dfrac{4}{100}$

84. $305(436)$

85. $403(522)$

86. $5(420 + 3)$

87. $3(550 + 2)$

Multiplication with Decimals

iStockPhoto/©Roberto Caucino

The distance around a circle is called the circumference. If you know the circumference of a bicycle wheel, and you ride the bicycle for one mile, you can calculate how many times the wheel has turned through one complete revolution. In this section we learn how to multiply decimal numbers, and this gives us the information we need to work with circles and their circumferences.

Before we introduce circumference, we need to back up and discuss multiplication with decimals. Suppose that during a half-price sale a calendar that usually sells for $6.42 is priced at $3.21. Therefore it must be true that

$$\frac{1}{2} \text{ of } 6.42 \text{ is } 3.21$$

But, because $\frac{1}{2}$ can be written as 0.5 and *of* translates to *multiply*, we can write this problem again as

$$0.5 \times 6.42 = 3.21$$

If we were to ignore the decimal points in this problem and simply multiply 5 and 642, the result would be 3,210. So, multiplication with decimal numbers is similar to multiplication with whole numbers. The difference lies in deciding where to place the decimal point in the answer. To find out how this is done, we can use fraction notation.

EXAMPLE 1   Change each decimal to a fraction and multiply:

0.5×0.3 To indicate multiplication we are using a $\times$ sign here instead of a dot so we won't confuse the decimal points with the multiplication symbol.

SOLUTION   Changing each decimal to a fraction and multiplying, we have

$$0.5 \times 0.3 = \frac{5}{10} \times \frac{3}{10} \qquad \text{Change to fractions}$$

$$= \frac{15}{100} \qquad \text{Multiply numerators and multiply denominators}$$

$$= 0.15 \qquad \text{Write the answer in decimal form}$$

The result is 0.15, which has two digits to the right of the decimal point.

What we want to do now is find a shortcut that will allow us to multiply decimals without first having to change each decimal number to a fraction. Let's look at another example.

EXAMPLE 2   Change each decimal to a fraction and multiply: 0.05×0.003

SOLUTION $0.05 \times 0.003 = \dfrac{5}{100} \times \dfrac{3}{1,000}$ *Change to fractions*

$$= \dfrac{15}{100,000}$$ *Multiply numerators and multiply denominators*

$$= 0.00015$$ *Write the answer in decimal form*

The result is 0.00015, which has a total of five digits to the right of the decimal point.

Looking over these first two examples, we can see that the digits in the result are just what we would get if we simply forgot about the decimal points and multiplied; that is, $3 \times 5 = 15$. The decimal point in the result is placed so that the total number of digits to its right is the same as the total number of digits to the right of both decimal points in the original two numbers. The reason this is true becomes clear when we look at the denominators after we have changed from decimals to fractions.

EXAMPLE 3   Multiply: 2.1×0.07

SOLUTION $2.1 \times 0.07 = 2\dfrac{1}{10} \times \dfrac{7}{100}$ *Change to fractions*

$$= \dfrac{21}{10} \times \dfrac{7}{100}$$

$$= \dfrac{147}{1,000}$$ *Multiply numerators and multiply denominators*

$$= 0.147$$ *Write the answer as a decimal*

Again, the digits in the answer come from multiplying $21 \times 7 = 147$. The decimal point is placed so that there are three digits to its right, because that is the total number of digits to the right of the decimal points in 2.1 and 0.07.

We summarize this discussion with a rule.

> **RULE:** To multiply two decimal numbers:
>
> **1.** Multiply as you would if the decimal points were not there.
>
> **2.** Place the decimal point in the answer so that the number of digits to its right is equal to the total number of digits to the right of the decimal points in the original two numbers in the problem.

EXAMPLE 4   How many digits will be to the right of the decimal point in the following product?

$$2.987 \times 24.82$$

SOLUTION   There are three digits to the right of the decimal point in 2.987 and two digits to the right in 24.82. Therefore, there will be $3 + 2 = 5$ digits to the right of the decimal point in their product.

EXAMPLE 5   Multiply: 3.05×4.36

SOLUTION   We can set this up as if it were a multiplication problem with whole numbers. We multiply and then place the decimal point in the correct position in the answer.

$$
\begin{array}{r}
3.05 \quad \longleftarrow \text{2 digits to the right of decimal point} \\
\times\,4.36 \quad \longleftarrow \text{2 digits to the right of decimal point} \\
\hline
1830 \\
915 \\
+\,12\,20 \\
\hline
13.2980
\end{array}
$$

The decimal point is placed so that there are $2 + 2 = 4$ digits to its right

As you can see, multiplying decimal numbers is just like multiplying whole numbers, except that we must place the decimal point in the result in the correct position.

Estimating

Look back to Example 5. We could have placed the decimal point in the answer by rounding the two numbers to the nearest whole number and then multiplying them. Because 3.05 rounds to 3 and 4.36 rounds to 4, and the product of 3 and 4 is 12, we estimate that the answer to 3.05×4.36 will be close to 12. We then place the decimal point in the product 132980 between the 3 and the 2 in order to make it into a number close to 12.

EXAMPLE 6   Estimate the answer to each of the following products.
a. 29.4×8.2 **b.** 68.5×172 **c.** $(6.32)^2$

SOLUTION   **a.** Because 29.4 is approximately 30 and 8.2 is approximately 8, we estimate this product to be about $30 \times 8 = 240$. (If we were to multiply 29.4 and 8.2, we would find the product to be exactly 241.08.)

b. Rounding 68.5 to 70 and 172 to 170, we estimate this product to be $70 \times 170 = 11{,}900$. (The exact answer is 11,782.) Note here that we do not always round the numbers to the nearest whole number when making estimates. The idea is to round to numbers that will be easy to multiply.

c. Because 6.32 is approximately 6 and $6^2 = 36$, we estimate our answer to be close to 36. (The actual answer is 39.9424.)

Combined Operations

We can use the rule for order of operations to simplify expressions involving decimal numbers and addition, subtraction, and multiplication.

EXAMPLE 7   Perform the indicated operations: $0.05(4.2 + 0.03)$

SOLUTION   We begin by adding inside the parentheses:

$$0.05(4.2 + 0.03) = 0.05(4.23) \qquad \text{Add}$$
$$= 0.2115 \qquad \text{Multiply}$$

Notice that we could also have used the distributive property first, and the result would be unchanged:

$$0.05(4.2 + 0.03) = 0.05(4.2) + 0.05(0.03) \qquad \text{Distributive property}$$
$$= 0.210 + 0.0015 \qquad \text{Multiply}$$
$$= 0.2115 \qquad \text{Add}$$

EXAMPLE 8   Simplify: $4.8 + 12(3.2)^2$

SOLUTION   According to the rule for order of operations, we must first evaluate the number with an exponent, then multiply, and finally add.

$$4.8 + 12(3.2)^2 = 4.8 + 12(10.24) \qquad (3.2)^2 = 10.24$$
$$= 4.8 + 122.88 \qquad \text{Multiply}$$
$$= 127.68 \qquad \text{Add}$$

Applications

EXAMPLE 9   Sally earns $6.32 for each of the first 36 hours she works in one week and $9.48 in overtime pay for each additional hour she works in the same week. How much money will she make if she works 42 hours in one week?

SOLUTION   The difference between 42 and 36 is 6 hours of overtime pay. The total amount of money she will make is

Pay for the first   Pay for the
36 hours   next 6 hours

$$6.32(36) + 9.48(6) = 227.52 + 56.88$$
$$= 284.40$$

She will make $284.40 for working 42 hours in one week.

> **Note**
>
> To estimate the answer to Example 10 before doing the actual calculations, we would do the following:
>
> $6(40) + 9(6) = 240 + 54 = 294$

GETTING READY FOR CLASS

After reading through the preceding section, respond in your own words and in complete sentences.

1. If you multiply 34.76 and 0.072, how many digits will be to the right of the decimal point in your answer?
2. To simplify the expression $0.053(9) + 67.42$, what would be the first step according to the rule for order of operations?
3. What is the purpose of estimating?
4. What are some applications of decimals that we use in our everyday lives?

Problem Set 4.3

Find each of the following products. (Multiply.)

1. 0.7
$\times\,0.4$

2. 0.8
$\times\,0.3$

3. 0.07
$\times\,0.4$

4. 0.8
$\times\,0.03$

5. 0.03
$\times\,0.09$

6. 0.07
$\times\,0.002$

7. 2.6(0.3)

8. 8.9(0.2)

9. 0.9
$\times\,0.88$

10. 0.8
$\times\,0.99$

11. 3.12
$\times\,0.005$

12. 4.69
$\times\,0.006$

13. 4.003
$\times\,6.07$

14. 7.0001
$\times\quad3.04$

15. 5(0.006)

16. 7(0.005)

17. 75.14
$\times\,2.5$

18. 963.8
$\times\,0.24$

19. 0.1
$\times\,0.02$

20. 0.3
$\times\,0.02$

21. 2.796(10)

22. 97.531(100)

23. 0.0043
$\times\,100$

24. 12.345
$\times\,1{,}000$

25. 49.94
$\times\,1{,}000$

26. 157.02
$\times\,10{,}000$

27. 987.654
$\times\,10{,}000$

28. 1.23
$\times\,100{,}000$

Perform the following operations according to the rule for order of operations.

29. $2.1(3.5 - 2.6)$

30. $5.4(9.9 - 6.6)$

31. $0.05(0.02 + 0.03)$

32. $0.04(0.07 + 0.09)$

33. $2.02(0.03 + 2.5)$

34. $4.04(0.05 + 6.6)$

35. $(2.1 + 0.03)(3.4 + 0.05)$

36. $(9.2 + 0.01)(3.5 + 0.03)$

37. $(2.1 - 0.1)(2.1 + 0.1)$

38. $(9.6 - 0.5)(9.6 + 0.5)$

39. $3.08 - 0.2(5 + 0.03)$

40. $4.09 + 0.5(6 + 0.02)$

41. $4.23 - 5(0.04 + 0.09)$

42. $7.89 - 2(0.31 + 0.76)$

43. $2.5 + 10(4.3)^2$

44. $3.6 + 15(2.1)^2$

45. $100(1 + 0.08)^2$

46. $500(1 + 0.12)^2$

47. $(1.5)^2 + (2.5)^2 + (3.5)^2$

48. $(1.1)^2 + (2.1)^2 + (3.1)^2$

Applying the Concepts

Solve each of the following word problems. Note that not all of the problems are solved by simply multiplying the numbers in the problems. Many of the problems involve addition and subtraction as well as multiplication.

49. Number Problem What is the product of 6 and the sum of 0.001 and 0.02?

50. Number Problem Find the product of 8 and the sum of 0.03 and 0.002.

51. **Number Problem** What does multiplying a decimal number by 100 do to the decimal point?

52. **Number Problem** What does multiplying a decimal number by 1,000 do to the decimal point?

53. **Home Mortgage** On a certain home mortgage, there is a monthly payment of $9.66 for every $1,000 that is borrowed. What is the monthly payment on this type of loan if $143,000 is borrowed?

54. **Caffeine Content** If 1 cup of regular coffee contains 105 milligrams of caffeine, how much caffeine is contained in 3.5 cups of coffee?

55. **Long-Distance Charges** If a phone company charges $0.45 for the first minute and $0.35 for each additional minute for a long-distance call, how much will a 20-minute long-distance call cost?

56. **Price of Gasoline** If gasoline costs $1.37 per gallon when you pay with a credit card, but $0.06 per gallon less if you pay with cash, how much do you save by filling up a 12-gallon tank and paying for it with cash?

57. **Car Rental** Suppose it costs $15 per day and $0.12 per mile to rent a car. What is the total bill if a car is rented for 2 days and is driven 120 miles?

58. **Car Rental** Suppose it costs $20 per day and $0.08 per mile to rent a car. What is the total bill if the car is rented for 2 days and is driven 120 miles?

59. **Wages** A man earns $5.92 for each of the first 36 hours he works in one week and $8.88 in overtime pay for each additional hour he works in the same week. How much money will he make if he works 45 hours in one week?

60. **Wages** A student earns $8.56 for each of the first 40 hours she works in one week and $12.84 in overtime pay for each additional hour she works in the same week. How much money will she make if she works 44 hours in one week?

61. **Rectangle** A rectangle has a width of 33.5 millimeters and a length of 254 millimeters. Find the area.

62. **Rectangle** A rectangle has a width of 2.56 inches and a length of 6.14 inches. Find the area.

63. **Rectangle** The logo on a business letter is rectangular. The rectangle has a width of 0.84 inches and a length of 1.41 inches. Find the area.

64. **Rectangle** A small sticky note is a rectangle. It has a width of 21.4 millimeters and a length of 35.8 millimeters. Find the area.

Getting Ready for the Next Section

To get ready for the next section, which covers division with decimals, we will review division and multiplication with whole numbers and fractions.

Perform each of the following calculations.

65. $3,758 \div 2$ 66. $9,900 \div 22$ 67. $50,032 \div 33$ 68. $90,902 \div 5$

69. $20\overline{)5,960}$ 70. $30\overline{)4,620}$ 71. 4×8.7 72. 5×6.7

73. 27×1.848 74. 35×32.54 75. $38\overline{)31,350}$ 76. $25\overline{)377,800}$

Division with Decimals

Suppose three friends go out for lunch and their total bill comes to $18.75. If they decide to split the bill equally, how much does each person owe? To find out, they will have to divide 18.75 by 3. If you think about the dollars and cents separately, you will see that each person owes $6.25. Therefore,

$$\$18.75 \div 3 = \$6.25$$

In this section we will find out how to do division with any combination of decimal numbers.

EXAMPLE 1   Divide: $5,974 \div 20$

Note

We can estimate the answer to Example 1 by rounding 5,974 to 6,000 and dividing by 20:

$$\frac{6,000}{20} = 300$$

SOLUTION

$$
\begin{array}{r}
298 \\
20\overline{)5,974} \\
\underline{4\ 0} \\
1\ 97 \\
\underline{1\ 80} \\
174 \\
\underline{160} \\
14
\end{array}
$$

In the past we have written this answer as $298\frac{14}{20}$ or, after reducing the fraction, $298\frac{7}{10}$. Because $\frac{7}{10}$ can be written as 0.7, we could also write our answer as 298.7. This last form of our answer is exactly the same result we obtain if we write 5,974 as 5,974.0 and continue the division until we have no remainder. Here is how it looks:

$$
\begin{array}{r}
298.7 \\
20\overline{)5,974.0} \\
\underline{4\ 0} \\
1\ 97 \\
\underline{1\ 80} \\
174 \\
\underline{160} \\
14\ 0 \\
\underline{14\ 0} \\
0
\end{array}
$$

Notice that we place the decimal point in the answer directly above the decimal point in the problem

Let's try another division problem. This time one of the numbers in the problem will be a decimal.

Note

We never need to make a mistake with division, because we can always check our results with multiplication.

EXAMPLE 2   Divide: $34.8 \div 4$

SOLUTION   We can use the ideas from Example 1 and divide as usual. The decimal point in the answer will be placed directly above the decimal point in the problem.

$$
\begin{array}{r}
8.7 \\
4\overline{)34.8} \\
\underline{32}\downarrow \\
2\,8 \\
\underline{2\,8} \\
0
\end{array}
\qquad
\begin{array}{r}
Check: \quad 8.7 \\
\times\quad 4 \\
\hline
34.8
\end{array}
$$

The answer is 8.7.

We can use these facts to write a rule for dividing decimal numbers.

RULE:

To divide a decimal by a whole number, we do the usual long division as if there were no decimal point involved. The decimal point in the answer is placed directly above the decimal point in the problem.

Here are some more examples to illustrate the procedure.

EXAMPLE 3   Divide: $49.896 \div 27$

SOLUTION

$$
\begin{array}{r}
1.848 \\
27\overline{)49.896} \\
\underline{27}\downarrow \\
22\,8 \\
\underline{21\,6}\downarrow \\
1\,29 \\
\underline{1\,08}\downarrow \\
216 \\
\underline{216} \\
0
\end{array}
$$

Check this result by multiplication:

$$
\begin{array}{r}
1.848 \\
\times\quad 27 \\
\hline
12\,936 \\
36\,96 \\
\hline
49.896
\end{array}
$$

We can write as many zeros as we choose after the rightmost digit in a decimal number without changing the value of the number. For example,

$$6.91 = 6.910 = 6.9100 = 6.91000$$

There are times when this can be very useful, as Example 4 shows.

EXAMPLE 4   Divide: $1,138.9 \div 35$

SOLUTION

```
            32.54
      35)1,138.90     Write 0 after the 9. It doesn't
         1 05          change the original number,
            88         but it gives us another digit
            70         to bring down.
           18 9     Check:    32.54
           17 5      ×          35
            1 40             162 70
            1 40             976 2
               0           1,138.90
```

Until now we have considered only division by whole numbers. Extending division to include division by decimal numbers is a matter of knowing what to do about the decimal point in the divisor.

EXAMPLE 5   Divide: $31.35 \div 3.8$

SOLUTION   In fraction form, this problem is equivalent to

$$\frac{31.35}{3.8}$$

If we want to write the divisor as a whole number, we can multiply the numerator and the denominator of this fraction by 10:

$$\frac{31.35 \times 10}{3.8 \times 10} = \frac{313.5}{38}$$

So, since this fraction is equivalent to the original fraction, our original division problem is equivalent to

```
            8.25
      38)313.50     Put 0 after the last digit
         304
          9 5
          7 6
          1 90
          1 90
             0
```

We can summarize division with decimal numbers by listing the following points, as illustrated in the first five examples.

> **Note**
>
> We do not always use the rules for rounding numbers to make estimates. For example, to estimate the answer in Example 5, $31.35 \div 3.8$, we can get a rough estimate of the answer by reasoning that 3.8 is close to 4 and 31.35 is close to 32. Therefore, our answer will be approximately $32 \div 4 = 8$.

Summary of Division with Decimals

1. We divide decimal numbers by the same process used in Chapter 1 to divide whole numbers. The decimal point in the answer is placed directly above the decimal point in the dividend.
2. We are free to write as many zeros after the last digit in a decimal number as we need.
3. If the divisor is a decimal, we can change it to a whole number by moving the decimal point to the right as many places as necessary so long as we move the decimal point in the dividend the same number of places.

EXAMPLE 6   Divide, and round the answer to the nearest hundredth:

$$0.3778 \div 0.25$$

SOLUTION   First, we move the decimal point two places to the right:

$$0.25.\overline{)37.78}$$

Note

Moving the decimal point two places in both the divisor and the dividend is justified like this:

$$\frac{0.3778 \times \mathbf{100}}{0.25 \times \mathbf{100}} = \frac{37.78}{25}$$

Then we divide, using long division:

$$
\begin{array}{r}
1.5112 \\
25\overline{)37.7800} \\
\underline{25} \\
12\,7 \\
\underline{12\,5} \\
28 \\
\underline{25} \\
30 \\
\underline{25} \\
50 \\
\underline{50} \\
0
\end{array}
$$

Rounding to the nearest hundredth, we have 1.51. We actually did not need to have this many digits to round to the hundredths column. We could have stopped at the thousandths column and rounded off.

EXAMPLE 7   Divide, and round to the nearest tenth: $17 \div 0.03$

SOLUTION   Because we are rounding to the nearest tenth, we will continue dividing until we have a digit in the hundredths column. We don't have to go any further to round to the tenths column.

$$
\begin{array}{r}
5\,66.66 \\
0.03.\overline{)17.00.00} \\
\underline{15} \\
2\,0 \\
\underline{1\,8} \\
20 \\
\underline{18} \\
2\,0 \\
\underline{1\,8} \\
20 \\
\underline{18} \\
2
\end{array}
$$

Rounding to the nearest tenth, we have 566.7.

Applications

EXAMPLE 8   If a man earning $5.26 an hour receives a paycheck for $170.95, how many hours did he work?

SOLUTION   To find the number of hours the man worked, we divide $170.95 by $5.26.

$$
\begin{array}{r}
32.5 \\
5.26.\overline{)170.95.0} \\
\underline{157\,8} \\
13\,15 \\
\underline{10\,52} \\
2\,63\,\,0 \\
\underline{2\,63\,\,0} \\
0
\end{array}
$$

The man worked 32.5 hours.

EXAMPLE 9   A telephone company charges $0.43 for the first minute and then $0.33 for each additional minute for a long-distance call. If a long-distance call costs $3.07, how many minutes was the call?

SOLUTION   To solve this problem we need to find the number of additional minutes for the call. To do so, we first subtract the cost of the first minute from the total cost, and then we divide the result by the cost of each additional minute. Without showing the actual arithmetic involved, the solution looks like this:

Total cost
of the call

Cost of the
first minute

$$\text{The number of additional minutes} = \frac{3.07 - 0.43}{0.33} = \frac{2.64}{0.33} = 8$$

Cost of each
additional minute

The call was 9 minutes long. (The number 8 is the number of additional minutes past the first minute.)

DESCRIPTIVE STATISTICS   *Grade Point Average*

I have always been surprised by the number of my students who have difficulty calculating their grade point average (GPA). During her first semester in college, my daughter, Amy, earned the following grades:

Class	Units	Grade
Algebra	5	B
Chemistry	4	C
English	3	A
History	3	B

When her grades arrived in the mail, she told me she had a 3.0 grade point average, because the A and C grades averaged to a B. I told her that her GPA was a little less than a 3.0. What do you think? Can you calculate her GPA? If not, you will be able to after you finish this section.

When you calculate your grade point average (GPA), you are calculating what is called a *weighted average.* To calculate your grade point average, you must first calculate the number of grade points you have earned in each class that you have completed. The number of grade points for a class is the product of the number of units the class is worth times the value of the grade received. The table below shows the value that is assigned to each grade.

Grade	Value
A	4
B	3
C	2
D	1
F	0

If you earn a B in a 4-unit class, you earn $4 \times 3 = 12$ grade points. A grade of C in the same class gives you $4 \times 2 = 8$ grade points. To find your grade point average for one term (a semester or quarter), you must add your grade points and divide that total by the number of units. Round your answer to the nearest hundredth.

EXAMPLE 10   Calculate Amy's grade point average using the information above.

SOLUTION   We begin by writing in two more columns, one for the value of each grade (4 for an A, 3 for a B, 2 for a C, 1 for a D, and 0 for an F), and another for the grade points earned for each class. To fill in the grade points column, we multiply the number of units by the value of the grade:

Class	Units	Grade	Value	Grade Points
Algebra	5	B	3	$5 \times 3 = 15$
Chemistry	4	C	2	$4 \times 2 = 8$
English	3	A	4	$3 \times 4 = 12$
History	3	B	3	$3 \times 3 = 9$
Total Units	15			Total Grade Points: 44

To find her grade point average, we divide 44 by 15 and round (if necessary) to the nearest hundredth:

$$\text{Grade point average} = \frac{44}{15} = 2.93$$

GETTING READY FOR CLASS

After reading through the preceding section, respond in your own words and in complete sentences.

1. The answer to the division problem in Example 1 is $298\frac{14}{20}$. Write this number in decimal notation.

2. In Example 4 we place a 0 at the end of a number without changing the value of the number. Why is the placement of this 0 helpful?

3. The expression $0.3778 \div 0.25$ is equivalent to the expression $37.78 \div 25$ because each number was multiplied by what?

4. Round 372.1675 to the nearest tenth.

Problem Set 4.4

Perform each of the following divisions.

1. $394 \div 20$ **2.** $486 \div 30$ **3.** $248 \div 40$ **4.** $372 \div 80$

5. $5\overline{)26}$ **6.** $8\overline{)36}$ **7.** $25\overline{)276}$ **8.** $50\overline{)276}$

9. $28.8 \div 6$ **10.** $15.5 \div 5$ **11.** $77.6 \div 8$ **12.** $31.48 \div 4$

13. $35\overline{)92.05}$ **14.** $26\overline{)146.38}$ **15.** $45\overline{)190.8}$ **16.** $55\overline{)342.1}$

17. $86.7 \div 34$ **18.** $411.4 \div 44$ **19.** $29.7 \div 22$ **20.** $488.4 \div 88$

21. $4.5\overline{)29.25}$ **22.** $3.3\overline{)21.978}$ **23.** $0.11\overline{)1.089}$ **24.** $0.75\overline{)2.40}$

25. $2.3\overline{)0.115}$ **26.** $6.6\overline{)0.198}$ **27.** $0.012\overline{)1.068}$ **28.** $0.052\overline{)0.23712}$

29. $1.1\overline{)2.42}$ **30.** $2.2\overline{)7.26}$

Carry out each of the following divisions only so far as needed to round the results to the nearest hundredth.

31. $26\overline{)35}$ **32.** $18\overline{)47}$ **33.** $3.3\overline{)56}$ **34.** $4.4\overline{)75}$

35. $0.1234 \div 0.5$ **36.** $0.543 \div 2.1$ **37.** $19 \div 7$ **38.** $16 \div 6$

39. $0.059\overline{)0.69}$ **40.** $0.048\overline{)0.49}$ **41.** $1.99 \div 0.5$ **42.** $0.99 \div 0.5$

43. $2.99 \div 0.5$ **44.** $3.99 \div 0.5$

Applying the Concepts

45. Hot Air Balloon Since the pilot of a hot air balloon can only control the balloon's altitude, he relies on the winds for travel. To ride on the jet streams, a hot air balloon must rise as high as 12 kilometers. Convert this to miles by dividing by 1.61. Round your answer to the nearest tenth of a mile.

46. Hot Air Balloon December and January are the best times for traveling in a hot-air balloon because the jet streams in the Northern Hemisphere are the strongest. They reach speeds of 400 kilometers per hour. Convert this to miles per hour by dividing by 1.61. Round to the nearest whole number.

47. Women's Golf The table below gives the top five money earners for the Ladies' Professional Golf Association (LPGA) in 2005, through September 19. Fill in the last column of the table by finding the average earning per tournament for each golfer. Round your answers to the nearest dollar.

Rank	Name	Number of Tournaments	Total Earnings	Average per Tournament
1.	Annika Sorenstam	14	$1,957,200	
2.	Paula Creamer	20	$1,332,254	
3.	Cristie Kerr	18	$1,297,864	
4.	Lorena Ochoa	18	$1,156,542	
5.	Jeong Jang	21	$950,709	

48. Men's Golf The table below gives the top five money earners for the men's Professional Golf Association (PGA) in 2005, through September 19. Fill in the last column of the table by finding the average earnings per tournament for each golfer. Round your answers to the nearest dollar.

Rank	Name	Number of Tournaments	Total Earnings	Average per Tournament
1.	Tiger Woods	18	$8,613,024	
2.	Vijay Singh	25	$7,463,503	
3.	Phil Mickelson	18	$5,609,025	
4.	David Toms	21	$3,656,213	
5.	Jim Furyk	22	$3,577,435	

49. Wages If a woman earns $33.90 for working 6 hours, how much does she earn per hour?

50. Wages How many hours does a person making $6.78 per hour have to work in order to earn $257.64?

51. Gas Mileage If a car travels 336 miles on 15 gallons of gas, how far will the car travel on 1 gallon of gas?

52. Gas Mileage If a car travels 392 miles on 16 gallons of gas, how far will the car travel on 1 gallon of gas?

53. Wages Suppose a woman earns $6.78 an hour for the first 36 hours she works in a week and then $10.17 an hour in overtime pay for each additional hour she works in the same week. If she makes $294.93 in one week, how many hours did she work overtime?

54. Wages Suppose a woman makes $286.08 in one week. If she is paid $5.96 an hour for the first 36 hours she works and then $8.94 an hour in overtime pay for each additional hour she works in the same week, how many hours did she work overtime that week?

55. Phone Bill Suppose a telephone company charges $0.41 for the first minute and then $0.32 for each additional minute for a long-distance call. If a long-distance call costs $2.33, how many minutes was the call?

56. Phone Bill Suppose a telephone company charges $0.45 for the first three minutes and then $0.29 for each additional minute for a long-distance call. If a long-distance call costs $2.77, how many minutes was the call?

Grade Point Average The following grades were earned by Steve during his first term in college. Use these data to answer Problems 57–60.

Class	Units	Grade
Basic mathematics	3	A
Health	2	B
History	3	B
English	3	C
Chemistry	4	C

57. Calculate Steve's GPA.

58. If his grade in chemistry had been a B instead of a C, by how much would his GPA have increased?

59. If his grade in health had been a C instead of a B, by how much would his grade point average have dropped?

60. If his grades in both English and chemistry had been B's, what would his GPA have been?

Calculator Problems Work each of the following problems on your calculator. If rounding is necessary, round to the nearest hundred thousandth.

61. $7 \div 9$

62. $11 \div 13$

63. $243 \div 0.791$

64. $67.8 \div 37.92$

65. $0.0503 \div 0.0709$

66. $429.87 \div 16.925$

Getting Ready for the Next Section

In the next section we will consider the relationship between fractions and decimals in more detail. The problems below review some of the material that is necessary to make a successful start in the next section.

Reduce to lowest terms.

67. $\dfrac{75}{100}$

68. $\dfrac{220}{1,000}$

69. $\dfrac{12}{18}$

70. $\dfrac{15}{30}$

71. $\dfrac{75}{200}$

72. $\dfrac{220}{2,000}$

73. $\dfrac{38}{100}$

74. $\dfrac{75}{1,000}$

Write each fraction as an equivalent fraction with denominator 10.

75. $\dfrac{3}{5}$

76. $\dfrac{1}{2}$

Write each fraction as an equivalent fraction with denominator 100.

77. $\dfrac{3}{5}$

78. $\dfrac{17}{20}$

Write each fraction as an equivalent fraction with denominator 15.

79. $\dfrac{4}{5}$

80. $\dfrac{2}{3}$

81. $\dfrac{4}{1}$

82. $\dfrac{2}{1}$

83. $\dfrac{6}{5}$

84. $\dfrac{7}{3}$

Divide.

85. $3 \div 4$

87. $7 \div 8$

86. $3 \div 5$

88. $3 \div 8$

Fractions and Decimals

If you are shopping for clothes and a store has a sale advertising $\frac{1}{3}$ off the regular price, how much can you expect to pay for a pair of pants that normally sells for $31.95? If the sale price of the pants is $22.30, have they really been marked down by $\frac{1}{3}$? To answer questions like these, we need to know how to solve problems that involve fractions and decimals together.

We begin this section by showing how to convert back and forth between fractions and decimals.

Converting Fractions to Decimals

You may recall that the notation we use for fractions can be interpreted as implying division. That is, the fraction $\frac{3}{4}$ can be thought of as meaning "3 divided by 4." We can use this idea to convert fractions to decimals.

EXAMPLE 1   Write $\frac{3}{4}$ as a decimal.

SOLUTION   Dividing 3 by 4, we have

$$
\begin{array}{r}
.75 \\
4\overline{)3.00} \\
2\,8\downarrow \\
\hline
20 \\
20 \\
\hline
0
\end{array}
$$

The fraction $\frac{3}{4}$ is equal to the decimal 0.75.

EXAMPLE 2   Write $\frac{7}{12}$ as a decimal correct to the thousandths column.

SOLUTION   Because we want the decimal to be rounded to the thousandths column, we divide to the ten thousandths column and round off to the thousandths column:

$$
\begin{array}{r}
.5833 \\
12\overline{)7.0000} \\
6\,0\downarrow \\
\hline
1\,00 \\
96\downarrow \\
\hline
40 \\
36\downarrow \\
\hline
40 \\
36 \\
\hline
4
\end{array}
$$

Rounding off to the thousandths column, we have 0.583. Because $\frac{7}{12}$ is not exactly the same as 0.583, we write

$$\frac{7}{12} \approx 0.583$$

where the symbol $\approx$ is read "is approximately."

If we wrote more zeros after 0.583 in Example 2, the pattern of 3's would continue for as many places as we could want. When we get a sequence of digits that repeat like this, 0.58333 . . . , we can indicate the repetition by writing

$$0.58\overline{3}$$ *The bar over the 3 indicates that the 3 repeats from there on*

EXAMPLE 3   Write $\frac{3}{11}$ as a decimal.

SOLUTION   Dividing 3 by 11, we have

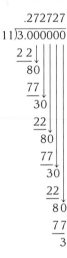

$$\begin{array}{r} .272727 \\ 11\overline{)3.000000} \\ 22 \\ \hline 80 \\ 77 \\ \hline 30 \\ 22 \\ \hline 80 \\ 77 \\ \hline 30 \\ 22 \\ \hline 80 \\ 77 \\ \hline 3 \end{array}$$

No matter how long we continue the division, the remainder will never be 0, and the pattern will continue. We write the decimal form of $\frac{3}{11}$ as $0.\overline{27}$, where

$$0.\overline{27} = 0.272727 \ldots$$ *The dots mean "and so on"*

Note

The bar over the 2 and the 7 in $0.\overline{27}$ is used to indicate that the pattern repeats itself indefinitely.

Converting Decimals to Fractions

To convert decimals to fractions, we take advantage of the place values we assigned to the digits to the right of the decimal point.

EXAMPLE 4   Write 0.38 as a fraction in lowest terms.

SOLUTION   0.38 is 38 hundredths, or

$$0.38 = \frac{38}{100}$$

$$= \frac{19}{50}$$ *Divide the numerator and the denominator by 2 to reduce to lowest terms*

The decimal 0.38 is equal to the fraction $\frac{19}{50}$.

We could check our work here by converting $\frac{19}{50}$ back to a decimal. We do this by dividing 19 by 50. That is,

$$
\begin{array}{r}
.38 \\
50\overline{)19.00} \\
\underline{15\ 0}\!\downarrow \\
4\ 00 \\
\underline{4\ 00} \\
0
\end{array}
$$

EXAMPLE 5   Convert 0.075 to a fraction.

SOLUTION   We have 75 thousandths, or

$$0.075 = \frac{75}{1,000}$$

$$= \frac{3}{40} \qquad \text{\textit{Divide the numerator and the denominator}} \\ \text{\textit{by 25 to reduce to lowest terms}}$$

EXAMPLE 6   Write 15.6 as a mixed number.

SOLUTION   Converting 0.6 to a fraction, we have

$$0.6 = \frac{6}{10} = \frac{3}{5} \qquad \text{\textit{Reduce to lowest terms}}$$

Since $0.6 = \frac{3}{5}$, we have $15.6 = 15\frac{3}{5}$.

Problems Containing Both Fractions and Decimals

We continue this section by working some problems that involve both fractions and decimals.

EXAMPLE 7   Simplify: $\frac{19}{50}(1.32 + 0.48)$

SOLUTION   In Example 4, we found that $0.38 = \frac{19}{50}$. Therefore we can rewrite the problem as

$$\frac{19}{50}(1.32 + 0.48) = 0.38(1.32 + 0.48) \qquad \text{\textit{Convert all numbers to decimals}}$$

$$= 0.38(1.80) \qquad \text{\textit{Add: 1.32 + 0.48}}$$
$$= 0.684 \qquad \text{\textit{Multiply: 0.38 $\times$ 1.80}}$$

EXAMPLE 8   Simplify: $\frac{1}{2} + (0.75)\left(\frac{2}{5}\right)$

SOLUTION   We could do this problem one of two different ways. First, we could convert all fractions to decimals and then simplify:

$$\frac{1}{2} + (0.75)\left(\frac{2}{5}\right) = 0.5 + 0.75(0.4) \qquad \text{\textit{Convert to decimals}}$$

$$= 0.5 + 0.300 \qquad \text{\textit{Multiply: 0.75 $\times$ 0.4}}$$
$$= 0.8 \qquad \text{\textit{Add}}$$

Or, we could convert 0.75 to $\frac{3}{4}$ and then simplify:

$$\frac{1}{2} + 0.75\left(\frac{2}{5}\right) = \frac{1}{2} + \frac{3}{4}\left(\frac{2}{5}\right) \qquad \textit{Convert decimals to fractions}$$

$$= \frac{1}{2} + \frac{3}{10} \qquad \textit{Multiply: } \frac{3}{4} \times \frac{2}{5}$$

$$= \frac{5}{10} + \frac{3}{10} \qquad \textit{The common denominator is 10}$$

$$= \frac{8}{10} \qquad \textit{Add numerators}$$

$$= \frac{4}{5} \qquad \textit{Reduce to lowest terms}$$

The answers are equivalent. That is, $0.8 = \frac{8}{10} = \frac{4}{5}$. Either method can be used with problems of this type.

EXAMPLE 9   Simplify: $\left(\frac{1}{2}\right)^3(2.4) + \left(\frac{1}{4}\right)^2(3.2)$

SOLUTION   This expression can be simplified without any conversions between fractions and decimals. To begin, we evaluate all numbers that contain exponents. Then we multiply. After that, we add.

$$\left(\frac{1}{2}\right)^3(2.4) + \left(\frac{1}{4}\right)^2(3.2) = \frac{1}{8}(2.4) + \frac{1}{16}(3.2) \qquad \textit{Evaluate exponents}$$

$$= 0.3 + 0.2 \qquad \textit{Multiply by } \tfrac{1}{8} \textit{ and } \tfrac{1}{16}$$
$$= 0.5 \qquad \textit{Add}$$

Applications

EXAMPLE 10   If a shirt that normally sells for $27.99 is on sale for $\frac{1}{3}$ off, what is the sale price of the shirt?

SOLUTION   To find out how much the shirt is marked down, we must find $\frac{1}{3}$ of 27.99. That is, we multiply $\frac{1}{3}$ and 27.99, which is the same as dividing 27.99 by 3.

$$\frac{1}{3}(27.99) = \frac{27.99}{3} = 9.33$$

The shirt is marked down $9.33. The sale price $9.33 less than the original price:

$$\text{Sale price} = 27.99 - 9.33 = 18.66$$

The sale price is $18.66. We also could have solved this problem by simply multiplying the original price by $\frac{2}{3}$, since, if the shirt is marked $\frac{1}{3}$ off, then the sale price must be $\frac{2}{3}$ of the original price. Multiplying by $\frac{2}{3}$ is the same as dividing by 3 and then multiplying by 2. The answer would be the same.

GETTING READY FOR CLASS

After reading through the preceding section, respond in your own words and in complete sentences.

1. To convert fractions to decimals, do we multiply or divide the numerator by the denominator?

2. The decimal 0.13 is equivalent to what fraction?

3. Write 36 thousandths in decimal form and in fraction form.

4. Explain how to write the fraction $\dfrac{84}{1,000}$ in lowest terms.

Problem Set 4.5

Each circle below is divided into 8 equal parts. The number below each circle indicates what fraction of the circle is shaded. Convert each fraction to a decimal.

1.

$\dfrac{1}{8}$

2.

$\dfrac{3}{8}$

3.

$\dfrac{5}{8}$

4.

$\dfrac{7}{8}$

Complete the following tables by converting each fraction to a decimal.

5.

Fraction	$\frac{1}{4}$	$\frac{2}{4}$	$\frac{3}{4}$	$\frac{4}{4}$
Decimal				

6.

Fraction	$\frac{1}{5}$	$\frac{2}{5}$	$\frac{3}{5}$	$\frac{4}{5}$	$\frac{5}{5}$
Decimal					

7.

Fraction	$\frac{1}{6}$	$\frac{2}{6}$	$\frac{3}{6}$	$\frac{4}{6}$	$\frac{5}{6}$	$\frac{6}{6}$
Decimal						

Convert each of the following fractions to a decimal.

8. $\dfrac{1}{2}$ **9.** $\dfrac{12}{25}$ **10.** $\dfrac{14}{25}$ **11.** $\dfrac{14}{32}$ **12.** $\dfrac{18}{32}$

Write each fraction as a decimal correct to the hundredths column.

13. $\dfrac{12}{13}$ **14.** $\dfrac{17}{19}$ **15.** $\dfrac{3}{11}$ **16.** $\dfrac{5}{11}$

17. $\dfrac{2}{23}$ **18.** $\dfrac{3}{28}$ **19.** $\dfrac{12}{43}$ **20.** $\dfrac{15}{51}$

Complete the following table by converting each decimal to a fraction.

21.

Decimal	0.125	0.250	0.375	0.500	0.625	0.750	0.875
Fraction							

22.

Decimal	0.1	0.2	0.3	0.4	0.5	0.6	0.7	0.8	0.9
Fraction									

Write each decimal as a fraction in lowest terms.

23. 0.15   **24.** 0.45   **25.** 0.08   **26.** 0.06   **27.** 0.375   **28.** 0.475

Write each decimal as a mixed number.

29. 5.6   **30.** 8.4   **31.** 5.06   **32.** 8.04   **33.** 1.22   **34.** 2.11

Simplify each of the following as much as possible, and write all answers as decimals.

35. $\frac{1}{2}(2.3 + 2.5)$ **36.** $\frac{3}{4}(1.8 + 7.6)$ **37.** $\dfrac{1.99}{\frac{1}{2}}$ **38.** $\dfrac{2.99}{\frac{1}{2}}$

39. $3.4 - \frac{1}{2}(0.76)$ **40.** $6.7 - \frac{1}{5}(0.45)$ **41.** $\frac{2}{5}(0.3) + \frac{3}{5}(0.3)$

42. $\frac{1}{8}(0.7) + \frac{3}{8}(0.7)$ **43.** $6\left(\frac{3}{5}\right)(0.02)$ **44.** $8\left(\frac{4}{5}\right)(0.03)$

45. $\frac{5}{8} + 0.35\left(\frac{1}{2}\right)$ **46.** $\frac{7}{8} + 0.45\left(\frac{3}{4}\right)$ **47.** $\left(\frac{1}{3}\right)^2(5.4) + \left(\frac{1}{2}\right)^3(3.2)$

48. $\left(\frac{1}{5}\right)^2(7.5) + \left(\frac{1}{4}\right)^2(6.4)$ **49.** $(0.25)^2 + \left(\frac{1}{4}\right)^2(3)$ **50.** $(0.75)^2 + \left(\frac{1}{4}\right)^2(7)$

Applying the Concepts

51. Price of Beef If each pound of beef costs $2.59, how much does $3\frac{1}{4}$ pounds cost?

52. Price of Gasoline What does it cost to fill a $15\frac{1}{2}$-gallon gas tank if the gasoline is priced at 129.9¢ per gallon?

53. Sale Price A dress that costs $57.99 is on sale for $\frac{1}{3}$ off. What is the sale price of the dress?

54. Sale Price A suit that normally sells for $121 is on sale for $\frac{1}{4}$ off. What is the sale price of the suit?

55. Perimeter of the Sierpinski Triangle The diagram shows one stage of what is known as the Sierpinski triangle. Each triangle in the diagram has three equal sides. The large triangle is made up of 4 smaller triangles. If each side of the large triangle is 2 inches, and each side of the smaller triangles is 1 inch, what is the perimeter of the shaded region?

56. Perimeter of the Sierpinski Triangle The diagram shows another stage of the Sierpinski triangle. Each triangle in the diagram has three equal sides. The largest triangle is made up of a number of smaller triangles. If each side of the large triangle is 2 inches, and each side of the smallest triangles is 0.5 inch, what is the perimeter of the shaded region?

57. Nutrition If 1 ounce of ground beef contains 50.75 calories and 1 ounce of halibut contains 27.5 calories, what is the difference in calories between a $4\frac{1}{2}$-ounce serving of ground beef and a $4\frac{1}{2}$-ounce serving of halibut?

58. Nutrition If a 1-ounce serving of baked potato contains 48.3 calories and a 1-ounce serving of chicken contains 24.6 calories, how many calories are in a meal of $5\frac{1}{4}$ ounces of chicken and a $3\frac{1}{3}$-ounce baked potato?

Taxi Ride Recently, the Texas Junior College Teachers Association annual conference was held in Austin. At that time a taxi ride in Austin was $1.25 for the first $\frac{1}{5}$ of a mile and $0.25 for each additional $\frac{1}{5}$ of a mile. The charge for a taxi to wait is $12.00 per hour. Use this information for Problems 59 through 62.

59. If the distance from one of the convention hotels to the airport is 7.5 miles, how much will it cost to take a taxi from that hotel to the airport?

60. If you were to tip the driver of the taxi in Problem 59 $1.50, how much would it cost to take a taxi from the hotel to the airport?

61. Suppose the distance from one of the hotels to one of the western dance clubs in Austin is 12.4 miles. If the fare meter in the taxi gives the charge for that trip as $16.50, is the meter working correctly?

62. Suppose that the distance from a hotel to the airport is 8.2 miles, and the ride takes 20 minutes. Is it more expensive to take a taxi to the airport or to just sit in the taxi?

Getting Ready for the Next Section

Expand and simplify.

63. 6^2

64. 8^2

65. 5^2

66. 10^2

67. 5^3

68. 2^5

69. 3^2

70. 2^3

71. $\left(\frac{1}{3}\right)^4$

72. $\left(\frac{3}{4}\right)^3$

73. $\left(\frac{5}{6}\right)^2$

74. $\left(\frac{3}{5}\right)^3$

75. $(0.5)^2$

76. $(0.1)^3$

77. $(1.2)^2$

78. $(2.1)^2$

79. $3^2 + 4^2$

80. $5^2 + 12^2$

81. $6^2 + 8^2$

82. $2^2 + 3^2$

Square Roots and the Pythagorean Theorem

Figure 1 shows the front view of the roof of a tool shed. How do we find the length d of the diagonal part of the roof? (Imagine that you are drawing the plans for the shed. Since the shed hasn't been built yet, you can't just measure the diagonal, but you need to know how long it will be so you can buy the correct amount of material to build the shed.)

There is a formula from geometry that gives the length d:

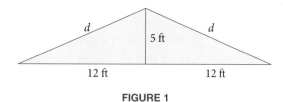

FIGURE 1

$$d = \sqrt{12^2 + 5^2}$$

where $\sqrt{}$ is called the *square root symbol.* If we simplify what is under the square root symbol, we have this:

$$d = \sqrt{144 + 25}$$
$$= \sqrt{169}$$

The expression $\sqrt{169}$ stands for the number we *square* to get 169. Because $13 \cdot 13 = 169$, that number is 13. Therefore the length d in our original diagram is 13 feet.

Here is a more detailed discussion of square roots. In Chapter 1, we did some work with exponents. In particular, we spent some time finding squares of numbers. For example, we considered expressions like this:

$$5^2 = 5 \cdot 5 = 25$$
$$7^2 = 7 \cdot 7 = 49$$
$$x^2 = x \cdot x$$

We say that "the square of 5 is 25" and "the square of 7 is 49." To square a number, we multiply it by itself. When we ask for the *square root* of a given number, we want to know what number we *square* in order to obtain the given number. We say that the square root of 49 is 7, because 7 is the number we square to get 49. Likewise, the square root of 25 is 5, because $5^2 = 25$. The symbol we use to denote square root is $\sqrt{}$, which is also called a *radical sign.* Here is the precise definition of square root.

Note

The square root we are describing here is actually the principal square root. There is another square root that is a negative number. We won't see it in this book, but, if you go on to take an algebra course, you will see it there.

DEFINITION

The *square root* of a positive number a, written $\sqrt{a}$, is the number we square to get a. In symbols:

If $\sqrt{a} = b$ then $b^2 = a$.

We list some common square roots in Table 1.

TABLE 1

Statement	In Words	Reason
$\sqrt{0} = 0$	The square root of 0 is 0	Because $0^2 = 0$
$\sqrt{1} = 1$	The square root of 1 is 1	Because $1^2 = 1$
$\sqrt{4} = 2$	The square root of 4 is 2	Because $2^2 = 4$
$\sqrt{9} = 3$	The square root of 9 is 3	Because $3^2 = 9$
$\sqrt{16} = 4$	The square root of 16 is 4	Because $4^2 = 16$
$\sqrt{25} = 5$	The square root of 25 is 5	Because $5^2 = 25$

Numbers like 1, 9, and 25, whose square roots are whole numbers, are called *perfect squares*. To find the square root of a perfect square, we look for the whole number that is squared to get the perfect square. The following examples involve square roots of perfect squares.

EXAMPLE 1 Simplify: $7\sqrt{64}$

SOLUTION The expression $7\sqrt{64}$ means 7 times $\sqrt{64}$. To simplify this expression, we write $\sqrt{64}$ as 8 and multiply:

$$7\sqrt{64} = 7 \cdot 8 = 56$$

We know $\sqrt{64} = 8$, because $8^2 = 64$.

EXAMPLE 2 Simplify: $\sqrt{9} + \sqrt{16}$

SOLUTION We write $\sqrt{9}$ as 3 and $\sqrt{16}$ as 4. Then we add:

$$\sqrt{9} + \sqrt{16} = 3 + 4 = 7$$

EXAMPLE 3 Simplify: $\sqrt{\dfrac{25}{81}}$

SOLUTION We are looking for the number we square (multiply times itself) to get $\frac{25}{81}$. We know that when we multiply two fractions, we multiply the numerators and multiply the denominators. Because $5 \cdot 5 = 25$ and $9 \cdot 9 = 81$, the square root of $\frac{25}{81}$ must be $\frac{5}{9}$:

$$\sqrt{\frac{25}{81}} = \frac{5}{9} \quad \text{because} \quad \left(\frac{5}{9}\right)^2 = \frac{5}{9} \cdot \frac{5}{9} = \frac{25}{81}$$

In Examples 4–6, we simplify each expression as much as possible.

EXAMPLE 4 Simplify: $12\sqrt{25} = 12 \cdot 5 = 60$

EXAMPLE 5 Simplify: $\sqrt{100} - \sqrt{36} = 10 - 6 = 4$

EXAMPLE 6 Simplify: $\sqrt{\dfrac{49}{121}} = \dfrac{7}{11}$ because $\left(\dfrac{7}{11}\right)^2 = \dfrac{7}{11} \cdot \dfrac{7}{11} = \dfrac{49}{121}$

So far in this section we have been concerned only with square roots of perfect squares. The next question is, "What about square roots of numbers that are not perfect squares, like $\sqrt{7}$, for example?" We know that

$$\sqrt{4} = 2 \quad \text{and} \quad \sqrt{9} = 3$$

And because 7 is between 4 and 9, $\sqrt{7}$ should be between $\sqrt{4}$ and $\sqrt{9}$. That is, $\sqrt{7}$ should be between 2 and 3. But what is it exactly? The answer is, we cannot write it exactly in decimal or fraction form. Because of this, it is called an *irrational number*. We can approximate it with a decimal, but we can never write it exactly with a decimal. Table 2 gives some decimal approximations for $\sqrt{7}$. The decimal approximations were obtained by using a calculator. We could continue the list to any accuracy we desired. However, we would never reach a number in decimal form whose square was exactly 7.

TABLE 2

APPROXIMATIONS FOR THE SQUARE ROOT OF 7

Accurate to the Nearest	The Square Root of 7 is	Check by Squaring
Tenth	$\sqrt{7} = 2.6$	$(2.6)^2 = 6.76$
Hundredth	$\sqrt{7} = 2.65$	$(2.65)^2 = 7.0225$
Thousandth	$\sqrt{7} = 2.646$	$(2.646)^2 = 7.001316$
Ten thousandth	$\sqrt{7} = 2.6458$	$(2.6458)^2 = 7.00025764$

EXAMPLE 7   Give a decimal approximation for the expression $5\sqrt{12}$ that is accurate to the nearest ten thousandth.

SOLUTION   Let's agree not to round to the nearest ten thousandth until we have first done all the calculations. Using a calculator, we find $\sqrt{12} \approx 3.4641016$. Therefore,

$$5\sqrt{12} \approx 5(3.4641016) \qquad \text{\small $\sqrt{12}$ on calculator}$$
$$= 17.320508 \qquad \text{\small Multiplication}$$
$$= 17.3205 \qquad \text{\small To the nearest ten thousandth}$$

EXAMPLE 8   Approximate $\sqrt{301} + \sqrt{137}$ to the nearest hundredth.

SOLUTION   Using a calculator to approximate the square roots, we have

$$\sqrt{301} + \sqrt{137} \approx 17.349352 + 11.704700 = 29.054052$$

To the nearest hundredth, the answer is 29.05.

EXAMPLE 9   Approximate $\sqrt{\dfrac{7}{11}}$ to the nearest thousandth.

SOLUTION   Because we are using calculators, we first change $\frac{7}{11}$ to a decimal and then find the square root:

$$\sqrt{\frac{7}{11}} \approx \sqrt{0.6363636} \approx 0.7977240$$

To the nearest thousandth, the answer is 0.798.

FACTS FROM GEOMETRY   *Pythagorean Theorem*

A **right triangle** is a triangle that contains a 90° (or right) angle. The longest side in a right triangle is called the **hypotenuse**, and we use the letter c to denote it. The two shorter sides are denoted by the letters a and b. The Pythagorean theorem states that the hypotenuse is the square root of the sum of the squares of the two shorter sides. In symbols:

$$c = \sqrt{a^2 + b^2}$$

EXAMPLE 10   Find the length of the hypotenuse in each right triangle.

a.

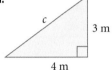

3 m

4 m

b.

5 in.

7 in.

SOLUTION   We apply the formula given above.

a.	**b.**
When $a = 3$ and $b = 4$:	When $a = 5$ and $b = 7$:
$c = \sqrt{3^2 + 4^2}$	$c = \sqrt{5^2 + 7^2}$
$ = \sqrt{9 + 16}$	$ = \sqrt{25 + 49}$
$ = \sqrt{25}$	$ = \sqrt{74}$
$c = 5$ meters	$c \approx 8.60$ inches

In part a, the solution is a whole number, whereas in part b, we must use a calculator to get 8.60 as an approximation to $\sqrt{74}$.

EXAMPLE 11   A ladder is leaning against the top of a 6-foot wall. If the bottom of the ladder is 8 feet from the wall, how long is the ladder?

SOLUTION   A picture of the situation is shown in Figure 2. We let c denote the length of the ladder. Applying the Pythagorean theorem, we have

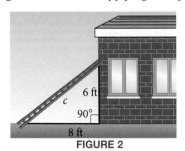

6 ft

c

90°

8 ft

FIGURE 2

$c = \sqrt{6^2 + 8^2}$

$ = \sqrt{36 + 64}$

$ = \sqrt{100}$

$ = 10$ feet

The ladder is 10 feet long.

FACTS FROM GEOMETRY   *The Spiral of Roots*

To visualize the square roots of the counting numbers, we can construct the spiral of roots, which we mentioned in the introduction to this chapter. To begin, we draw two line segments, each of length 1, at right angles to each other. Then we use the Pythagorean theorem to find the length of the diagonal. Figure 3 illustrates:

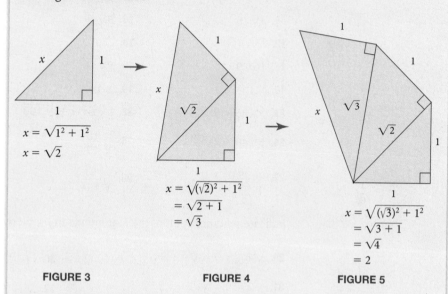

$x = \sqrt{1^2 + 1^2}$
$x = \sqrt{2}$

$x = \sqrt{(\sqrt{2})^2 + 1^2}$
$\;\; = \sqrt{2 + 1}$
$\;\; = \sqrt{3}$

$x = \sqrt{(\sqrt{3})^2 + 1^2}$
$\;\; = \sqrt{3 + 1}$
$\;\; = \sqrt{4}$
$\;\; = 2$

FIGURE 3   **FIGURE 4**   **FIGURE 5**

Next, we construct a second triangle by connecting a line segment of length 1 to the end of the first diagonal so that the angle formed is a right angle. We find the length of the second diagonal using the Pythagorean theorem. Figure 4 illustrates this procedure. As we continue to draw new triangles by connecting line segments of length 1 to the end of each previous diagonal, so that the angle formed is a right angle, the spiral of roots begins to appear (see Figure 5).

GETTING READY FOR CLASS

After reading through the preceding section, respond in your own words and in complete sentences.

1. Which number is larger, the square of 10 or the square root of 10?

2 Give a definition for the square root of a number.

3. What two numbers will the square root of 20 fall between?

4. What is the Pythagorean theorem?

Problem Set 4.6

Find each of the following square roots without using a calculator.

1. $\sqrt{64}$ **2.** $\sqrt{100}$ **3.** $\sqrt{81}$ **4.** $\sqrt{49}$

5. $\sqrt{36}$ **6.** $\sqrt{144}$ **7.** $\sqrt{25}$ **8.** $\sqrt{169}$

Simplify each of the following expressions without using a calculator.

9. $3\sqrt{25}$ **10.** $9\sqrt{49}$ **11.** $6\sqrt{64}$

12. $11\sqrt{100}$ **13.** $15\sqrt{9}$ **14.** $8\sqrt{36}$

15. $16\sqrt{9}$ **16.** $9\sqrt{16}$ **17.** $\sqrt{49} + \sqrt{64}$

18. $\sqrt{1} + \sqrt{0}$ **19.** $\sqrt{16} - \sqrt{9}$ **20.** $\sqrt{25} - \sqrt{4}$

21. $3\sqrt{25} + 9\sqrt{49}$ **22.** $6\sqrt{64} + 11\sqrt{100}$ **23.** $15\sqrt{9} - 9\sqrt{16}$

24. $7\sqrt{49} - 2\sqrt{4}$ **25.** $\sqrt{\dfrac{16}{49}}$ **26.** $\sqrt{\dfrac{100}{121}}$

27. $\sqrt{\dfrac{36}{64}}$ **28.** $\sqrt{\dfrac{81}{144}}$

Indicate whether each of the statements in Problems 29–32 is *True* or *False*.

29. $\sqrt{4} + \sqrt{9} = \sqrt{4 + 9}$ **30.** $\sqrt{\dfrac{16}{25}} = \dfrac{\sqrt{16}}{\sqrt{25}}$

31. $\sqrt{25 \cdot 9} = \sqrt{25} \cdot \sqrt{9}$ **32.** $\sqrt{100} - \sqrt{36} = \sqrt{100 - 36}$

Find the length of the hypotenuse in each right triangle. Round to the nearest hundredth, if rounding is necessary.

33.

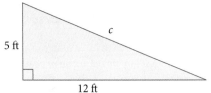

6 in. c 8 in.

34.

5 yd c 5 yd

35.

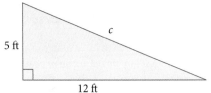

5 ft c 12 ft

36.

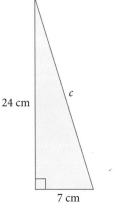

24 cm c 7 cm

37.

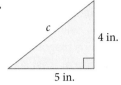

c

4 in.

5 in.

38.

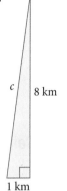

c

6 ft

6 ft

39.

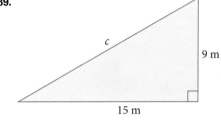

c

9 m

15 m

40.

c

8 km

1 km

Applying the Concepts

41. Geometry One end of a wire is attached to the top of a 24-foot pole; the other end of the wire is anchored to the ground 18 feet from the bottom of the pole. If the pole makes an angle of 90° with the ground, find the length of the wire.

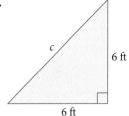

24 ft

90°

18 ft

42. Geometry Two children are trying to cross a stream. They want to use a log that goes from one bank to the other. If the left bank is 5 feet higher than the right bank and the stream is 12 feet wide, how long must a log be to just barely reach?

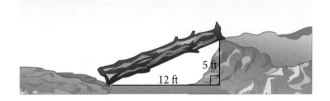

5 ft

12 ft

43. Geometry A ladder is leaning against the top of a 15-foot wall. If the bottom of the ladder is 20 feet from the wall, how long is the ladder?

44. Geometry A wire from the top of a 24-foot pole is fastened to the ground by a stake that is 10 feet from the bottom of the pole. How long is the wire?

45. Spiral of Roots Construct your own spiral of roots by using a ruler. Draw the first triangle by using two 1-inch lines. The first diagonal will have a length of $\sqrt{2}$ inches. Each new triangle will be formed by drawing a 1-inch line segment at the end of the previous diagonal so that the angle formed is 90°. Draw your spiral until you have at least six right triangles.

46. Spiral of Roots Construct a spiral of roots by using line segments of length 2 inches. The length of the first diagonal will be $2\sqrt{2}$ inches. The length of the second diagonal will be $2\sqrt{3}$ inches. What will be the length of the third diagonal?

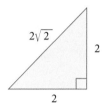

Calculator Problems

Use a calculator to work problems 47 through 68.

Approximate each of the following square roots to the nearest ten thousandth.

47. $\sqrt{1.25}$ **48.** $\sqrt{12.5}$ **49.** $\sqrt{125}$ **50.** $\sqrt{1250}$

Approximate each of the following expressions to the nearest hundredth.

51. $2\sqrt{3}$ **52.** $3\sqrt{2}$ **53.** $5\sqrt{5}$ **54.** $5\sqrt{3}$

55. $\dfrac{\sqrt{3}}{3}$ **56.** $\dfrac{\sqrt{2}}{2}$ **57.** $\sqrt{\dfrac{1}{3}}$ **58.** $\sqrt{\dfrac{1}{2}}$

Approximate each of the following expressions to the nearest thousandth.

59. $\sqrt{12} + \sqrt{75}$ **60.** $\sqrt{18} + \sqrt{50}$ **61.** $\sqrt{87}$ **62.** $\sqrt{68}$

63. $2\sqrt{3} + 5\sqrt{3}$ **64.** $3\sqrt{2} + 5\sqrt{2}$ **65.** $7\sqrt{3}$ **66.** $8\sqrt{2}$

67. Lighthouse Problem The higher you are above the ground, the farther you can see. If your view is unobstructed, then the distance in miles that you can see from h feet above the ground is given by the formula

$$d = \sqrt{\dfrac{3h}{2}}$$

The following figure shows a lighthouse with a door and windows at various heights. The preceding formula can be used to find the distance to the ocean horizon from these heights. Use the formula and a calculator to complete the following table. Round your answers to the nearest whole number.

Height h (feet)	Distance d (miles)
10	
50	
90	
130	
170	
190	

68. Pendulum Problem The time (in seconds) it takes for the pendulum on a clock to swing through one complete cycle is given by the formula

$$T = \frac{11}{7}\sqrt{\frac{L}{2}}$$

where L is the length (in feet) of the pendulum. Use this formula and a calculator to complete the following table. Round your answers to the nearest hundredth.

Length L (feet)	Time T (seconds)
1	
2	
3	
4	
5	
6	

L ft

T sec
out and back

Chapter 4 Summary

Place Value [4.1]

1. The number 4.123 in words is "four and one hundred twenty-three thousandths."

The place values for the first five places to the right of the decimal point are

Decimal Point	Tenths	Hundredths	Thousandths	Ten Thousandths	Hundred Thousandths
.	$\frac{1}{10}$	$\frac{1}{100}$	$\frac{1}{1,000}$	$\frac{1}{10,000}$	$\frac{1}{100,000}$

Rounding Decimals [4.1]

2. 357.753 rounded to the nearest

Tenth: 357.8

Ten: 360

If the digit in the column to the right of the one we are rounding to is 5 or more, we add 1 to the digit in the column we are rounding to; otherwise, we leave it alone. We then replace all digits to the right of the column we are rounding to with zeros if they are to the left of the decimal point; otherwise, we simply delete them.

Addition and Subtraction with Decimals [4.2]

3.
```
   3.400
  25.060
+  0.347
-------
  28.807
```

To add (or subtract) decimal numbers, we align the decimal points and add (or subtract) as if we were adding (or subtracting) whole numbers. The decimal point in the answer goes directly below the decimal points in the problem.

Multiplication with Decimals [4.3]

4. If we multiply 3.49×5.863, there will be a total of $2 + 3 = 5$ digits to the right of the decimal point in the answer.

To multiply two decimal numbers, we multiply as if the decimal points were not there. The decimal point in the product has as many digits to the right as there are total digits to the right of the decimal points in the two original numbers.

Division with Decimals [4.4]

5.
```
          1.39
   2.5.)3.4.75
        2 5
        ---
          9 7
          7 5
          ---
          2 25
          2 25
          ----
             0
```

To begin a division problem with decimals, we make sure that the divisor is a whole number. If it is not, we move the decimal point in the divisor to the right as many places as it takes to make it a whole number. We must then be sure to move the decimal point in the dividend the same number of places to the right. Once the divisor is a whole number, we divide as usual. The decimal point in the answer is placed directly above the decimal point in the dividend.

Changing Fractions to Decimals [4.5]

To change a fraction to a decimal, we divide the numerator by the denominator.

6. $\frac{4}{15} = 0.2\overline{6}$ because

$$
\begin{array}{r}
.266 \\
15\overline{)4.000} \\
\underline{3\ 0} \\
1\ 00 \\
\underline{90} \\
100 \\
\underline{90} \\
10
\end{array}
$$

Changing Decimals to Fractions [4.5]

To change a decimal to a fraction, we write the digits to the right of the decimal point over the appropriate power of 10.

7. $0.781 = \dfrac{781}{1,000}$

Square Roots [4.6]

The square root of a positive number a, written $\sqrt{a}$, is the number we square to get a.

8. $\sqrt{49} = 7$ because
$7^2 = 7 \cdot 7 = 49$

Pythagorean Theorem [4.6]

In any right triangle, the length of the longest side (the hypotenuse) is equal to the square root of the sum of the squares of the two shorter sides.

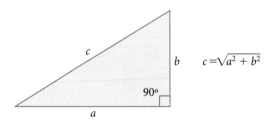

Chapter 4 Test Form A

These problems are all taken from examples in your text. Work each problem and check your answers. If you have made a mistake, work the problem again. If you cannot get the correct answer after two tries, look up the correct solution in your text.

1. Round 0.00346 to the nearest ten thousandth.

2. Write each of the following decimals in words.

    a. 0.4   **b.** 0.04   **c.** 0.004

3. Add: $8 + 0.002 + 3.1 + 0.04$

4. Subtract: $5.9 - 3.0814$

5. Multiply: 3.05×4.36

6. Perform the indicated operations: $0.05(4.2 + 0.03)$

7. Divide: $34.8 \div 4$

8. Divide: $31.35 \div 3.8$

9. Write $\dfrac{3}{11}$ as a decimal.

10. Write 0.38 as a fraction in lowest terms.

11. Simplify: $\dfrac{19}{50}(1.32 + 0.48)$

12. Simplify: $\sqrt{9} + \sqrt{16}$

13. Simplify: $\sqrt{100} - \sqrt{36}$

14. Simplify: $\sqrt{\dfrac{25}{81}}$

15. Approximate $5\sqrt{12}$ to the nearest ten thousandth.

Chapter Test, Form B

For an alternate, more comprehensive, chapter test, go to MathTV.com and select the test and summary for this chapter of the textbook. Click the worksheet labeled Chapter 4 Test, Form B to download it.

Ratio and Proportion

iStockphoto.com © Summit's Peak

Every gallon of gasoline you burn produces 20 pounds of carbon dioxide.
www.fueleconomy.gov

Carbon dioxide (CO_2) is one of the greenhouse gases. Greenhouse gases trap the sun's energy within the Earth's atmosphere, contributing to global climate change.

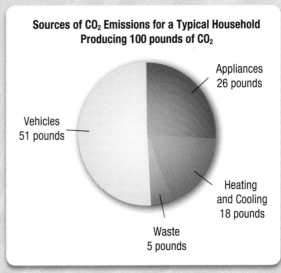

Sources of CO_2 Emissions for a Typical Household Producing 100 pounds of CO_2

Appliances 26 pounds

Vehicles 51 pounds

Heating and Cooling 18 pounds

Waste 5 pounds

As you can see from the pie chart, cars contribute more CO_2 to the environment than all other human contributors combined. Later in this chapter, we will show that each gallon of gasoline you burn in your car produces 20 pounds of CO_2.

According to the website www.carbonfootprint.com, a Carbon Footprint is a measure of the impact our activities have on the environment in terms of the amount of greenhouse gases we produce. It is measured in units of carbon dioxide. We will use the carbon footprint graphic (in the margin) when we are working problems involving greenhouse gases and carbon dioxide.

BigStockPhoto.com © Kathy Kifer

Study Skills

A great deal of your success in this class will depend on attending lectures and taking good, useful notes. They are many different styles and methods of note taking. In a math class, it's important to understand and follow your instructor's lecture. Here are some things to keep in mind about note taking.

Find a Method That Works for You If you can't read or understand your notes from a previous lecture, something is wrong. Your notes need to be legible and clear when you review them later.

Imitate Success Copy down everything your instructor writes on the board, especially any worked examples. If your teacher took the time to write it for the class, it's probably important. You should also look around the class at notes other students are taking. There will be a variety of styles, accuracy and thoroughness. Choose a method that is clear and understandable to you.

Take Notes with Sharing in Mind What if your best friend was in class with you and needed to share your notes for a particular day? You would probably do a better job, write neater, and use different color pens and pencils, mimicking your instructor's work. Work on taking these kinds of notes at each class meeting.

Review With Your Book Open After class, sit for 15 or 20 minutes and review your notes with your book open to the topic you went over in class. Is the material in the textbook similar to your notes? Did the teacher leave some things out or perhaps introduce material not covered in this book? In either case, this is something to confirm with a classmate.

Ratios

Introduction

Recently, *USA Today* gave the average student/teacher ratio for public and private schools for a recent school year. For public schools, the average ratio is 15.6 students for every 1 teacher, while the average ratio for private schools is 13.2 students for every 1 teacher. Generally, in education, lower student/teacher ratios are associated with higher-quality education. There are many other factors that contribute to quality education as well.

©iStockPhoto.com

The *ratio* of two numbers is a way of comparing them. If we say that the ratio of two numbers is 2 to 1, then the first number is twice as large as the second number. For example, if there are 10 men and 5 women enrolled in a math class, then the ratio of men to women is 10 to 5. Because 10 is twice as large as 5, we can also say that the ratio of men to women is 2 to 1.

We can define the ratio of two numbers in terms of fractions.

DEFINITION

The **ratio** of two numbers is a fraction, where the first number in the ratio is the numerator and the second number in the ratio is the denominator. *In symbols:*

If a and b are any two numbers, then the ratio of a to b is $\dfrac{a}{b}$. $\quad (b \neq 0)$

We handle ratios the same way we handle fractions. For example, when we said that the ratio of 10 men to 5 women was the same as the ratio 2 to 1, we were actually saying

$$\frac{10}{5} = \frac{2}{1} \qquad \textit{Reducing to lowest terms}$$

Because we have already studied fractions in detail, much of the introductory material on ratios will seem like review.

EXAMPLE 1   Express the ratio of 16 to 48 as a fraction in lowest terms.

SOLUTION   Because the ratio is 16 to 48, the numerator of the fraction is 16 and the denominator is 48:

$$\frac{16}{48} = \frac{1}{3} \qquad \textit{In lowest terms}$$

Notice that the first number in the ratio becomes the numerator of the fraction, and the second number in the ratio becomes the denominator.

EXAMPLE 2   Give the ratio of $\frac{2}{3}$ to $\frac{4}{9}$ as a fraction in lowest terms.

SOLUTION   We begin by writing the ratio of $\frac{2}{3}$ to $\frac{4}{9}$ as a complex fraction. The numerator is $\frac{2}{3}$, and the denominator is $\frac{4}{9}$. Then we simplify.

$$\dfrac{\dfrac{2}{3}}{\dfrac{4}{9}} = \frac{2}{3} \cdot \frac{9}{4} \qquad \textit{Division by } \frac{4}{9} \textit{ is the same as multiplication by } \frac{9}{4}$$

$$= \frac{18}{12} \qquad \textit{Multiply}$$

$$= \frac{3}{2} \qquad \textit{Reduce to lowest terms}$$

EXAMPLE 3   Write the ratio of 0.08 to 0.12 as a fraction in lowest terms.

SOLUTION   When the ratio is in reduced form, it is customary to write it with whole numbers and not decimals. For this reason we multiply the numerator and the denominator of the ratio by 100 to clear it of decimals. Then we reduce to lowest terms.

$$\frac{0.08}{0.12} = \frac{0.08 \times 100}{0.12 \times 100} \qquad \textit{Multiply the numerator and the denominator by 100 to clear the ratio of decimals}$$

$$= \frac{8}{12} \qquad \textit{Multiply}$$

$$= \frac{2}{3} \qquad \textit{Reduce to lowest terms}$$

Note

Another symbol used to denote ratio is the colon (:). The ratio of, say, 5 to 4 can be written as 5:4. Although we will not use it here, this notation is fairly common.

Table 1 shows several more ratios and their fractional equivalents. Notice that in each case the fraction has been reduced to lowest terms. Also, the ratio that contains decimals has been rewritten as a fraction that does not contain decimals.

TABLE 1

Ratio	Fraction	Fraction in Lowest Terms
25 to 35	$\frac{25}{35}$	$\frac{5}{7}$
35 to 25	$\frac{35}{25}$	$\frac{7}{5}$
8 to 2	$\frac{8}{2}$	$\frac{4}{1}$ We can also write this as just 4.
$\frac{1}{4}$ to $\frac{3}{4}$	$\dfrac{\frac{1}{4}}{\frac{3}{4}}$	$\frac{1}{3}$ because $\dfrac{\frac{1}{4}}{\frac{3}{4}} = \frac{1}{4} \cdot \frac{4}{3} = \frac{1}{3}$
0.6 to 1.7	$\frac{0.6}{1.7}$	$\frac{6}{17}$ because $\frac{0.6 \times 10}{1.7 \times 10} = \frac{6}{17}$

EXAMPLE 4   During a game, a basketball player makes 12 out of the 18 free throws he attempts. Write the ratio of the number of free throws he makes to the number of free throws he attempts as a fraction in lowest terms.

SOLUTION   Because he makes 12 out of 18, we want the ratio 12 to 18, or

$$\frac{12}{18} = \frac{2}{3}$$

Because the ratio is 2 to 3, we can say that, in this particular game, he made 2 out of every 3 free throws he attempted.

EXAMPLE 5   A solution of alcohol and water contains 15 milliliters of water and 5 milliliters of alcohol. Find the ratio of alcohol to water, water to alcohol, water to total solution, and alcohol to total solution. Write each ratio as a fraction and reduce to lowest terms.

SOLUTION   There are 5 milliliters of alcohol and 15 milliliters of water, so there are 20 milliliters of solution (alcohol + water). The ratios are as follows:

The ratio of alcohol to water is 5 to 15, or

$$\frac{5}{15} = \frac{1}{3} \qquad \text{In lowest terms}$$

The ratio of water to alcohol is 15 to 5, or

$$\frac{15}{5} = \frac{3}{1} \qquad \text{In lowest terms}$$

The ratio of water to total solution is 15 to 20, or

$$\frac{15}{20} = \frac{3}{4} \qquad \text{In lowest terms}$$

The ratio of alcohol to total solution is 5 to 20, or

$$\frac{5}{20} = \frac{1}{4} \qquad \text{In lowest terms}$$

EXAMPLE 6   The website fueleconomy.gov provides carbon footprint scores for cars. The carbon footprint is a measure of the tons per year of CO_2 produced by driving the car. Use the carbon footprints below to find the ratio of the CO_2 output of a Hummer to that of a Honda Civic. Then change the ratio to a decimal rounded to the nearest tenth.

Car	Carbon Footprint
Honda Civic	6.3
Hummer H3 4WD	12.2

SOLUTION The ratio of the carbon footprint of the Hummer to that of the Civic is

$$\frac{12.2}{6.3} = \frac{12.2 \times 10}{6.3 \times 10}$$ *Multiply the numerator and denominator by 10 to clear the ratio of decimals*

$$= \frac{122}{63}$$ *Multiply*

This ratio is in lowest terms (no reducing is possible)

To convert to a decimal, we divide 122 by 63 and round to the nearest tenth.

$$\frac{122}{63} \approx 1.9$$

GETTING READY FOR CLASS

After reading through the preceding section, respond in your own words and in complete sentences.

1. In your own words, write a definition for the ratio of two numbers.

2. What does a ratio compare?

3. What are some different ways of using mathematics to write the ratio of a to b?

4. When will the ratio of two numbers be a complex fraction?

Problem Set 5.1

Write each of the following ratios as a fraction in lowest terms. None of the answers should contain decimals.

1. 8 to 6

2. 6 to 8

3. 64 to 12

4. 12 to 64

5. 100 to 250

6. 250 to 100

7. 13 to 26

8. 36 to 18

9. $\frac{3}{4}$ to $\frac{1}{4}$

10. $\frac{5}{8}$ to $\frac{3}{8}$

11. $\frac{7}{3}$ to $\frac{6}{3}$

12. $\frac{9}{5}$ to $\frac{11}{5}$

13. $\frac{6}{5}$ to $\frac{6}{7}$

14. $\frac{5}{3}$ to $\frac{5}{8}$

15. $2\frac{1}{2}$ to $3\frac{1}{2}$

16. $5\frac{1}{4}$ to $1\frac{3}{4}$

17. $2\frac{2}{3}$ to $\frac{5}{3}$

18. $\frac{1}{2}$ to $3\frac{1}{2}$

19. 0.05 to 0.15

20. 0.21 to 0.03

21. 0.3 to 3

22. 0.5 to 10

23. 1.2 to 10

24. 6.4 to 0.8

Use the figures to the right to answer the following questions.

25. a. What is the ratio of shaded squares to nonshaded squares?

   b. What is the ratio of shaded squares to total squares?

   c. What is the ratio of nonshaded squares to total squares?

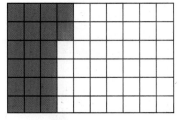

26. a. What is the ratio of shaded squares to non-shaded squares?

   b. What is the ratio of shaded squares to total squares?

   c. What is the ratio of nonshaded squares to total squares?

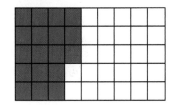

Applying the Concepts

27. Family Budget A family of four budgeted the amounts shown below for some of their monthly bills.

   a. What is the ratio of the rent to the food bill?

   b. What is the ratio of the gas bill to the food bill?

   c. What is the ratio of the utilities bills to the food bill?

   d. What is the ratio of the rent to the utilities bills?

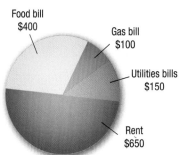

Food bill
$400

Gas bill
$100

Utilities bills
$150

Rent
$650

28. Nutrition One cup of breakfast cereal was found to contain the nutrients shown here.

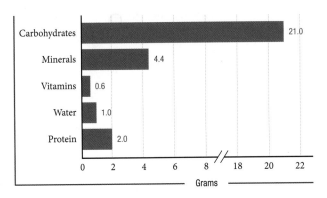

a. Find the ratio of water to protein.
b. Find the ratio of carbohydrates to protein.
c. Find the ratio of vitamins to minerals.
d. Find the ratio of protein to vitamins and minerals.

29. Carbon Footprints The carbon footprints (tons per year of CO_2 produced) shown are from the website fueleconomy.gov. Find the ratios of the carbon footprints for the following cars.

Car	Carbon Footprint
Chevy Malibu Hybrid	6.8
Ford Mustang	9.6
Nissan Pathfinder 2WD	10.8
Toyota Corolla	6.3

a. Chevrolet Malibu Hybrid to Toyota Corolla
b. Nissan Pathfinder to Ford Mustang
c. Toyota Corolla to Nissan Pathfinder
d. Ford Mustang to Chevrolet Malibu Hybrid

30. Profit and Revenue The following bar chart shows the profit and revenue of the Baby Steps Shoe Company each quarter for one year.

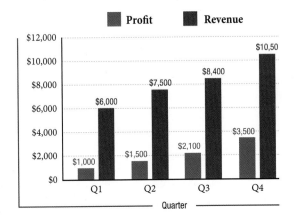

Find the ratio of revenue to profit for each of the following quarters. Write your answers in lowest terms.

a. Q1   **b.** Q2   **c.** Q3   **d.** Q4

e. Find the ratio of revenue to profit for the entire year.

31. Geometry In the diagram below, *AC* represents the length of the line segment that starts at *A* and ends at *C*. From the diagram we see that *AC* = 8.

a. Find the ratio of *BC* to *AC*.

b. What is the length *AE*?

c. Find the ratio of *DE* to *AE*.

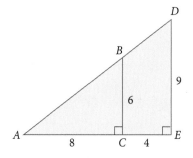

Calculator Problems

Write each of the following ratios as a fraction, and then use a calculator to change the fraction to a decimal. Round all decimal answers to the nearest hundredth. Do not reduce fractions.

Number of Students The total number of students attending a community college in the Midwest is 4,722. Of these students, 2,314 are male and 2,408 are female.

32. Give the ratio of males to females as a fraction and as a decimal.

33. Give the ratio of females to males as a fraction and as a decimal.

34. Give the ratio of males to total number of students as a fraction and as a decimal.

35. Give the ratio of total number of students to females as a fraction and as a decimal.

Getting Ready for the Next Section

The following problems review material from a previous section. Reviewing these problems will help you with the next section.

Write as a decimal.

36. $\dfrac{90}{5}$ **37.** $\dfrac{120}{3}$ **38.** $\dfrac{125}{2}$ **39.** $\dfrac{2}{10}$ **40.** $\dfrac{1.23}{2}$

41. $\dfrac{1.39}{2}$ **42.** $\dfrac{88}{0.5}$ **43.** $\dfrac{1.99}{0.5}$ **44.** $\dfrac{46}{0.25}$ **45.** $\dfrac{92}{0.25}$

Divide. Round answers to the nearest thousandth.

46. $0.48 \div 5.5$ **47.** $0.75 \div 11.5$ **48.** $2.19 \div 46$ **49.** $1.25 \div 50$

Rates and Unit Pricing

Here is the first paragraph of an article that appeared in *USA Today* in 2003.

Culture Clash

Dannon recently shrank its 8-ounce cup of yogurt by 25% to 6 ounces, but cut its suggested retail price by only 20% from 89 cents to 71 cents, which would raise the unit price a penny an ounce, 9%, to 12 cents. At the Hoboken store, which charges more than Dannon's suggested prices, the unit price went from 12 cents to 13 cents with the size change.

DANNON YOGURT		
	Old	New
Size	8 ounces	6 ounces
Container cost	88 cents	72 cents
Price per ounce	11 cents	12 cents
Price difference per ounce: 9%		

In this section we cover material that will give you a better understanding of the information in this article. We start this section with a discussion of rates, then we move on to unit pricing.

Whenever a ratio compares two quantities that have different units (and neither unit can be converted to the other), then the ratio is called a *rate*. For example, if we were to travel 120 miles in 3 hours, then our average rate of speed expressed as the ratio of miles to hours would be

$$\frac{120 \text{ miles}}{3 \text{ hours}} = \frac{40 \text{ miles}}{1 \text{ hour}}$$

Divide the numerator and the denominator by 3 to reduce to lowest terms

The ratio $\frac{40 \text{ miles}}{1 \text{ hour}}$ can be expressed as

$$40 \frac{\text{miles}}{\text{hour}} \quad \text{or} \quad 40 \text{ miles/hour} \quad \text{or} \quad 40 \text{ miles per hour}$$

A rate is expressed in simplest form when the numerical part of the denominator is 1. To accomplish this we use division.

EXAMPLE 1   A train travels 125 miles in 2 hours. What is the train's rate in miles per hour?

SOLUTION   The ratio of miles to hours is

$$\frac{125 \text{ miles}}{2 \text{ hours}} = 62.5 \frac{\text{miles}}{\text{hour}} \qquad \text{Divide 125 by 2}$$

$$= 62.5 \text{ miles per hour}$$

If the train travels 125 miles in 2 hours, then its average rate of speed is 62.5 miles per hour.

EXAMPLE 2   A car travels 90 miles on 5 gallons of gas. Give the ratio of miles to gallons as a rate in miles per gallon.

SOLUTION   The ratio of miles to gallons is

$$\frac{90 \text{ miles}}{5 \text{ gallons}} = 18 \frac{\text{miles}}{\text{gallon}}$$ *Divide 90 by 5*

$$= 18 \text{ miles/gallon}$$

The gas mileage of the car is 18 miles per gallon.

Unit Pricing

One kind of rate that is very common is *unit pricing*. Unit pricing is the ratio of price to quantity when the quantity is one unit. Suppose a 1-liter bottle of a certain soft drink costs $1.19, whereas a 2-liter bottle of the same drink costs $1.39. Which is the better buy? That is, which has the lower price per liter?

$$\frac{\$1.19}{1 \text{ liter}} = \$1.19 \text{ per liter}$$

$$\frac{\$1.39}{2 \text{ liters}} = \$0.695 \text{ per liter}$$

The unit price for the 1-liter bottle is $1.19 per liter, whereas the unit price for the 2-liter bottle is 69.5¢ per liter. The 2-liter bottle is a better buy.

EXAMPLE 3   A supermarket sells low-fat milk in three different containers at the following prices:

| 1 gallon | ½ gallon | 1 quart | (1 quart = ¼ gallon) |
| $3.59 | $1.99 | $1.29 | |

Give the unit price in dollars per gallon for each one.

SOLUTION   Because 1 quart $= \frac{1}{4}$ gallon, we have

1-gallon container $$\frac{\$3.59}{1 \text{ gallon}} = \frac{\$3.59}{1 \text{ gallon}} = \$3.59 \text{ per gallon}$$

½-gallon container $$\frac{\$1.99}{\frac{1}{2} \text{ gallon}} = \frac{\$1.99}{0.5 \text{ gallon}} = \$3.98 \text{ per gallon}$$

1-quart container $$\frac{\$1.29}{1 \text{ quart}} = \frac{\$1.29}{0.25 \text{ gallon}} = \$5.16 \text{ per gallon}$$

The 1-gallon container has the lowest unit price, whereas the 1-quart container has the highest unit price.

GETTING READY FOR CLASS

After reading through the preceding section, respond in your own words and in complete sentences.

1. A rate is a special type of ratio. In your own words, explain what a rate is.
2. When is a rate written in simplest terms?
3. What is *unit pricing*?
4. Give some examples of rates **not** found in your textbook.

SPOTLIGHT ON SUCCESS   *Student Instructor Stefanie*

Never confuse a single defeat with a final defeat.
—F. Scott Fitzgerald

The idea that has worked best for my success in college, and more specifically in my math courses, is to stay positive and be resilient. I have learned that a 'bad' grade doesn't make me a failure; if anything it makes me strive to do better. That is why I never let a bad grade on a test or even in a class get in the way of my overall success.

By sticking with this positive attitude, I have been able to achieve my goals. My grades have never represented how well I know the material. This is because I have struggled with test anxiety and it has consistently lowered my test scores in a number of courses. However, I have not let it defeat me. When I applied to graduate school, I did not meet the grade requirements for my top two schools, but that did not stop me from applying.

One school asked that I convince them that my knowledge of mathematics was more than my grades indicated. If I had let my grades stand in the way of my goals, I wouldn't have been accepted to both of my top two schools, and will be attending one of them in the Fall, on my way to becoming a mathematics teacher.

Problem Set 5.2

1. **Miles/Hour** A car travels 220 miles in 4 hours. What is the rate of the car in miles per hour?

2. **Miles/Hour** A train travels 360 miles in 5 hours. What is the rate of the train in miles per hour?

3. **Kilometers/Hour** It takes a car 3 hours to travel 252 kilometers. What is the rate in kilometers per hour?

4. **Kilometers/Hour** In 6 hours an airplane travels 4,200 kilometers. What is the rate of the airplane in kilometers per hour?

5. **Gallons/Second** The flow of water from a water faucet can fill a 3-gallon container in 15 seconds. Give the ratio of gallons to seconds as a rate in gallons per second.

6. **Gallons/Minute** A 225-gallon drum is filled in 3 minutes. What is the rate in gallons per minute?

7. **Liters/Minute** A gas tank which can hold a total of 56 liters contains only 8 liters of gas when the driver stops to refuel. If it takes 4 minutes to fill up the tank, what is the rate in liters per minute?

8. **Liters/Hour** The gas tank on a car holds 60 liters of gas. At the beginning of a 6-hour trip, the tank is full. At the end of the trip, it contains only 12 liters. What is the rate at which the car uses gas in liters per hour?

9. **Miles/Gallon** A car travels 95 miles on 5 gallons of gas. Give the ratio of miles to gallons as a rate in miles per gallon.

10. **Miles/Gallon** On a 384-mile trip, an economy car uses 8 gallons of gas. Give this as a rate in miles per gallon.

11. **Miles/Liter** The gas tank on a car has a capacity of 75 liters. On a full tank of gas, the car travels 325 miles. What is the gas mileage in miles per liter?

12. **Miles/Liter** A car pulling a trailer can travel 105 miles on 70 liters of gas. What is the gas mileage in miles per liter?

13. **Cents/Ounce** A 6-ounce can of frozen orange juice costs 96¢. Give the unit price in cents per ounce.

14. **Cents/Liter** A 2-liter bottle of root beer costs $1.25. Give the unit price in cents per liter.

15. **Cents/Ounce** A 20-ounce package of frozen peas is priced at 99¢. Give the unit price in cents per ounce.

16. **Dollars/Pound** A 4-pound bag of cat food costs $8.12. Give the unit price in dollars per pound.

17. **Best Buy** Find the unit price in cents per diaper for each of the brands shown below. Round to the nearest tenth of a cent. Which is the better buy?

Dry Baby   *Happy Baby*

36 Diapers, $12.49   38 Diapers, $11.99

18. **Best Buy** Find the unit price in cents per pill for each of the brands shown below. Round to the nearest tenth of a cent. Which is the better buy?

Relief   *New Life*

100 Pills, $5.99   225 Pills, $13.96

19. **Carbon Footprint** A car produces 38.5 tons of CO_2 over a 5-year period. Find its carbon footprint (tons per year of CO_2)

20. **Pounds/Gallon** A car uses 5 gallons of gas on a trip and produces 101 pounds of carbon dioxide. Find the amount of CO_2 per gallon produced by the car.

21. **Cents/Day** If a 15-day supply of vitamins costs $1.62, what is the price in cents per day?

22. **Miles/Hour** A car travels 675.4 miles in $12\frac{1}{2}$ hours. Give the rate in miles per hour to the nearest hundredth.

23. **Miles/Gallon** A truck's odometer reads 15,208.3 at the beginning of a trip and 15,336.7 at the end of the trip. If the trip takes 13.8 gallons of gas, what is the gas mileage in miles per gallon? (Round to the nearest tenth.)

24. **Miles/Hour** At the beginning of a trip, the odometer on a car read 32,567.2 miles. At the end of the trip, it read 32,741.8 miles. If the trip took $4\frac{1}{4}$ hours, what was the rate of the car in miles per hour to the nearest tenth?

Hourly Wages Jane has a job at the local Marcy's department store. The graph shows how much Jane earns for working 8 hours per day for 5 days.

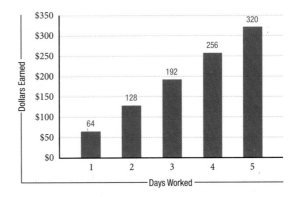

25. What is her daily rate of pay? (Assume she works 8 hours per day.)
26. What is her weekly rate of pay? (Assume she works 5 days per week.)
27. What is her annual rate of pay? (Assume she works 50 weeks per year.)
28. What is her hourly rate of pay? (Assume she works 8 hours per day.)

Getting Ready for the Next Section

Solve each equation by finding a number to replace n with that will make the equation a true statement.

29. $2 \cdot n = 12$ **30.** $3 \cdot n = 27$ **31.** $6 \cdot n = 24$ **32.** $8 \cdot n = 16$

33. $20 = 5 \cdot n$ **34.** $35 = 7 \cdot n$ **35.** $650 = 10 \cdot n$ **36.** $630 = 7 \cdot n$

Solving Equations by Division

In Chapter 1 we solved equations like $3 \cdot n = 12$ by finding a number with which to replace n that would make the equation a true statement. The solution for the equation $3 \cdot n = 12$ is $n = 4$, because

$$
\begin{aligned}
\text{when} \qquad & n = 4 \\
\text{the equation} \qquad & 3 \cdot n = 12 \\
\text{becomes} \qquad & 3 \cdot 4 = 12 \\
\text{or} \qquad & 12 = 12 \qquad \text{A true statement}
\end{aligned}
$$

The problem with this method of solving equations is that we have to guess at the solution and then check it in the equation to see if it works. In this section we will develop a method of solving equations like $3 \cdot n = 12$ that does not require any guessing.

In Chapter 2 we simplified expressions such as

$$\frac{2 \cdot 2 \cdot 3 \cdot 5 \cdot 7}{2 \cdot 5}$$

by dividing out any factors common to the numerator and the denominator. For example:

$$\frac{2 \cdot 2 \cdot 3 \cdot 5 \cdot 7}{2 \cdot 5} = 2 \cdot 3 \cdot 7 = 42$$

The same process works with expressions that have variables for some of their factors. For example, the expression

$$\frac{2 \cdot n \cdot 7 \cdot 11}{n \cdot 11}$$

can be simplified by dividing out the factors common to the numerator and the denominator—namely, n and 11:

$$\frac{2 \cdot n \cdot 7 \cdot 11}{n \cdot 11} = 2 \cdot 7 = 14$$

EXAMPLE 1   Divide the expression $5 \cdot n$ by 5.

SOLUTION   Applying the method above, we have:

$$5 \cdot n \text{ divided by 5 is } \frac{5 \cdot n}{5} = n$$

If you are having trouble understanding this process because there is a variable involved, consider what happens when we divide 6 by 2 and when we divide 6 by 3. Because $6 = 2 \cdot 3$, when we divide by 2 we get 3. Like this:

$$\frac{6}{2} = \frac{2 \cdot 3}{2} = 3$$

When we divide by 3, we get 2:

$$\frac{6}{3} = \frac{2 \cdot 3}{3} = 2$$

EXAMPLE 2   Divide $7 \cdot y$ by 7.

SOLUTION   Dividing by 7, we have:

$$7 \cdot y \text{ divided by 7 is } \frac{7 \cdot y}{7} = y$$

We can use division to solve equations such as $3 \cdot n = 12$. Notice that the left side of the equation is $3 \cdot n$. The equation is solved when we have just n, instead of $3 \cdot n$, on the left side and a number on the right side. That is, we have solved the equation when we have rewritten it as

$$n = \text{a number}$$

We can accomplish this by dividing *both* sides of the equation by 3:

$$\frac{3 \cdot n}{3} = \frac{12}{3} \qquad \text{Divide both sides by 3}$$
$$n = 4$$

Because 12 divided by 3 is 4, the solution to the equation is $n = 4$, which we know to be correct from our discussion at the beginning of this section. Notice that it would be incorrect to divide just the left side by 3 and not the right side also. *Whenever we divide one side of an equation by a number, we must also divide the other side by the same number.*

EXAMPLE 3   Solve the equation $7 \cdot y = 42$ for y by dividing both sides by 7.

SOLUTION   Dividing both sides by 7, we have:

$$\frac{7 \cdot y}{7} = \frac{42}{7}$$
$$y = 6$$

We can check our solution by replacing y with 6 in the original equation:

when	$y = 6$
the equation	$7 \cdot y = 42$
becomes	$7 \cdot 6 = 42$
or	$42 = 42$ A true statement

EXAMPLE 4   Solve for a: $30 = 5 \cdot a$

SOLUTION   Our method of solving equations by division works regardless of which side the variable is on. In this case, the right side is $5 \cdot a$, and we would like it to be just a. Dividing both sides by 5, we have:

$$\frac{30}{5} = \frac{5 \cdot a}{5}$$
$$6 = a$$

The solution is $a = 6$. (If 6 is a, then a is 6.)

We can write our solutions as improper fractions, mixed numbers, or decimals. Let's agree to write our answers as either whole numbers, proper fractions, or mixed numbers unless otherwise stated.

Note

The choice of the letter we use for the variable is not important. The process works just as well with y as it does with n. The letters used for variables in equations are most often the letters a, n, x, y, or z.

Note

In the last chapter of this book, we will devote a lot of time to solving equations. For now, we are concerned only with equations that can be solved by division.

GETTING READY FOR CLASS

After reading through the preceding section, respond in your own words and in complete sentences.

1. In your own words, explain what a solution to an equation is.

2. What number results when you simplify $\dfrac{2 \cdot n \cdot 7 \cdot 11}{n \cdot 11}$?

3. What is the result of dividing $7 \cdot y$ by 7?

4. Explain how division is used to solve the equation $30 = 5 \cdot a$.

Problem Set 5.3

Simplify each of the following expressions by dividing out any factors common to the numerator and the denominator and then simplifying the result.

1. $\dfrac{3 \cdot 5 \cdot 5 \cdot 7}{3 \cdot 5}$

2. $\dfrac{2 \cdot 2 \cdot 3 \cdot 5 \cdot 7}{2 \cdot 5 \cdot 7}$

3. $\dfrac{2 \cdot n \cdot 3 \cdot 3 \cdot 5}{n \cdot 5}$

4. $\dfrac{3 \cdot 5 \cdot n \cdot 7 \cdot 7}{3 \cdot n \cdot 7}$

5. $\dfrac{2 \cdot 2 \cdot n \cdot 7 \cdot 11}{2 \cdot n \cdot 11}$

6. $\dfrac{3 \cdot n \cdot 7 \cdot 13 \cdot 17}{n \cdot 13 \cdot 17}$

7. $\dfrac{9 \cdot n}{9}$

8. $\dfrac{8 \cdot a}{8}$

9. $\dfrac{4 \cdot y}{4}$

10. $\dfrac{7 \cdot x}{7}$

Solve each of the following equations by dividing both sides by the appropriate number. Be sure to show the division in each case.

11. $4 \cdot n = 8$ **12.** $2 \cdot n = 8$ **13.** $5 \cdot x = 35$ **14.** $7 \cdot x = 35$

15. $3 \cdot y = 21$ **16.** $7 \cdot y = 21$ **17.** $6 \cdot n = 48$ **18.** $16 \cdot n = 48$

19. $5 \cdot a = 40$ **20.** $10 \cdot a = 40$ **21.** $3 \cdot x = 6$ **22.** $8 \cdot x = 40$

23. $2 \cdot y = 2$ **24.** $2 \cdot y = 12$ **25.** $3 \cdot a = 18$ **26.** $4 \cdot a = 4$

27. $5 \cdot n = 25$ **28.** $9 \cdot n = 18$ **29.** $6 = 2 \cdot x$ **30.** $56 = 7 \cdot x$

31. $42 = 6 \cdot n$ **32.** $30 = 5 \cdot n$ **33.** $4 = 4 \cdot y$ **34.** $90 = 9 \cdot y$

35. $63 = 7 \cdot y$ **36.** $3 = 3 \cdot y$ **37.** $2 \cdot n = 7$ **38.** $4 \cdot n = 10$

39. $6 \cdot x = 21$ **40.** $7 \cdot x = 8$ **41.** $5 \cdot a = 12$ **42.** $8 \cdot a = 13$

43. $4 = 7 \cdot y$ **44.** $3 = 9 \cdot y$ **45.** $10 = 13 \cdot y$ **46.** $9 = 11 \cdot y$

47. $12 \cdot x = 30$ **48.** $16 \cdot x = 56$ **49.** $21 = 14 \cdot n$ **50.** $48 = 20 \cdot n$

Getting Ready for the Next Section

Reduce.

51. $\dfrac{6}{8}$ **52.** $\dfrac{17}{34}$

Multiply.

53. $3(0.4)$ **54.** $\dfrac{2}{3} \cdot 6$

Divide.

55. $65 \div 10$ **56.** $1.2 \div 8$

Proportions

Millions of people are turning to the Internet to view music videos of their favorite musician. Many Web sites offer different sizes of video based on the speed of a user's Internet connection. Even though the figures below are not the same size, their sides are proportional. Later in this chapter we will use proportions to find the unknown height in the larger figure.

h 320

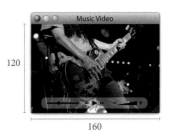

120   160

In this section we will solve problems using proportions. As you will see later in this chapter, proportions can model a number of everyday applications.

DEFINITION

A statement that two ratios are equal is called a ***proportion***. If $\frac{a}{b}$ and $\frac{c}{d}$ are two equal ratios, then the statement

$$\frac{a}{b} = \frac{c}{d}$$

is called a proportion.

Each of the four numbers in a proportion is called a *term* of the proportion. We number the terms of a proportion as follows:

First term $\longrightarrow \dfrac{a}{b} = \dfrac{c}{d} \longleftarrow$ Third term
Second term $\longrightarrow \phantom{\dfrac{a}{b}} \longleftarrow$ Fourth term

The first and fourth terms of a proportion are called the *extremes,* and the second and third terms of a proportion are called the *means.*

Means $\longrightarrow \dfrac{a}{b} = \dfrac{c}{d} \longleftarrow$ Extremes

EXAMPLE 1   In the proportion $\frac{3}{4} = \frac{6}{8}$, name the four terms, the means, and the extremes.

SOLUTION   The terms are numbered as follows:

First term = 3   Third term = 6
Second term = 4   Fourth term = 8

The means are 4 and 6; the extremes are 3 and 8.

The only additional thing we need to know about proportions is the following property.

Fundamental Property of Proportions

In any proportion, the product of the extremes is equal to the product of the means. In symbols, it looks like this:

$$\text{If } \frac{a}{b} = \frac{c}{d} \quad \text{then} \quad ad = bc \quad \text{for } b \neq 0 \text{ and } a \neq 0$$

EXAMPLE 2   Verify the fundamental property of proportions for the following proportions.

$$\textbf{a. } \frac{3}{4} = \frac{6}{8} \qquad \textbf{b. } \frac{17}{34} = \frac{1}{2}$$

SOLUTION   We verify the fundamental property by finding the product of the means and the product of the extremes in each case.

Proportion	Product of the Means	Product of the extremes
a. $\frac{3}{4} = \frac{6}{8}$	$4 \cdot 6 = 24$	$3 \cdot 8 = 24$
b. $\frac{17}{34} = \frac{1}{2}$	$34 \cdot 1 = 34$	$17 \cdot 2 = 34$

For each proportion the product of the means is equal to the product of the extremes.

We can use the fundamental property of proportions, along with a property we encountered in Section 5.3, to solve an equation that has the form of a proportion.

A Note on Multiplication Previously, we have used a multiplication dot to indicate multiplication, both with whole numbers and with variables. A more compact form for multiplication involving variables is simply to leave out the dot. That is, $5 \cdot y = 5y$ and $10 \cdot x \cdot y = 10xy$.

EXAMPLE 3   Solve for x.

$$\frac{2}{3} = \frac{4}{x}$$

SOLUTION   Applying the fundamental property of proportions, we have

$$\text{If} \quad \frac{2}{3} = \frac{4}{x}$$

$$\text{then} \quad 2 \cdot x = 3 \cdot 4 \qquad \text{\textit{The product of the extremes equals the product of the means}}$$

$$2x = 12 \qquad \text{\textit{Multiply}}$$

> **Note**
>
> In some of these problems you will be able to see what the solution is just by looking the problem over. In those cases it is still best to show all the work involved in solving the proportion. It is good practice for the more difficult problems.

The result is an equation. We know from Section 5.3 that we can divide both sides of an equation by the same nonzero number without changing the solution to the equation. In this case we divide both sides by 2 to solve for x:

$$2x = 12$$

$$\frac{2x}{2} = \frac{12}{2} \qquad \text{Divide both sides by 2}$$

$$x = 6 \qquad \text{Simplify each side}$$

The solution is 6. We can check our work by using the fundamental property of proportions:

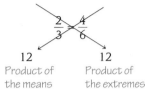

$$\begin{array}{cc} 12 & 12 \\ \text{Product of} & \text{Product of} \\ \text{the means} & \text{the extremes} \end{array}$$

Because the product of the means and the product of the extremes are equal, our work is correct.

EXAMPLE 4   Solve for y: $\dfrac{5}{y} = \dfrac{10}{13}$

SOLUTION   We apply the fundamental property and solve as we did in Example 3:

$$\text{If} \qquad \frac{5}{y} = \frac{10}{13}$$

$$\text{then} \qquad 5 \cdot 13 = y \cdot 10 \qquad \begin{array}{l}\text{The product of the extremes equals} \\ \text{the product of the means}\end{array}$$

$$65 = 10y \qquad \text{Multiply } 5 \cdot 13$$

$$\frac{65}{10} = \frac{10\,y}{10} \qquad \text{Divide both sides by 10}$$

$$6.5 = y \qquad 65 \div 10 = 6.5$$

The solution is 6.5. We could check our result by substituting 6.5 for y in the original proportion and then finding the product of the means and the product of the extremes.

EXAMPLE 5   Find n if $\dfrac{n}{3} = \dfrac{0.4}{8}$.

SOLUTION   We proceed as we did in the previous two examples:

$$\text{If} \qquad \frac{n}{3} = \frac{0.4}{8}$$

$$\text{then} \qquad n \cdot 8 = 3(0.4) \qquad \begin{array}{l}\text{The product of the extremes equals} \\ \text{the product of the means}\end{array}$$

$$8n = 1.2 \qquad 3(0.4) = 1.2$$

$$\frac{8n}{8} = \frac{1.2}{8} \qquad \text{Divide both sides by 8}$$

$$n = 0.15 \qquad 1.2 \div 8 = 0.15$$

The missing term is 0.15.

EXAMPLE 6   Solve for x: $\dfrac{\frac{2}{3}}{5} = \dfrac{x}{6}$

SOLUTION   We begin by multiplying the means and multiplying the extremes:

If $\dfrac{\frac{2}{3}}{5} = \dfrac{x}{6}$

then $\dfrac{2}{3} \cdot 6 = 5 \cdot x$ *The product of the extremes equals the product of the means*

$4 = 5 \cdot x$ $\frac{2}{3} \cdot 6 = 4$

$\dfrac{4}{5} = \dfrac{5 \cdot x}{5}$ *Divide both sides by 5*

$\dfrac{4}{5} = x$

The missing term is $\dfrac{4}{5}$, or 0.8.

EXAMPLE 7   Solve $\dfrac{b}{15} = 2$.

SOLUTION   Since the number 2 can be written as the ratio of 2 to 1, we can write this equation as a proportion, and then solve as we have in the examples above.

$\dfrac{b}{15} = 2$

$\dfrac{b}{15} = \dfrac{2}{1}$ *Write 2 as a ratio*

$b \cdot 1 = 15 \cdot 2$ *Product of the extremes equals the product of the means*

$b = 30$

The procedure for finding a missing term in a proportion is always the same. We first apply the fundamental property of proportions to find the product of the extremes and the product of the means. Then we solve the resulting equation.

GETTING READY FOR CLASS

After reading through the preceding section, respond in your own words and in complete sentences.

1. In your own words, give a definition of a *proportion*.

2. In the proportion $\dfrac{2}{5} = \dfrac{4}{x}$, name the means and the extremes.

3. State the Fundamental Property of Proportions in words and in symbols.

4. For the proportion $\dfrac{2}{5} = \dfrac{4}{x}$, find the product of the means and the product of the extremes.

Problem Set 5.4

For each of the following proportions, name the means, name the extremes, and show that the product of the means is equal to the product of the extremes.

1. $\dfrac{1}{3} = \dfrac{5}{15}$ **2.** $\dfrac{6}{12} = \dfrac{1}{2}$ **3.** $\dfrac{10}{25} = \dfrac{2}{5}$ **4.** $\dfrac{5}{8} = \dfrac{10}{16}$

5. $\dfrac{\frac{1}{3}}{\frac{1}{2}} = \dfrac{4}{6}$ **6.** $\dfrac{2}{\frac{1}{4}} = \dfrac{4}{\frac{1}{2}}$ **7.** $\dfrac{0.5}{5} = \dfrac{1}{10}$ **8.** $\dfrac{0.3}{1.2} = \dfrac{1}{4}$

Find the missing term in each of the following proportions. Set up each problem like the examples in this section. For problems 30–36 write your answers in decimal form. For the other problems, write your answers as fractions in lowest terms.

9. $\dfrac{2}{5} = \dfrac{4}{x}$ **10.** $\dfrac{3}{8} = \dfrac{9}{x}$ **11.** $\dfrac{1}{y} = \dfrac{5}{12}$ **12.** $\dfrac{2}{y} = \dfrac{6}{10}$

13. $\dfrac{x}{4} = \dfrac{3}{8}$ **14.** $\dfrac{x}{5} = \dfrac{7}{10}$ **15.** $\dfrac{5}{9} = \dfrac{x}{2}$ **16.** $\dfrac{3}{7} = \dfrac{x}{3}$

17. $\dfrac{3}{7} = \dfrac{3}{x}$ **18.** $\dfrac{2}{9} = \dfrac{2}{x}$ **19.** $\dfrac{x}{2} = 7$ **20.** $\dfrac{x}{3} = 10$

21. $\dfrac{\frac{1}{2}}{y} = \dfrac{\frac{1}{3}}{12}$ **22.** $\dfrac{\frac{2}{3}}{y} = \dfrac{\frac{1}{3}}{5}$ **23.** $\dfrac{n}{12} = \dfrac{\frac{1}{4}}{\frac{1}{2}}$ **24.** $\dfrac{n}{10} = \dfrac{\frac{3}{5}}{\frac{3}{8}}$

25. $\dfrac{10}{20} = \dfrac{20}{n}$ **26.** $\dfrac{8}{4} = \dfrac{4}{n}$ **27.** $\dfrac{x}{10} = \dfrac{10}{2}$ **28.** $\dfrac{x}{12} = \dfrac{12}{48}$

29. $\dfrac{y}{12} = 9$ **30.** $\dfrac{y}{16} = 0.75$ **31.** $\dfrac{0.4}{1.2} = \dfrac{1}{x}$ **32.** $\dfrac{5}{0.5} = \dfrac{20}{x}$

33. $\dfrac{0.3}{0.18} = \dfrac{n}{0.6}$ **34.** $\dfrac{0.01}{0.1} = \dfrac{n}{10}$ **35.** $\dfrac{0.5}{x} = \dfrac{1.4}{0.7}$ **36.** $\dfrac{0.3}{x} = \dfrac{2.4}{0.8}$

37. $\dfrac{168}{324} = \dfrac{56}{x}$ **38.** $\dfrac{280}{530} = \dfrac{112}{x}$ **39.** $\dfrac{429}{y} = \dfrac{858}{130}$ **40.** $\dfrac{573}{y} = \dfrac{2{,}292}{316}$

41. $\dfrac{n}{39} = \dfrac{533}{507}$ **42.** $\dfrac{n}{47} = \dfrac{1{,}003}{799}$ **43.** $\dfrac{756}{903} = \dfrac{x}{129}$ **44.** $\dfrac{321}{1{,}128} = \dfrac{x}{376}$

Getting Ready for the Next Section

Divide.

45. $360 \div 18$ **46.** $2{,}700 \div 6$

Multiply.

47. $3.5(85)$ **48.** $4.75(105)$

Solve each equation.

49. $\dfrac{x}{10} = \dfrac{270}{6}$ **50.** $\dfrac{x}{45} = \dfrac{8}{18}$ **51.** $\dfrac{x}{25} = \dfrac{4}{20}$ **52.** $\dfrac{x}{3.5} = \dfrac{85}{1}$

Introduction

Model railroads continue to be as popular today as they ever have been. One of the first things model railroaders ask each other is what scale they work with. The scale of a model train indicates its size relative to a full-size train. Each scale is associated with a ratio and a fraction, as shown in the table and bar chart below. An HO scale model train has a ratio of 1 to 87, meaning it is $\frac{1}{87}$ as large as an actual train.

iStockPhoto/©Victor Maffe

Scale	Ratio	As a Fraction
LGB	1 to 22.5	$\frac{1}{22.5}$
#1	1 to 32	$\frac{1}{32}$
O	1 to 43.5	$\frac{1}{43.5}$
S	1 to 64	$\frac{1}{64}$
HO	1 to 87	$\frac{1}{87}$
TT	1 to 120	$\frac{1}{120}$

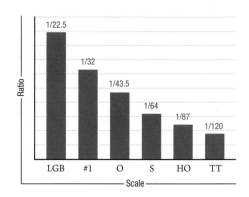

How long is an actual boxcar that has an HO scale model 5 inches long? In this section we will solve this problem using proportions.

Proportions can be used to solve a variety of word problems. The examples that follow show some of these word problems. In each case we will translate the word problem into a proportion and then solve the proportion using the method developed in this chapter.

EXAMPLE 1   A woman drives her car 270 miles in 6 hours. If she continues at the same rate, how far will she travel in 10 hours?

SOLUTION   We let x represent the distance traveled in 10 hours. Using x, we translate the problem into the following proportion:

$$\text{Miles} \longrightarrow \frac{x}{10} = \frac{270}{6} \longleftarrow \text{Miles}$$
$$\text{Hours} \longrightarrow \qquad\qquad \longleftarrow \text{Hours}$$

iStockPhoto/ ©Yakov Stavchansky

Notice that the two ratios in the proportion compare the same quantities. That is, both ratios compare miles to hours. In words this proportion says:

$$x \text{ miles is to } 10 \text{ hours as } 270 \text{ miles is to } 6 \text{ hours}$$

$$\frac{x}{10} = \frac{270}{6}$$

Next, we solve the proportion.

$$x \cdot 6 = 10 \cdot 270$$
$$x \cdot 6 = 2{,}700 \qquad \text{\small\color{gray}{10 · 270 = 2,700}}$$
$$\frac{x \cdot \cancel{6}}{\cancel{6}} = \frac{2{,}700}{6} \qquad \text{\small\color{gray}{Divide both sides by 6}}$$
$$x = 450 \text{ miles} \qquad \text{\small\color{gray}{2,700 ÷ 6 = 450}}$$

If the woman continues at the same rate, she will travel 450 miles in 10 hours.

EXAMPLE 2   A baseball player gets 8 hits in the first 18 games of the season. If he continues at the same rate, how many hits will he get in 45 games?

SOLUTION   We let x represent the number of hits he will get in 45 games. Then

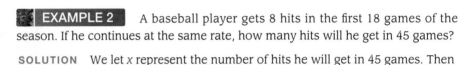

$$x \text{ is to } 45 \text{ as } 8 \text{ is to } 18$$

$$\text{Hits} \longrightarrow \frac{x}{45} = \frac{8}{18} \longleftarrow \text{Hits}$$
$$\text{Games} \longrightarrow \quad\quad \longleftarrow \text{Games}$$

Notice again that the two ratios are comparing the same quantities, hits to games. We solve the proportion as follows:

$$18x = 360 \qquad \text{\small\color{gray}{45 · 8 = 360}}$$
$$\frac{\cancel{18} x}{\cancel{18}} = \frac{360}{18} \qquad \text{\small\color{gray}{Divide both sides by 18}}$$
$$x = 20 \qquad \text{\small\color{gray}{360 ÷ 18 = 20}}$$

If he continues to hit at the rate of 8 hits in 18 games, he will get 20 hits in 45 games.

EXAMPLE 3   A solution contains 4 milliliters of alcohol and 20 milliliters of water. If another solution is to have the same ratio of milliliters of alcohol to milliliters of water and must contain 25 milliliters of water, how much alcohol should it contain?

SOLUTION We let x represent the number of milliliters of alcohol in the second solution. The problem translates to

x milliliters *is to* 25 milliliters *as* 4 milliliters *is to* 20 milliliters

$$\text{Alcohol} \longrightarrow \frac{x}{25} = \frac{4}{20} \longleftarrow \text{Alcohol} \atop \text{Water}$$

$$20x = 100 \qquad\qquad 25 \cdot 4 = 100$$

$$\frac{20x}{20} = \frac{100}{20} \qquad\qquad \text{Divide both sides by 20}$$

$$x = 5 \text{ milliliters of alcohol} \qquad 100 \div 20 = 5$$

EXAMPLE 4   The scale on a map indicates that 1 inch on the map corresponds to an actual distance of 85 miles. Two cities are 3.5 inches apart on the map. What is the actual distance between the two cities?

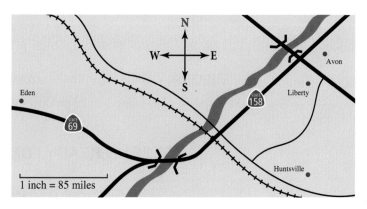

1 inch = 85 miles

SOLUTION We let x represent the actual distance between the two cities. The proportion is

$$\text{Miles} \longrightarrow \frac{x}{3.5} = \frac{85}{1} \longleftarrow \text{Miles} \atop \text{Inches}$$

$$x \cdot 1 = 3.5(85)$$

$$x = 297.5 \text{ miles}$$

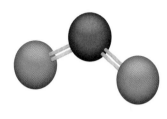

EXAMPLE 5   One gallon of gasoline weighs 6.3 pounds, of which 5.5 pounds is carbon. The carbon is combined with hydrogen in gasoline. When gasoline is burned, the carbon and hydrogen separate, and the carbon recombines with oxygen from air to form carbon dioxide. The atomic weight of oxygen is 16. Show that burning 1 gallon of gasoline produces 20.2 pounds of carbon dioxide.

SOLUTION   First we find the ratio of the weight of carbon to the weight of the whole molecule in carbon dioxide.

$$\text{Atomic weight of carbon} = 12$$

$$\text{Atomic weight of carbon dioxide} = 12 + 16 + 16$$

$$\text{Ratio of weight of carbon to weight of carbon dioxide} = \frac{12}{44} = \frac{3}{11}$$

Next, since the weight of carbon in one gallon of gasoline is 5.5 pounds, if we let x = the weight of carbon dioxide produced by burning one gallon of gasoline, we have

$$\begin{array}{l}\text{weight of carbon} \longrightarrow \dfrac{3}{11} = \dfrac{5.5}{x} \longleftarrow \text{weight of carbon} \\ \text{weight of carbon dioxide} \longrightarrow \phantom{\dfrac{3}{11} = \dfrac{5.5}{x}} \longleftarrow \text{weight of carbon dioxide}\end{array}$$

$$3x = 11(5.5) \qquad \text{Extremes/means property}$$

$$\frac{\cancel{3}x}{\cancel{3}} = \frac{11(5.5)}{3} \qquad \text{Divide both sides by 3}$$

$$x = 20.2 \qquad \text{To the nearest tenth}$$

Each gallon of gasoline burned produces 20.2 pounds of carbon dioxide.

GETTING READY FOR CLASS

After reading through the preceding section, respond in your own words and in complete sentences.

1. Give an example, not found in the book, of a proportion problem you may encounter.
2. Write a word problem for the proportion $\frac{2}{5} = \frac{4}{x}$.
3. What does it mean to translate a word problem into a proportion?
4. Name some jobs that may frequently require solving proportion problems.

Problem Set 5.5

Solve each of the following word problems by translating the statement into a proportion. Be sure to show the proportion used in each case.

1. **Distance** A woman drives her car 235 miles in 5 hours. At this rate how far will she travel in 7 hours?

2. **Distance** An airplane flies 1,260 miles in 3 hours. How far will it fly in 5 hours?

3. **Basketball** A basketball player scores 162 points in 9 games. At this rate how many points will he score in 20 games?

4. **Football** In the first 4 games of the season, a football team scores a total of 68 points. At this rate how many points will the team score in 11 games?

5. **Mixture** A solution contains 8 pints of antifreeze and 5 pints of water. How many pints of water must be added to 24 pints of antifreeze to get a solution with the same concentration?

6. **Nutrition** If 10 ounces of a certain breakfast cereal contains 3 ounces of sugar, how many ounces of sugar does 25 ounces of the same cereal contain?

7. **Map Reading** The scale on a map indicates that 1 inch corresponds to an actual distance of 95 miles. Two cities are 4.5 inches apart on the map. What is the actual distance between the two cities?

8. **Map Reading** A map is drawn so that every 2.5 inches on the map corresponds to an actual distance of 100 miles. If the actual distance between two cities is 350 miles, how far apart are they on the map?

9. **Farming** A farmer knows that of every 50 eggs his chickens lay, only 45 will be marketable. If his chickens lay 1,000 eggs in a week, how many of them will be marketable?

10. **Manufacturing** Of every 17 parts manufactured by a certain machine, only 1 will be defective. How many parts were manufactured by the machine if 8 defective parts were found?

Model Trains In the introduction to this section we indicated that the size of a model train relative to an actual train is referred to as its scale. Each scale is associated with a ratio as shown in the table. For example, an HO model train has a ratio of 1 to 87, meaning it is $\frac{1}{87}$ as large as an actual train.

Scale	Ratio
LGB	1 to 22.5
#1	1 to 32
O	1 to 43.5
S	1 to 64
HO	1 to 87
TT	1 to 120

11. **Length of a Boxcar** How long is an actual boxcar that has an HO scale model 5 inches long? Give your answer in inches, then divide by 12 to give the answer in feet.

12. **Length of a Flatcar** How long is an actual flatcar that has an LGB scale model 24 inches long? Give your answer in feet.

13. **Travel Expenses** A traveling salesman figures it costs 21¢ for every mile he drives his car. How much does it cost him a week to drive his car if he travels 570 miles a week?

14. **Travel Expenses** A family plans to drive their car during their annual vacation. The car can go 350 miles on a tank of gas, which is 18 gallons of gas. The vacation they have planned will cover 1,785 miles. How many gallons of gas will that take?

15. **Nutrition** A 6-ounce serving of grapefruit juice contains 159 grams of water. How many grams of water are in 10 ounces of grapefruit juice?

16. **Nutrition** If 100 grams of ice cream contains 13 grams of fat, how much fat is in 250 grams of ice cream?

17. **Travel Expenses** If a car travels 378.9 miles on 50 liters of gas, how many liters of gas will it take to go 692 miles if the car travels at the same rate? (Round to the nearest tenth.)

18. **Nutrition** If 125 grams of peas contains 26 grams of carbohydrates, how many grams of carbohydrates does 375 grams of peas contain?

19. **Elections** During a recent election, 47 of every 100 registered voters in a certain city voted. If there were 127,900 registered voters in that city, how many people voted?

20. **Map Reading** The scale on a map is drawn so that 4.5 inches corresponds to an actual distance of 250 miles. If two cities are 7.25 inches apart on the map, how many miles apart are they? (Round to the nearest tenth.)

Getting Ready for the Next Section

Simplify.

21. $\dfrac{320}{160}$ **22.** $21 \cdot 105$ **23.** $2{,}205 \div 15$ **24.** $\dfrac{48}{24}$

Solve each equation.

25. $\dfrac{x}{5} = \dfrac{28}{7}$ **26.** $\dfrac{x}{4} = \dfrac{6}{3}$ **27.** $\dfrac{x}{21} = \dfrac{105}{15}$ **28.** $\dfrac{b}{15} = 2$

Similar Figures

This 8-foot-high bronze sculpture "Cellarman" in Napa, California, is an exact replica of the smaller, 12-inch sculpture. Both pieces are the product of artist Tim Lloyd of Arroyo Grande, California.

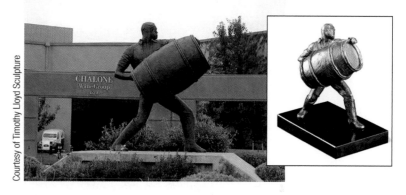

Courtesy of Timothy Lloyd Sculpture

In mathematics, when two or more objects have the same shape, but are different sizes, we say they are similar. If two figures are similar, then their corresponding sides are proportional.

In order to give more details on what we mean by corresponding sides of similar figures, it is helpful to label the parts of a triangle as shown in the margin.

Similar Triangles

Labeling Triangles

One way to label the important parts of a triangle is to label the vertices with capital letters and the sides with lower-case letters.

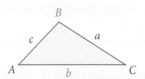

Notice that side *a* is opposite vertex *A*, side *b* is opposite vertex *B*, and side *c* is opposite vertex *C*. Also, because each vertex is the vertex of one of the angles of the triangle, we refer to the three interior angles as *A*, *B*, and *C*.

Two triangles that have the same shape are similar when their corresponding sides are proportional, or have the same ratio. The triangles below are similar.

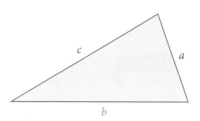

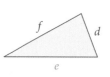

Corresponding Sides	**Ratio**
side *a* corresponds with side *d*	$\dfrac{a}{d}$
side *b* corresponds with side *e*	$\dfrac{b}{e}$
side *c* corresponds with side *f*	$\dfrac{c}{f}$

Because their corresponding sides are proportional, we write

$$\frac{a}{d} = \frac{b}{e} = \frac{c}{f}$$

EXAMPLE 1   The two triangles below are similar. Find side x.

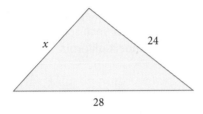

SOLUTION   To find the length x, we set up a proportion of equal ratios. The ratio of x to 5 is equal to the ratio of 24 to 6 and to the ratio of 28 to 7. Algebraically we have

$$\frac{x}{5} = \frac{24}{6} \quad \text{and} \quad \frac{x}{5} = \frac{28}{7}$$

We can solve either proportion to get our answer. The first gives us

$$6x = 5 \cdot 24 \qquad \text{Extremes/means property}$$
$$6x = 120 \qquad 5 \cdot 24 = 120$$
$$\frac{\cancel{6}x}{\cancel{6}} = \frac{120}{6} \qquad \text{Divide both sides by 6}$$
$$x = 20 \qquad 120 \div 6 = 20$$

Other Similar Figures

When one shape or figure is either a reduced or enlarged copy of the same shape or figure, we consider them similar. For example, video viewed over the Internet was once confined to a small "postage stamp" size. Now it is common to see larger video over the Internet. Although the width and height has increased, the shape of the video has not changed.

EXAMPLE 2   The width and height of the two video clips are proportional. Find the height, h, in pixels of the larger video window.

Note

A pixel is the smallest dot made on a computer monitor. Many computer monitors have a width of 800 pixels and a height of 600 pixels.

SOLUTION   We write our proportion as the ratio of the height of the new video to the height of the old video is equal to the ratio of the width of the new video to the width of the old video:

$$\frac{h}{120} = \frac{320}{160}$$

$160h = 120 \cdot 320$ Extremes/means property

$160h = 38{,}400$ $120 \cdot 320 = 38{,}400$

$h = 240$ Divide both sides by 160

The height of the larger video is 240 pixels.

Drawing Similar Figures

EXAMPLE 3   Draw a triangle similar to triangle *ABC*, if *AC* is proportional to *DF*. Make *E* the third vertex of the new triangle.

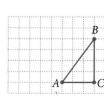
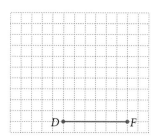

SOLUTION   We see that *AC* is 3 units in length and *BC* has a length of 4 units. Since *AC* is proportional to *DF*, which has a length of 6 units, we set up a proportion to find the length *EF*.

$$\frac{EF}{BC} = \frac{DF}{AC}$$

$$\frac{EF}{4} = \frac{6}{3}$$

$3EF = 24$ Extremes/means property

$EF = 8$ Divide both sides by 3

Now we can draw *EF* with a length of 8 units, then complete the triangle by drawing line *DE*.

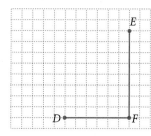
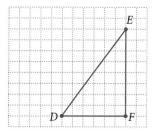

We have drawn triangle *DEF* similar to triangle *ABC*.

Applications

EXAMPLE 4   A building casts a shadow of 105 feet while a 21-foot flag-pole casts a shadow that is 15 feet. Find the height of the building.

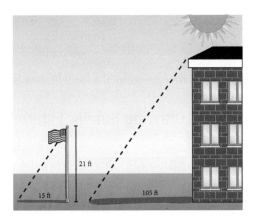

21 ft

15 ft   105 ft

SOLUTION   The figure shows both the building and the flagpole, along with their respective shadows. From the figure it is apparent that we have two similar tri-angles. Letting x = the height of the building, we have

$$\frac{x}{21} = \frac{105}{15}$$

$15x = 2205$ Extremes/means property

$x = 147$ Divide both sides by 15

The height of the building is 147 feet.

The Violin Family   The instruments in the violin family include the bass, cello, viola, and violin. These instruments can be considered similar figures because the entire length of each instrument is proportional to its body length.

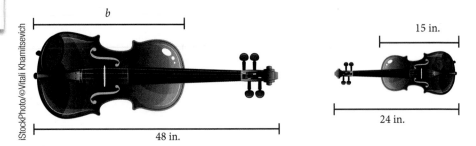

iStockPhoto/©Vitali Khamitsevich

b

48 in.

15 in.

24 in.

EXAMPLE 5   The entire length of a violin is 24 inches, while the body length is 15 inches. Find the body length of a cello if the entire length is 48 inches.

SOLUTION   Let's let b equal the body length of the cello, and set up the proportion.

$$\frac{b}{15} = \frac{48}{24}$$

$24b = 720$ Extremes/means property

$b = 30$ Divide both sides by 24

The body length of a cello is 30 inches.

GETTING READY FOR CLASS

After reading through the preceding section, respond in your own words and in complete sentences.

1. What are similar figures?

2. How do we know if corresponding sides of two triangles are proportional?

3. When labeling a triangle ABC, how do we label the sides?

4. How are proportions used when working with similar figures?

SPOTLIGHT ON SUCCESS   *Student Instructor Aaron*

Sometimes you have to take a step back in order to get a running start forward.
—Anonymous

As a high school senior I was encouraged to go to college immediately after graduating. I earned good grades in high school and I knew that I would have a pretty good group of schools to pick from. Even though I felt like "more school" was not quite what I wanted, the counselors had so much faith and had done this process so many times that it was almost too easy to get the applications out. I sent out applications to schools I knew I could get into and a "dream school."

One night in my email inbox there was a letter of acceptance from my dream school. There was just one problem with getting into this school. It was going to be difficult and I still had senioritis. Going into my first quarter of college was as exciting and difficult as I knew it would be. But after my first quarter I could see that this was not the time for me to be here. I was interested in the subject matter but I could not find my motivating purpose like I had in high school. Instead of dropping out completely, I decided a community college would be a good way for me to stay on track. Without necessarily knowing my direction, I could take the general education classes and get those out of the way while figuring out exactly what and where I felt a good place for me to be.

Now I know what I want to go to school for and the next time I walk onto a four year campus it will be on my terms with my reasons for being there driving me to succeed. I encourage everyone to continue school after high school, even if you have no clue as to what you want to study. There are always stepping stones, like community colleges, that can help you get a clearer picture of what you want to strive for.

Problem Set 5.6

In problems 1–4, for each pair of similar triangles, set up a proportion in order to find the unknown.

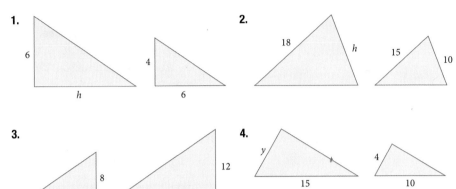

1.

6

h

4

6

2.

18

h

15

10

3.

8

y

12

21

4.

y

15

4

10

In problems 5–10, for each pair of similar figures, set up a proportion in order to find the unknown.

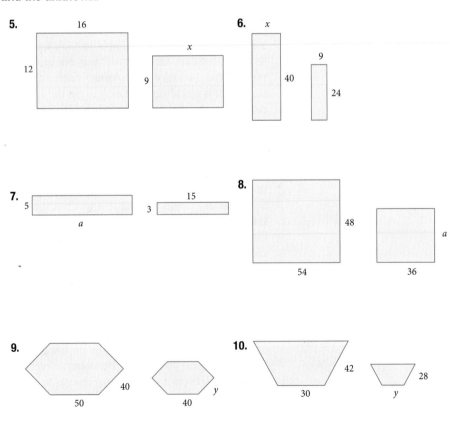

5.

16

12

x

9

6.

x

40

9

24

7.

5

a

3

15

8.

48

54

a

36

9.

40

50

40

y

10.

42

30

28

y

For each problem, draw a figure on the grid on the right that is similar to the given figure.

11. *AC* is proportional to *DF*.

12. *AB* is proportional to *DE*.

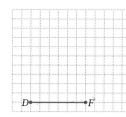

13. *BC* is proportional to *EF*.

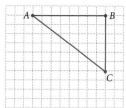

14. *AC* is proportional to *DF*.

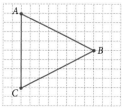

15. *DC* is proportional to *HG*.

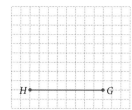

16. *AD* is proportional to *EH*.

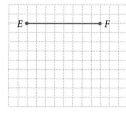
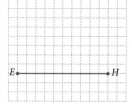

17. *AB* is proportional to *FG*.

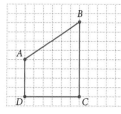

18. *BC* is proportional to *FG*.

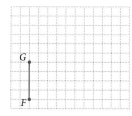

Applying the Concepts

19. Length of a Bass The entire length of a violin is 24 inches, while its body length is 15 inches. The bass is an instrument proportional to the violin. If the total length of a bass is 72 inches, find its body length.

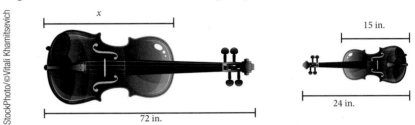

20. Length of an Instrument The entire length of a violin is 24 inches, while the body length is 15 inches. Another instrument proportional to the violin has a body length of 25 inches. What is the total length of this instrument?

21. Length of a Viola The entire length of a cello is 48 inches, while its body length is 30 inches. The viola is an instrument proportional to the cello. If the total length of a viola is 26 inches, find its body length.

22. Length of an Instrument The entire length of a cello is 48 inches, while its body length is 30 inches. Another instrument proportional to the cello has a body length of 35 inches. What is the total length of this instrument?

23. Video Resolution A new graphics card can increase the resolution of a computer's monitor. Suppose a monitor has a horizontal resolution of 800 pixels and a vertical resolution of 600 pixels. By adding a new graphics card, the resolutions remain in the same proportions, but the horizontal resolution increases to 1,280 pixels. What is the new vertical resolution?

24. Screen Resolution The display of a 20″ computer monitor is proportional to that of a 23″ monitor. A 20″ monitor has a horizontal resolution of 1,680 pixels and a vertical resolution of 1,050 pixels. If a 23″ monitor has a horizontal resolution of 1,920 pixels, what is its vertical resolution?

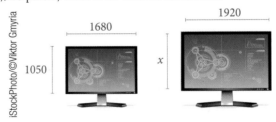

25. Screen Resolution The display of a 20″ computer monitor is proportional to that of a 17″ monitor. A 20″ monitor has a horizontal resolution of 1,680 pixels and a vertical resolution of 1,050 pixels. If a 17″ monitor has a vertical resolution of 900 pixels, what is its horizontal resolution?

26. Video Resolution A new graphics card can increase the resolution of a computer's monitor. Suppose a monitor has a horizontal resolution of 640 pixels and a vertical resolution of 480 pixels. By adding a new graphics card, the resolutions remain in the same proportions, but the vertical resolution increases to 786 pixels. What is the new horizontal resolution?

27. Height of a Tree A tree casts a shadow 38 feet long, while a 6-foot man casts a shadow 4 feet long. How tall is the tree?

28. Height of a Building A building casts a shadow 128 feet long, while a 24-foot flagpole casts a shadow 32 feet long. How tall is the building?

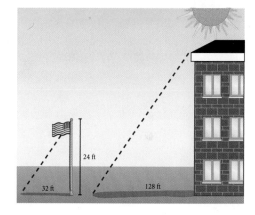

29. Height of a Child A water tower is 36 feet tall and casts a shadow 54 feet long, while a child casts a shadow 6 feet long. How tall is the child?

30. Height of a Truck A clock tower is 36 feet tall and casts a shadow 30 feet long, while a large truck next to the tower casts a shadow 15 feet long. How tall is the truck?

Chapter 5 Summary

Ratio [5.1]

1. The ratio of 6 to 8 is $\frac{6}{8}$ which can be reduced to $\frac{3}{4}$

The ratio of a to b is $\frac{a}{b}$. The ratio of two numbers is a way of comparing them using fraction notation.

Rates [5.2]

2. If a car travels 150 miles in 3 hours, then the ratio of miles to hours is considered a rate:

$$\frac{150 \text{ miles}}{3 \text{ hours}} = 50 \frac{\text{miles}}{\text{hour}}$$
$$= 50 \text{ miles per hour}$$

A ratio that compares two different quantities, like miles and hours, gallons and seconds, etc., is called a *rate*.

Unit Pricing [5.2]

3. If a 10-ounce package of frozen peas costs 69¢, then the price per ounce, or unit price, is

$$\frac{69 \text{ cents}}{10 \text{ ounces}} = 6.9 \frac{\text{cents}}{\text{ounce}}$$
$$= 6.9 \text{ cents per ounce}$$

The *unit price* of an item is the ratio of price to quantity when the quantity is one unit.

Solving Equations by Division [5.3]

4. Solve: $5 \cdot x = 40$

$$5 \cdot x = 40$$
$$\frac{5 \cdot x}{5} = \frac{40}{5} \quad \text{Divide both sides by 5}$$
$$x = 8 \qquad 40 \div 5 = 8$$

Dividing both sides of an equation by the same number will not change the solution to the equation. For example, the equation $5 \cdot x = 40$ can be solved by dividing both sides by 5.

Proportion [5.4]

5. The following is a proportion:

$$\frac{6}{8} = \frac{3}{4}$$

A proportion is an equation that indicates that two ratios are equal.

The numbers in a proportion are called *terms* and are numbered as follows:

$$\text{First term} \longrightarrow \frac{a}{b} = \frac{c}{d} \longleftarrow \text{Third term}$$
$$\text{Second term} \qquad \qquad \qquad \longleftarrow \text{Fourth term}$$

The first and fourth terms are called the *extremes*. The second and third terms are called the *means*.

$$\text{Means} \longrightarrow \frac{a}{b} = \frac{c}{d} \longleftarrow \text{Extremes}$$

Fundamental Property of Proportions [5.4]

In any proportion the product of the extremes is equal to the product of the means. In symbols,

$$\text{If} \quad \frac{a}{b} = \frac{c}{d} \quad \text{then} \quad ad = bc$$

Finding an Unknown Term in a Proportion [5.4]

6. Find x: $\dfrac{2}{5} = \dfrac{8}{x}$

$2 \cdot x = 5 \cdot 8$

$2 \cdot x = 40$

$\dfrac{2 \cdot x}{2} = \dfrac{40}{2}$

$x = 20$

To find the unknown term in a proportion, we apply the fundamental property of proportions and solve the equation that results by dividing both sides by the number that is multiplied by the unknown. For instance, if we want to find the unknown in the proportion

$$\frac{2}{5} = \frac{8}{x}$$

we use the fundamental property of proportions to set the product of the extremes equal to the product of the means.

Using Proportions to Find Unknown Length with Similar Figures [5.6]

7. Find x.

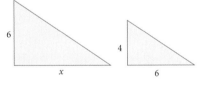

$\dfrac{4}{6} = \dfrac{6}{x}$

$36 = 4x$

$9 = x$

Two triangles that have the same shape are similar when their corresponding sides are proportional, or have the same ratio. The triangles below are similar.

| **Corresponding Sides** | **Ratio** |

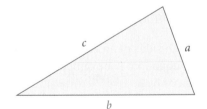   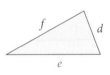

side a corresponds with side d $\dfrac{a}{d}$

side b corresponds with side e $\dfrac{b}{e}$

side c corresponds with side f $\dfrac{c}{f}$

Because their corresponding sides are proportional, we write

$$\frac{a}{d} = \frac{b}{e} = \frac{c}{f}$$

Chapter 5 Test Form A

These problems are all taken from examples in your text. Work each problem and check your answers. If you have made a mistake, work the problem again. If you cannot get the correct answer after two tries, look up the correct solution in your text.

1. Write the ratio 16 to 48 as a fraction in lowest terms.

2. Give the ratio $\frac{2}{3}$ to $\frac{4}{9}$ as a fraction in lowest terms.

3. A train travels 125 miles in 2 hours. What is the train's rate in miles per hour?

4. A car travels 90 miles on 5 gallons of gas. Give the ratio of miles to gallon as a rate in miles per gallon.

5. Solve for a: $30 = 5a$

6. Divide the expression $5n$ by 5.

7. Solve for b: $\frac{b}{15} = 2$

8. Solve for n: $\frac{n}{3} = \frac{0.4}{8}$

9. Solve for y: $\frac{5}{y} = \frac{10}{13}$

10. A woman drives 270 miles in 6 hours. If she continues at the same rate, how far will she travel in 10 hours?

11. The scale on the map indicates that 1 inch on the map corresponds to an actual distance of 85 miles. Two cities are 3.5 inches apart on the map. What is the actual distance between the two cities?

12. A building casts a shadow of 105 feet while a 21-foot flagpole casts a shadow that is 15 feet. Find the height of the building.

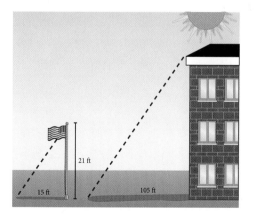

13. Two triangles are similar, one has sides x, 24, and 28. The second has sides 5, 6, and 7 respectfully. Find side x.

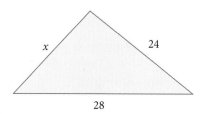

Chapter Test, Form B

For an alternate, more comprehensive, chapter test, go to MathTV.com and select the test and summary for this chapter of the textbook. Click the worksheet labeled Chapter 5 Test, Form B to download it.

Percent

iStockphoto.com © Nathan Jones

I n preceding chapters we have used various methods to compare quantities. For example, we have used the unit price of items to find the best buy. In this chapter we will study percent. When we use percent, we are comparing everything to 100, because percent means "per hundred."

The bar chart in Figure 1 shows the production costs for each of the first four *Star Wars* movies. As you can see, the largest increase in production costs occurred between *Return of the Jedi*, in 1983, and *The Phantom Menace*, in 1999. To see how this increase compares with the other increases, we look at the percent increase in production costs. The bar chart in Figure 2 shows the percent increase in these costs from each *Star Wars* movie to the next.

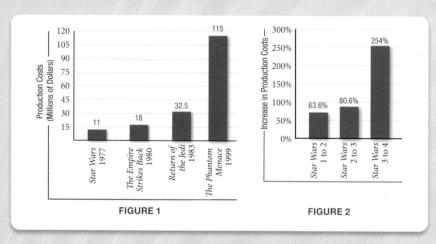

FIGURE 1   **FIGURE 2**

Because percent means "per hundred," comparing increases (and decreases) using percent gives us a way of standardizing our comparisons.

Study Skills

This is the last chapter in which we will mention study skills. You know by now what works best for you and what you have to do to achieve your goals for this course. From now on, it is simply a matter of sticking with the things that work for you and avoiding the things that do not. It seems simple, but as with anything that takes effort, it is up to you to see that you maintain the skills that get you where you want to be in the course.

If you intend to take more classes in mathematics and want to ensure your success in those classes, then you can work toward this goal: *Become the type of student who can learn mathematics on his or her own.* Most people who have degrees in mathematics were students who could learn mathematics on their own. This doesn't mean that you have to learn it all on your own; it simply means that if you have to, you can learn it on your own. Attaining this goal gives you independence and puts you in control of your success in any math class you take.

Percents, Decimals, and Fractions

The size of categories in the pie chart below are given as percents. The whole pie chart is represented by 100%. In general, 100% of something is the whole thing.

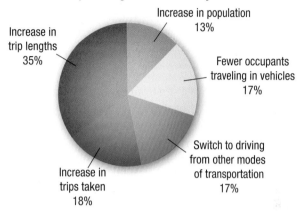

Factors producing more traffic today

Increase in trip lengths 35%

Increase in population 13%

Fewer occupants traveling in vehicles 17%

Switch to driving from other modes of transportation 17%

Increase in trips taken 18%

In this section we will look at the meaning of percent. To begin, we learn to change decimals to percents and percents to decimals.

The Meaning of Percent

Percent means "per hundred." Writing a number as a percent is a way of comparing the number with 100. For example, the number 42% (the % symbol is read "percent") is the same as 42 one-hundredths. That is:

$$42\% = \frac{42}{100}$$

Percents are really fractions (or ratios) with denominator 100.

EXAMPLE 1 Write each percent as a fraction with a denominator of 100.
    a. 33%   **b.** 6%   **c.** 160%

SOLUTION   **a.** $33\% = \dfrac{33}{100}$

            b. $6\% = \dfrac{6}{100}$

            c. $160\% = \dfrac{160}{100}$

If you are wondering if we could reduce some of these fractions further, the answer is yes. We have not done so because the point of this example is that every percent can be written as a fraction with denominator 100.

Changing Percents to Decimals

To change a percent to a decimal number, we simply use the meaning of percent.

EXAMPLE 2   Change 35.2% to a decimal.

SOLUTION   We drop the % symbol and write 35.2 over 100.

$$35.2\% = \frac{35.2}{100}$$ *Use the meaning of % to convert to a fraction with denominator 100*

$$= 0.352$$ *Divide 35.2 by 100*

We see from Example 2 that 35.2% is the same as the decimal 0.352. The result is that the % symbol has been dropped and the decimal point has been moved two places to the *left*. Because % always means "per hundred," we will always end up moving the decimal point two places to the left when we change percents to decimals. Because of this, we can write the following rule.

> **RULE:**
>
> To change a percent to a decimal, drop the % symbol and move the decimal point two places to the *left*.

EXAMPLE 3   Write each percent as a decimal.

     a. 37%   **b.** 68%   **c.** 120%   **d.** 0.8%

SOLUTION   We drop the % symbol and move the decimal point to the left, two places

     a. 37% = 0.37

Decimal point originally here   **b.** 68% = 0.68   Decimal point moved to here

     c. 120% = 1.20

     d. 0.8% = 0.008

EXAMPLE 4   A typical cortisone cream is 0.5% hydrocortisone. Writing this number as a decimal we have

$$0.5\% = 0.005$$

Changing Decimals to Percents

Now we want to do the opposite of what we just did in Examples 2–4. We want to change decimals to percents. We know that 42% written as a decimal is 0.42, which means that in order to change 0.42 back to a percent, we must move the decimal point two places to the *right* and use the % symbol:

$$0.42 = 42\%$$ Notice that we don't show the new decimal point if it is at the end of the number

> **RULE:**
>
> To change a decimal to a percent, we move the decimal point two places to the *right* and use the % symbol.

EXAMPLE 5   Write each decimal as a percent.

a. 0.27   **b.** 4.89   **c.** 0.5   **d.** 0.09

SOLUTION   **a.** 0.27 = 27%

b. 4.89 = 489%

c. 0.5 = 0.50 = 50%   Notice here that we put a 0 after the 5 so we can move the decimal point two places to the right

d. 0.09 = 09% = 9%   Notice that we can drop the 0 at the left without changing the value of the number

EXAMPLE 6   A softball player has a batting average of 0.650. As a percent, this number is 0.650 = 65.0%.

   As you can see from the examples above, percent is just a way of comparing numbers to 100. To multiply decimals by 100, we move the decimal point two places to the right. To divide by 100, we move the decimal point two places to the left. Because of this, it is a fairly simple procedure to change percents to decimals and decimals to percents.

Changing Percents to Fractions

To change a percent to a fraction, drop the % symbol and write the original number over 100.

EXAMPLE 7   The pie chart shows who pays health care bills. Change each percent to a fraction.

SOLUTION   In each case, we drop the percent symbol and write the number over 100. Then we reduce to lowest terms if possible.

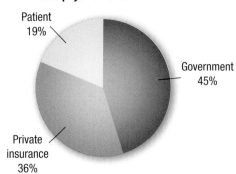

Who pays health care bills

Patient 19%

Government 45%

Private insurance 36%

$$19\% = \frac{19}{100} \qquad 45\% = \frac{45}{100} = \frac{9}{20} \qquad 36\% = \frac{36}{100} = \frac{9}{25}$$

$$\underset{\text{reduce}}{\uparrow} \qquad\qquad\qquad \underset{\text{reduce}}{\uparrow}$$

EXAMPLE 8   Change 4.5% to a fraction in lowest terms.

SOLUTION   We begin by writing 4.5 over 100:

$$4.5\% = \frac{4.5}{100}$$

We now multiply the numerator and the denominator by 10 so the numerator will be a whole number:

$$\frac{4.5}{100} = \frac{4.5 \times 10}{100 \times 10} \qquad \text{Multiply the numerator and the denominator by 10}$$

$$= \frac{45}{1,000}$$

$$= \frac{9}{200} \qquad \text{Reduce to lowest terms}$$

EXAMPLE 9   Change $32\frac{1}{2}\%$ to a fraction in lowest terms.

SOLUTION   Writing $32\frac{1}{2}\%$ over 100 produces a complex fraction. We change $32\frac{1}{2}$ to an improper fraction and simplify:

$$32\frac{1}{2}\% = \frac{32\frac{1}{2}}{100}$$

$$= \frac{\frac{65}{2}}{100} \qquad \text{Change } 32\frac{1}{2} \text{ to the improper fraction } \frac{65}{2}$$

$$= \frac{65}{2} \times \frac{1}{100} \qquad \text{Dividing by 100 is the same as multiplying by } \frac{1}{100}$$

$$= \frac{5 \cdot 13 \cdot 1}{2 \cdot 5 \cdot 20} \qquad \text{Multiplication}$$

$$= \frac{13}{40} \qquad \text{Reduce to lowest terms}$$

Note that we could have changed our original mixed number to a decimal first and then changed to a fraction:

$$32\frac{1}{2}\% = 32.5\% = \frac{32.5}{100} = \frac{32.5 \times 10}{100 \times 10} = \frac{325}{1000} = \frac{5 \cdot 5 \cdot 13}{5 \cdot 5 \cdot 40} = \frac{13}{40}$$

The result is the same in both cases.

Changing Fractions to Percents

To change a fraction to a percent, we can change the fraction to a decimal and then change the decimal to a percent.

EXAMPLE 10   Suppose the price your bookstore pays for your textbook is $\frac{7}{10}$ of the price you pay for your textbook. Write $\frac{7}{10}$ as a percent.

SOLUTION   We can change $\frac{7}{10}$ to a decimal by dividing 7 by 10:

$$\begin{array}{r} 0.7 \\ 10\overline{)7.0} \\ \underline{7\,0} \\ 0 \end{array}$$

We then change the decimal 0.7 to a percent by moving the decimal point two places to the *right* and using the % symbol:

$$0.7 = 70\%$$

You may have noticed that we could have saved some time by simply writing $\frac{7}{10}$ as an equivalent fraction with denominator 100. That is:

$$\frac{7}{10} = \frac{7 \cdot 10}{10 \cdot 10} = \frac{70}{100} = 70\%$$

This is a good way to convert fractions like $\frac{7}{10}$ to percents. It works well for fractions with denominators of 2, 4, 5, 10, 20, 25, and 50, because they are easy to change to fractions with denominators of 100.

EXAMPLE 11   Change $\frac{3}{8}$ to a percent.

SOLUTION   We write $\frac{3}{8}$ as a decimal by dividing 3 by 8. We then change the decimal to a percent by moving the decimal point two places to the right and using the % symbol.

$$\frac{3}{8} = 0.375 = 37.5\%$$

$$\begin{array}{r} .375 \\ 8\overline{)3.000} \\ \underline{2\,4} \\ 60 \\ \underline{56} \\ 40 \\ \underline{40} \\ 0 \end{array}$$

EXAMPLE 12   Change $\frac{5}{12}$ to a percent.

SOLUTION   We begin by dividing 5 by 12:

$$
\begin{array}{r}
.4166 \\
12\overline{)5.0000} \\
\underline{4\,8} \\
20 \\
\underline{12} \\
80 \\
\underline{72} \\
80 \\
\underline{72} \\
8
\end{array}
$$

Because the 6's repeat indefinitely, we can use mixed number notation to write

$$\frac{5}{12} = 0.41\overline{6} = 41\frac{2}{3}\%$$

Or, rounding, we can write

$$\frac{5}{12} = 41.7\% \qquad \textit{To the nearest tenth of a percent}$$

Note

When rounding off, let's agree to round off to the nearest thousandth and then move the decimal point. Our answers in percent form will then be accurate to the nearest tenth of a percent, as in Example 12.

EXAMPLE 13   Change $2\frac{1}{2}$ to a percent.

SOLUTION   We first change to a decimal and then to a percent:

$$2\frac{1}{2} = 2.5$$

$$= 250\%$$

Table 1 lists some of the most commonly used fractions and decimals and their equivalent percents.

TABLE 1

Fraction	Decimal	Percent
$\frac{1}{2}$	0.5	50%
$\frac{1}{4}$	0.25	25%
$\frac{3}{4}$	0.75	75%
$\frac{1}{3}$	$0.\overline{3}$	$33\frac{1}{3}\%$
$\frac{2}{3}$	$0.\overline{6}$	$66\frac{2}{3}\%$
$\frac{1}{5}$	0.2	20%
$\frac{2}{5}$	0.4	40%
$\frac{3}{5}$	0.6	60%
$\frac{4}{5}$	0.8	80%

GETTING READY FOR CLASS

After reading through the preceding section, respond in your own words and in complete sentences.

1. What is the relationship between the word *percent* and the number 100?

2. Explain in words how you would change 25% to a decimal.

3. Explain in words how you would change 25% to a fraction.

4. After reading this section you know that $\frac{1}{2}$, 0.5, and 50% are equivalent. Show mathematically why this is true.

Problem Set 6.1

Write each percent as a fraction with denominator 100.

1. 20%   **2.** 40%   **3.** 60%   **4.** 80%

5. 24%   **6.** 48%   **7.** 65%   **8.** 35%

Change each percent to a decimal.

9. 23%   **10.** 34%   **11.** 92%   **12.** 87%

13. 9%   **14.** 7%   **15.** 3.4%   **16.** 5.8%

17. 6.34%   **18.** 7.25%   **19.** 0.9%   **20.** 0.6%

Change each decimal to a percent.

21. 0.23   **22.** 0.34   **23.** 0.92   **24.** 0.87

25. 0.45   **26.** 0.54   **27.** 0.03   **28.** 0.04

29. 0.6   **30.** 0.9   **31.** 0.8   **32.** 0.5

33. 0.27   **34.** 0.62   **35.** 1.23   **36.** 2.34

Change each percent to a fraction in lowest terms.

37. 60%   **38.** 40%   **39.** 75%   **40.** 25%

41. 4%   **42.** 2%   **43.** 26.5%   **44.** 34.2%

45. 71.87%   **46.** 63.6%   **47.** 0.75%   **48.** 0.45%

49. $6\frac{1}{4}$%   **50.** $5\frac{1}{4}$%   **51.** $33\frac{1}{3}$%   **52.** $66\frac{2}{3}$%

Change each fraction or mixed number to a percent.

53. $\frac{1}{2}$ **54.** $\frac{1}{4}$ **55.** $\frac{3}{4}$ **56.** $\frac{2}{3}$

57. $\frac{1}{3}$ **58.** $\frac{1}{5}$ **59.** $\frac{4}{5}$ **60.** $\frac{1}{6}$

61. $\frac{7}{8}$ **62.** $\frac{1}{8}$ **63.** $\frac{7}{50}$ **64.** $\frac{9}{25}$

65. $3\frac{1}{4}$ **66.** $2\frac{1}{8}$ **67.** $1\frac{1}{2}$ **68.** $1\frac{3}{4}$

69. Change $\frac{21}{43}$ to the nearest tenth of a percent.

70. Change $\frac{36}{49}$ to the nearest tenth of a percent.

Applying the Concepts

71. **Physiology** The human body is between 50% and 75% water. Write each of these percents as a decimal.

72. **Alcohol Consumption** In the United States, 2.7% of those over 15 years of age drink more than 6.3 ounces of alcohol per day. In France, the same figure is 9%. Write each of these percents as a decimal.

73. **Paying Bills** According to Pew Research, a non-political organization that provides information on the issues, attitudes and trends shaping America, most people still pay their monthly bills by check.

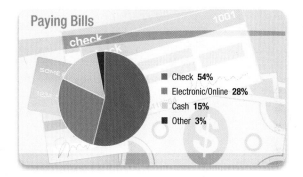

Paying Bills
- Check **54%**
- Electronic/Online **28%**
- Cash **15%**
- Other **3%**

   a. Convert each percent to a fraction.
   b. Convert each percent to a decimal.
   c. About how many times more likely are you to pay a bill with a check than by electronic or online methods?

74. **Commuting** The pie chart here shows how we, as a nation, commute to work. Change each percent to a fraction in lowest terms.

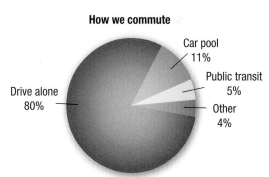

How we commute

Car pool 11%
Public transit 5%
Other 4%
Drive alone 80%

75. **Nutrition** Although, nutritionally, breakfast is the most important meal of the day, only $\frac{1}{5}$ of the people in the United States consistently eat breakfast. What percent of the population is this?

76. **Children in School** In Belgium, 96% of all children between 3 and 6 years of age go to school. In Sweden, the same figure is only 25%. In the United States, the figure is 60%. Write each of these percents as a fraction in lowest terms.

77. Student Enrollment The pie chart shown here is from Chapter 2. Change each fraction to a percent.

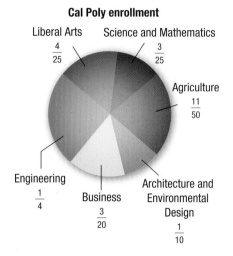

Cal Poly enrollment

Liberal Arts $\frac{4}{25}$ Science and Mathematics $\frac{3}{25}$

Agriculture $\frac{11}{50}$

Engineering $\frac{1}{4}$ Business $\frac{3}{20}$ Architecture and Environmental Design $\frac{1}{10}$

78. Mothers The chart shows the percentage of women who continue working after having a baby. Using the chart, convert the percentage for the following years to a decimal.

a. 1997   **b.** 2000   **c.** 2003

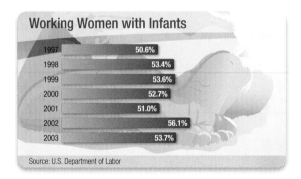

Working Women with Infants

1997	50.6%
1998	53.4%
1999	53.6%
2000	52.7%
2001	51.0%
2002	56.1%
2003	53.7%

Source: U.S. Department of Labor

Calculator Problems

Use a calculator to write each fraction as a decimal, and then change the decimal to a percent. Round all answers to the nearest tenth of a percent.

79. $\frac{29}{37}$ **80.** $\frac{18}{83}$ **81.** $\frac{6}{51}$ **82.** $\frac{8}{95}$ **83.** $\frac{236}{327}$ **84.** $\frac{568}{732}$

85. Women in the Military During World War II, $\frac{1}{12}$ of the Soviet armed forces were women, whereas today only $\frac{1}{450}$ of the Russian armed forces are women. Change both fractions to percents (to the nearest tenth of a percent).

86. Number of Teachers The ratio of the number of teachers to the number of students in secondary schools in Japan is 1 to 17. In the United States, the ratio is 1 to 19. Write each of these ratios as a fraction and then as a percent. Round to the nearest tenth of a percent.

Getting Ready for the Next Section

Multiply.

87. 0.25(74)

88. 0.15(63)

89. 0.435(25)

90. 0.635(45)

Divide. Round the answers to the nearest thousandth, if necessary.

91. $\dfrac{21}{42}$

92. $\dfrac{21}{84}$

93. $\dfrac{25}{0.4}$

94. $\dfrac{31.9}{78}$

Solve for n.

95. $42n = 21$

96. $25 = 0.40n$

Basic Percent Problems

The American Dietetic Association (ADA) recommends eating foods in which the number of calories from fat is less than 30% of the total number of calories. Foods that satisfy this requirement are considered healthy foods. Is the nutrition label shown below from a food that the ADA would consider healthy? This is the type of question we will be able to answer after we have worked through the examples in this section.

Nutrition Facts

Serving Size 1/2 cup (65g)
Servings Per Container: 8

Amount Per Serving

Calories 150   Calories from fat 90

% Daily Value*

Total Fat 10g	**16%**
Saturated Fat 6g	**32%**
Cholesterol 35mg	**12%**
Sodium 30mg	**1%**
Total Carbohydrate 14g	**5%**
Dietary Fiber 0g	**0%**
Sugars 11g	
Protein 2g	

Vitamin A 6%	•	Vitamin C 0%
Calcium 6%	•	Iron 0%

*Percent Daily Values are based on a 2,000 calorie diet.

FIGURE 1   Nutrition label from vanilla ice cream

This section is concerned with three kinds of word problems that are associated with percents. Here is an example of each type:

Type A: What number is 15% of 63?
Type B: What percent of 42 is 21?
Type C: 25 is 40% of what number?

Solving Percent Problems Using Equations

The first method we use to solve all three types of problems involves translating the sentences into equations and then solving the equations. The following translations are used to write the sentences as equations:

English	Mathematics
is	=
of	· (multiply)
a number	n
what number	n
what percent	n

The word *is* always translates to an = sign, the word *of* almost always means multiply, and the number we are looking for can be represented with a letter, such as n or x.

EXAMPLE 1   What number is 15% of 63?

SOLUTION   We translate the sentence into an equation as follows:

What number is 15% of 63?
$$n = 0.15 \cdot 63$$

To do arithmetic with percents, we have to change to decimals. That is why 15% is rewritten as 0.15. Solving the equation, we have

$$n = 0.15 \cdot 63$$
$$n = 9.45$$

15% of 63 is 9.45

EXAMPLE 2   What percent of 42 is 21?

SOLUTION   We translate the sentence as follows:

What percent of 42 is 21?
$$n \cdot 42 = 21$$

We solve for n by dividing both sides by 42.

$$\frac{n \cdot 42}{42} = \frac{21}{42}$$

$$n = \frac{21}{42}$$

$$n = 0.50$$

Because the original problem asked for a percent, we change 0.50 to a percent:

$$n = 50\%$$

21 is 50% of 42

EXAMPLE 3   25 is 40% of what number?

SOLUTION   Following the procedure from the first two examples, we have

25 is 40% of what number?
$$25 = 0.40 \cdot n$$

Again, we changed 40% to 0.40 so we can do the arithmetic involved in the problem. Dividing both sides of the equation by 0.40, we have

$$\frac{25}{0.40} = \frac{0.40 \cdot n}{0.40}$$

$$\frac{25}{0.40} = n$$

$$62.5 = n$$

25 is 40% of 62.5

As you can see, all three types of percent problems are solved in a similar manner. We write *is* as =, *of* as ·, and *what number* as *n*. The resulting equation is then solved to obtain the answer to the original question.

EXAMPLE 4   What number is 43.5% of 25?

$$n = 0.435 \cdot 25$$

$$n = 10.9 \qquad \textit{Rounded to the nearest tenth}$$

10.9 is 43.5% of 25

EXAMPLE 5   What percent of 78 is 31.9?

$$n \cdot 78 = 31.9$$

$$\frac{n \cdot 78}{78} = \frac{31.9}{78}$$

$$n = \frac{31.9}{78}$$

$$n = 0.409 \qquad \textit{Rounded to the nearest thousandth}$$

$$n = 40.9\%$$

40.9% of 78 is 31.9

EXAMPLE 6   34 is 29% of what number?

$$34 = 0.29 \cdot n$$

$$\frac{34}{0.29} = \frac{0.29 \cdot n}{0.29}$$

$$\frac{34}{0.29} = n$$

$$117.2 = n \qquad \textit{Rounded to the nearest tenth}$$

34 is 29% of 117.2

EXAMPLE 7   As we mentioned in the introduction to this section, the American Dietetic Association recommends eating foods in which the number of calories from fat is less than 30% of the total number of calories. According to the nutrition label below, what percent of the total number of calories are fat calories?

SOLUTION   To solve this problem, we must write the question in the form of one of the three basic percent problems shown in Examples 1–6. Because there are 90 calories from fat and a total of 150 calories, we can write the question this way: 90 is what percent of 150?

Nutrition Facts	
Serving Size 1/2 cup (65g)	
Servings Per Container: 8	
Amount Per Serving	
Calories 150	Calories from fat 90
	% Daily Value*
Total Fat 10g	**16%**
Saturated Fat 6g	**32%**
Cholesterol 35mg	**12%**
Sodium 30mg	**1%**
Total Carbohydrate 14g	**5%**
Dietary Fiber 0g	**0%**
Sugars 11g	
Protein 2g	
Vitamin A 6% • Vitamin C 0%	
Calcium 6% • Iron 0%	
*Percent Daily Values are based on a 2,000 calorie diet.	

FIGURE 2   Nutrition label from vanilla ice cream

Now that we have written the question in the form of one of the basic percent problems, we simply translate it into an equation. Then we solve the equation.

90 *is what percent of* 150?

$$90 = n \cdot 150$$

$$\frac{90}{150} = n$$

$$n = 0.60 = 60\%$$

The number of calories from fat in this package of ice cream is 60% of the total number of calories. Thus the ADA would not consider this to be a healthy food.

Solving Percent Problems Using Proportions

We can look at percent problems in terms of proportions also. For example, we know that 24% is the same as $\frac{24}{100}$, which reduces to $\frac{6}{25}$. That is

$$\frac{24}{100} = \frac{6}{25}$$

24 is to 100   as   6 is to 25

We can illustrate this visually with boxes of proportional lengths:

In general, we say

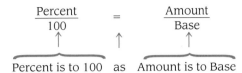

$$\frac{\text{Percent}}{100} = \frac{\text{Amount}}{\text{Base}}$$

Percent is to 100   as   Amount is to Base

EXAMPLE 8   What number is 15% of 63?

SOLUTION   This is the same problem we worked in Example 1. We let n be the number in question. We reason that n will be smaller than 63 because it is only 15% of 63. The base is 63 and the amount is n. We compare n to 63 as we compare 15 to 100. Our proportion sets up as follows:

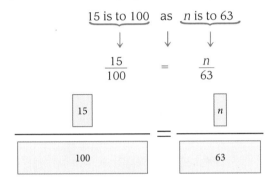

15 is to 100   as n is to 63

$$\frac{15}{100} = \frac{n}{63}$$

Solving the proportion, we have

$$15 \cdot 63 = 100n \qquad \text{Extremes/means property}$$
$$945 = 100n \qquad \text{Simplify the left side}$$
$$9.45 = n \qquad \text{Divide each side by 100}$$

This gives us the same result we obtained in Example 1.

EXAMPLE 9   What percent of 42 is 21?

SOLUTION   This is the same problem we worked in Example 2. We let n be the percent in question. The amount is 21 and the base is 42. We compare n to 100 as we compare 21 to 42. Here is our reasoning and proportion:

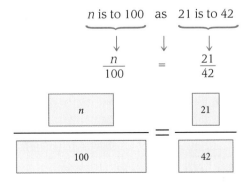

n is to 100   as   21 is to 42

$$\frac{n}{100} = \frac{21}{42}$$

Solving the proportion, we have

$$42n = 21 \cdot 100 \qquad \text{Extremes/means property}$$
$$42n = 2{,}100 \qquad \text{Simplify the right side}$$
$$n = 50 \qquad \text{Divide each side by 42}$$

Since n is a percent, our answer is 50%, giving us the same result we obtained in Example 2.

EXAMPLE 10   25 is 40% of what number?

SOLUTION   This is the same problem we worked in Example 3. We let n be the number in question. The base is n and the amount is 25. We compare 25 to n as we compare 40 to 100. Our proportion sets up as follows:

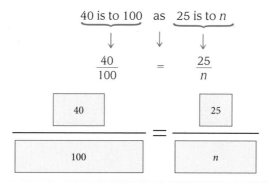

$$40 \text{ is to } 100 \quad \text{as} \quad 25 \text{ is to } n$$

$$\frac{40}{100} = \frac{25}{n}$$

$$\frac{40}{100} = \frac{25}{n}$$

Solving the proportion, we have

$$40 \cdot n = 25 \cdot 100 \qquad \text{Extremes/means property}$$
$$40 \cdot n = 2{,}500 \qquad \text{Simplify the right side}$$
$$n = 62.5 \qquad \text{Divide each side by 40}$$

So, 25 is 40% of 62.5, which is the same result we obtained in Example 3.

Note

When you work the problems in the problem set, use whichever method you like, unless your instructor indicates that you are to use one method instead of the other.

GETTING READY FOR CLASS

After reading through the preceding section, respond in your own words and in complete sentences.

1. When we translate a sentence such as "What number is 15% of 63?" into symbols, what does each of the following translate to?
   a. is   **b.** of   **c.** what number

2. Look at Example 1 in your text and answer the question below.

   The number 9.45 is what percent of 63?

3. Show that the answer to the question below is the same as the answer to the question in Example 2 of your text.

   The number 21 is what percent of 42?

4. If 21 is 50% of 42, then 21 is what percent of 84?

Problem Set 6.2

Solve each of the following problems.

1. What number is 25% of 32?

2. What number is 10% of 80?

3. What number is 20% of 120?

4. What number is 15% of 75?

5. What number is 54% of 38?

6. What number is 72% of 200?

7. What number is 11% of 67?

8. What number is 2% of 49?

9. What percent of 24 is 12?

10. What percent of 80 is 20?

11. What percent of 50 is 5?

12. What percent of 20 is 4?

13. What percent of 36 is 9?

14. What percent of 70 is 14?

15. What percent of 8 is 6?

16. What percent of 16 is 9?

17. 32 is 50% of what number?

18. 16 is 20% of what number?

19. 10 is 20% of what number?

20. 11 is 25% of what number?

21. 37 is 4% of what number?

22. 90 is 80% of what number?

23. 8 is 2% of what number?

24. 6 is 3% of what number?

The following problems can be solved by the same method you used in Problems 1–24.

25. What is 6.4% of 87?

26. What is 10% of 102?

27. 25% of what number is 30?

28. 10% of what number is 22?

29. 28% of 49 is what number?

30. 97% of 28 is what number?

31. 27 is 120% of what number?

32. 24 is 150% of what number?

33. 65 is what percent of 130?

34. 26 is what percent of 78?

35. What is 0.4% of 235,671?

36. What is 0.8% of 721,423?

37. 4.89% of 2,000 is what number?

38. 3.75% of 4,000 is what number?

39. Write a basic percent problem, the solution to which can be found by solving the equation $n = 0.25(350)$.

40. Write a basic percent problem, the solution to which can be found by solving the equation $n = 0.35(250)$.

41. Write a basic percent problem, the solution to which can be found by solving the equation $n \cdot 24 = 16$.

42. Write a basic percent problem, the solution to which can be found by solving the equation $n \cdot 16 = 24$.

43. Write a basic percent problem, the solution to which can be found by solving the equation $46 = 0.75 \cdot n$.

44. Write a basic percent problem, the solution to which can be found by solving the equation $75 = 0.46 \cdot n$.

Applying the Concepts

Nutrition For each nutrition label in Problems 45–48, find what percent of the total number of calories comes from fat calories. Then indicate whether the label is from a food considered healthy by the American Dietetic Association. Round to the nearest tenth of a percent if necessary.

45. Spaghetti

Nutrition Facts

Serving Size 2 oz. (56g per 1/8 of pkg) dry
Servings Per Container: 8

Amount Per Serving

Calories 210	Calories from fat 10

	% Daily Value*
Total Fat 1g	2%
Saturated Fat 0g	0%
Polyunsaturated Fat 0.5g	
Monounsaturated Fat 0g	
Cholesterol 0mg	0%
Sodium 0mg	0%
Total Carbohydrate 42g	14%
Dietary Fiber 2g	7%
Sugars 3g	
Protein 7g	

Vitamin A 0%	•	Vitamin C 0%
Calcium 0%	•	Iron 10%
Thiamin 30%	•	Riboflavin 10%
Niacin 15%	•	

*Percent Daily Values are based on a 2,000 calorie diet

46. Canned Italian tomatoes

Nutrition Facts

Serving Size 1/2 cup (121g)
Servings Per Container: about 3 1/2

Amount Per Serving

Calories 25	Calories from fat 0

	% Daily Value*
Total Fat 0g	0%
Saturated Fat 0g	0%
Cholesterol 0mg	0%
Sodium 300mg	12%
Potassium 145mg	4%
Total Carbohydrate 4g	2%
Dietary Fiber 1g	4%
Sugars 4g	
Protein 1g	

Vitamin A 20%	•	Vitamin C 15%
Calcium 4%	•	Iron 15%

*Percent Daily Values are based on a 2,000 calorie diet.

47. Shredded Romano cheese

Nutrition Facts

Serving Size 2 tsp (5g)
Servings Per Container: 34

Amount Per Serving

Calories 20	Calories from fat 10

	% Daily Value*
Total Fat 1.5g	2%
Saturated Fat 1g	5%
Cholesterol 5mg	2%
Sodium 70mg	3%
Total Carbohydrate 0g	0%
Fiber 0g	0%
Sugars 0g	
Protein 2g	

Vitamin A 0%	•	Vitamin C 0%
Calcium 4%	•	Iron 0%

*Percent Daily Values are based on a 2,000 calorie diet.

48. Tortilla chips

Nutrition Facts

Serving Size 1 oz (28g/About 12 chips)
Servings Per Container: about 2

Amount Per Serving

Calories 140	Calories from fat 60

	% Daily Value*
Total Fat 7g	1%
Saturated Fat 1g	6%
Cholesterol 0mg	0%
Sodium 170mg	7%
Total Carbohydrate 18g	6%
Dietary Fiber 1g	4%
Sugars less than 1g	
Protein 2g	

Vitamin A 0%	•	Vitamin C 0%
Calcium 4%	•	Iron 2%

*Percent Daily Values are based on a 2,000 calorie diet.

Getting Ready for the Next Section

Solve each equation.

49. $96 = n \cdot 120$

50. $2{,}400 = 0.48 \cdot n$

51. $114 = 150n$

52. $3{,}360 = 0.42n$

53. What number is 80% of 60?

54. What number is 25% of 300?

General Applications of Percent 6.3

As you know from watching television and reading the newspaper, we encounter percents in many situations in everyday life. A 1995 newspaper article discussing the effects of a cholesterol-lowering drug stated that the drug in question "lowered levels of LDL cholesterol by an average of 35%." As we progress through this chapter, we will become more and more familiar with percent, and as a result, we will be better equipped to understand statements like the one above concerning cholesterol.

In this section we continue our study of percent by doing more of the translations that were introduced in Section 6.2. The better you are at working the problems in Section 6.2, the easier it will be for you to get started on the problems in this section.

EXAMPLE 1 On a 120-question test, a student answered 96 correctly. What percent of the problems did the student work correctly?

SOLUTION We have 96 correct answers out of a possible 120. The problem can be restated as

$$96 \text{ is what percent of } 120?$$

$$96 = n \cdot 120$$

$$\frac{96}{120} = \frac{n \cdot 120}{120} \qquad \text{Divide both sides by 120}$$

$$n = \frac{96}{120} \qquad \text{Switch the left and right sides of the equation}$$

$$n = 0.80 \qquad \text{Divide 96 by 120}$$

$$= 80\% \qquad \text{Rewrite as a percent}$$

When we write a test score as a percent, we are comparing the original score to an equivalent score on a 100-question test. That is, 96 correct out of 120 is the same as 80 correct out of 100.

EXAMPLE 2 How much HCl (hydrochloric acid) is in a 60-milliliter bottle that is marked 80% HCl?

SOLUTION If the bottle is marked 80% HCl, that means 80% of the solution is HCl and the rest is water. Because the bottle contains 60 milliliters, we can restate the question as:

$$What \text{ is } 80\% \text{ of } 60?$$

$$n = 0.80 \cdot 60$$

$$n = 48$$

HCL 80%
60 ml

There are 48 milliliters of HCl in 60 milliliters of 80% HCl solution.

EXAMPLE 3   If 48% of the students in a certain college are female and there are 2,400 female students, what is the total number of students in the college?

SOLUTION   We restate the problem as:

2,400 *is* 48% *of what number?*

$$2,400 = 0.48 \cdot n$$

$$\frac{2,400}{0.48} = \frac{0.48 \cdot n}{0.48} \qquad \text{Divide both sides by 0.48}$$

$$n = \frac{2,400}{0.48} \qquad \text{Switch the left and right sides of the equation}$$

$$n = 5,000$$

There are 5,000 students.

EXAMPLE 4   If 25% of the students in elementary algebra courses receive a grade of A, and there are 300 students enrolled in elementary algebra this year, how many students will receive A's?

SOLUTION   After reading the question a few times, we find that it is the same as this question:

What number is 25% *of* 300?

$$n = 0.25 \cdot 300$$

$$n = 75$$

Thus, 75 students will receive A's in elementary algebra.

Almost all application problems involving percents can be restated as one of the three basic percent problems we listed in Section 6.2. It takes some practice before the restating of application problems becomes automatic. You may have to review Section 6.2 and Examples 1–4 above several times before you can translate word problems into mathematical expressions yourself.

GETTING READY FOR CLASS

After reading through the preceding section, respond in your own words and in complete sentences.

1. On the test mentioned in Example 1, how many questions would the student have answered correctly if she answered 40% of the questions correctly?

2. If the bottle in example 2 contained 30 milliliters instead of 60, what would the answer be?

3. In Example 3, how many of the students were male?

4. How many of the students mentioned in Example 4 received a grade lower than A?

Solve each of the following problems by first restating it as one of the three basic percent problems of Section 6.2. In each case, be sure to show the equation.

1. **Test Scores** On a 120-question test a student answered 84 correctly. What percent of the problems did the student work correctly?

2. **Test Scores** An engineering student answered 81 questions correctly on a 90-question trigonometry test. What percent of the questions did she answer correctly? What percent were answered incorrectly?

3. **Basketball** A basketball player made 63 out of 75 free throws. What percent is this?

4. **Family Budget** A family spends $450 every month on food. If the family's income each month is $1,800, what percent of the family's income is spent on food?

5. **Chemistry** How much HCl (hydrochloric acid) is in a 60-milliliter bottle that is marked 75% HCl?

HCL 75%
60 ml

6. **Chemistry** How much acetic acid is in a 5-liter container of acetic acid and water that is marked 80% acetic acid? How much is water?

7. **Farming** A farmer owns 28 acres of land. Of the 28 acres, only 65% can be farmed. How many acres are available for farming? How many are not available for farming?

8. **Number of Students** Of the 420 students enrolled in a basic math class, only 30% are first-year students. How many are first-year students? How many are not?

9. **Number of Students** If 48% of the students in a certain college are female and there are 1,440 female students, what is the total number of students in the college?

10. **Mixture Problem** A solution of alcohol and water is 80% alcohol. The solution is found to contain 32 milliliters of alcohol. How many milliliters total (both alcohol and water) are in the solution?

Alcohol
80%

11. **Number of Graduates** Suppose 60% of the graduating class in a certain high school goes on to college. If 240 students from this graduating class are going on to college, how many students are there in the graduating class?

12. **Defective Parts** In a shipment of airplane parts, 3% are known to be defective. If 15 parts are found to be defective, how many parts are in the shipment?

13. **Number of Students** There are 3,200 students at our school. If 52% of them are female, how many female students are there at our school?

14. **Number of Students** In a certain school, 75% of the students in first-year chemistry have had algebra. If there are 300 students in first-year chemistry, how many of them have had algebra?

15. **Population** In a city of 32,000 people, there are 10,000 people under 25 years of age. What percent of the population is under 25 years of age?

16. **Number of Students** If 45 people enrolled in a psychology course but only 35 completed it, what percent of the students completed the course? (Round to the nearest tenth of a percent.)

Calculator Problems

The following problems are similar to Problems 1–16. They should be set up the same way. Then the actual calculations should be done on a calculator.

17. **Number of People** Of 7,892 people attending an outdoor concert in Los Angeles, 3,972 are over 18 years of age. What percent is this? (Round to the nearest whole-number percent.)

18. **Manufacturing** A car manufacturer estimates that 25% of the new cars sold in one city have defective engine mounts. If 2,136 new cars are sold in that city, how many will have defective engine mounts?

19. **Population** The map shows the most populated cities in the United States. If the population of New York City is about 42% of the state's population, what is the approximate population of the state?

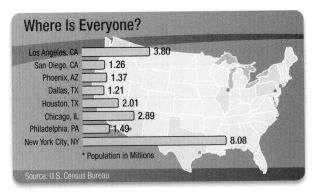

Where Is Everyone?

City	Population*
Los Angeles, CA	3.80
San Diego, CA	1.26
Phoenix, AZ	1.37
Dallas, TX	1.21
Houston, TX	2.01
Chicago, IL	2.89
Philadelphia, PA	1.49
New York City, NY	8.08

* Population in Millions

Source: U.S. Census Bureau

20. **Prom** The graph shows how much girls plan to spend on the prom. If 5,086 girls were surveyed, how many are planning on spending less than $200 on the prom? Round to the nearest whole number.

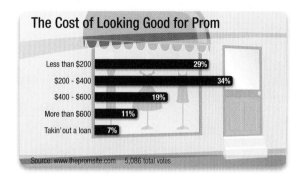

The Cost of Looking Good for Prom

Less than $200	29%
$200 - $400	34%
$400 - $600	19%
More than $600	11%
Takin' out a loan	7%

Source: www.thepromsite.com   5,086 total votes

Getting Ready for the Next Section

Multiply.

21. 0.06(550)   **22.** 0.06(625)   **23.** 0.03(289,500)   **24.** 0.03(115,900)

Divide. Write your answers as decimals.

25. 5.44 ÷ 0.04   **26.** 4.35 ÷ 0.03   **27.** 19.80 ÷ 396   **28.** 11.82 ÷ 197

29. $\dfrac{1,836}{0.12}$ **30.** $\dfrac{115}{0.1}$ **31.** $\dfrac{90}{600}$ **32.** $\dfrac{105}{750}$

Sales Tax and Commission

To solve the problems in this section, we will first restate them in terms of the problems we have already learned how to solve.

Sales Tax

EXAMPLE 1   Suppose the sales tax rate in Mississippi is 6% of the purchase price. If the price of a used refrigerator is $550, how much sales tax must be paid?

SOLUTION   Because the sales tax is 6% of the purchase price, and the purchase price is $550, the problem can be restated as:

What is 6% of $550?

We solve this problem, as we did in Section 6.2, by translating it into an equation:

What is 6% of $550?
$$n = 0.06 \cdot 550$$
$$n = 33$$

The sales tax is $33. The total price of the refrigerator would be

Purchase price		Sales tax		Total price
↓		↓		↓
$550	+	$33	=	$583

> **Note**
>
> In Example 1, the *sales tax rate* is 6%, and the *sales tax* is $33. In most everyday communications, people say "The sales tax is 6%," which is incorrect. The 6% is the tax *rate*, and the $33 is the actual tax.

EXAMPLE 2   Suppose the sales tax rate is 4%. If the sales tax on a 10-speed bicycle is $5.44, what is the purchase price, and what is the total price of the bicycle?

SOLUTION   We know that 4% of the purchase price is $5.44. We find the purchase price first by restating the problem as:

$5.44 *is* 4% *of what number?*
$$5.44 = 0.04 \cdot n$$

We solve the equation by dividing both sides by 0.04:

$$\frac{5.44}{0.04} = \frac{0.04 \cdot n}{0.04} \qquad \text{Divide both sides by 0.04}$$

$$n = \frac{5.44}{0.04} \qquad \text{Switch the left and right sides of the equation}$$

$$n = 136 \qquad \text{Divide}$$

The purchase price is $136. The total price is the sum of the purchase price and the sales tax.

Purchase price	= $136.00
Sales tax	= + 5.44
Total price	= $141.44

EXAMPLE 3   Suppose the purchase price of a stereo system is $396 and the sales tax is $19.80. What is the sales tax rate?

SOLUTION   We restate the problem as:

19.80 *is what percent of* $396?

$$19.80 = n \cdot 396$$

To solve this equation, we divide both sides by 396:

$$\frac{19.80}{396} = \frac{n \cdot 396}{396} \qquad \text{Divide both sides by 396}$$

$$n = \frac{19.80}{396} \qquad \text{Switch the left and right sides of the equation}$$

$$n = 0.05 \qquad \text{Divide}$$

$$n = 5\% \qquad 0.05 = 5\%$$

The sales tax rate is 5%.

Commission

Many salespeople work on a *commission* basis. That is, their earnings are a percentage of the amount they sell. The *commission rate* is a percent, and the actual commission they receive is a dollar amount.

EXAMPLE 4   A real estate agent gets 3% of the price of each house she sells. If she sells a house for $289,500, how much money does she earn?

SOLUTION   The commission is 3% of the price of the house, which is $289,500. We restate the problem as:

What is 3% *of* $289,500?

$$n = 0.03 \cdot 289,500$$

$$n = 8,685$$

The commission is $8,685.

EXAMPLE 5   Suppose a car salesperson's commission rate is 12%. If the commission on one of the cars is $1,836, what is the purchase price of the car?

SOLUTION   12% of the sales price is $1,836. The problem can be restated as:

12% *of what number is* $1,836?

$$0.12 \cdot n = 1,836$$

$$\frac{0.12 \cdot n}{0.12} = \frac{1,836}{0.12} \qquad \text{Divide both sides by 0.12}$$

$$n = 15,300$$

The car sells for $15,300.

EXAMPLE 6   If the commission on a $600 dining room set is $90, what is the commission rate?

SOLUTION   The commission rate is a percentage of the selling price. What we want to know is:

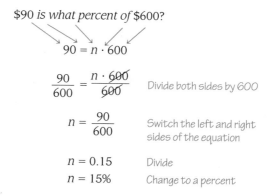

90 *is what percent of* 600?

$$90 = n \cdot 600$$

$$\frac{90}{600} = \frac{n \cdot 600}{600} \qquad \text{Divide both sides by 600}$$

$$n = \frac{90}{600} \qquad \text{Switch the left and right sides of the equation}$$

$$n = 0.15 \qquad \text{Divide}$$

$$n = 15\% \qquad \text{Change to a percent}$$

The commission rate is 15%.

GETTING READY FOR CLASS

After reading through the preceding section, respond in your own words and in complete sentences.

1. Explain the difference between the sales tax and the sales tax rate.

2. Rework Example 1 using a sales tax rate of 7% instead of 6%.

3. Suppose the bicycle in Example 2 was purchased in California, where the sales tax rate in 2004 was 7.5%. How much more would the bicycle have cost?

4. Suppose the car salesperson in Example 5 receives a commission of $3,672. Assuming the same commission rate of 12%, how much does this car sell for?

Problem Set 6.4

These problems should be solved by the method shown in this section. In each case show the equation needed to solve the problem. Write neatly, and show your work.

1. **Sales Tax** Suppose the sales tax rate in Mississippi is 7% of the purchase price. If a new food processor sells for $750, how much is the sales tax?

2. **Sales Tax** If the sales tax rate is 5% of the purchase price, how much sales tax is paid on a television that sells for $980?

3. **Sales Tax and Purchase Price** Suppose the sales tax rate in Michigan is 6%. How much is the sales tax on a $45 concert ticket? What is the total price?

4. **Sales Tax and Purchase Price** Suppose the sales tax rate in Hawaii is 4%. How much sales tax is charged on a new car if the purchase price is $16,400? What is the total price?

5. **Total Price** The sales tax rate is 4%. If the sales tax on a 10-speed bicycle is $6, what is the purchase price? What is the total price?

6. **Total Price** The sales tax on a new microwave oven is $30. If the sales tax rate is 5%, what is the purchase price? What is the total price?

7. **Tax Rate** Suppose the purchase price of a dining room set is $450. If the sales tax is $22.50, what is the sales tax rate?

8. **Tax Rate** If the purchase price of a bottle of California wine is $24 and the sales tax is $1.50, what is the sales tax rate?

9. **Commission** A real estate agent has a commission rate of 3%. If a piece of property sells for $94,000, what is her commission?

10. **Commission** A tire salesperson has a 12% commission rate. If he sells a set of radial tires for $400, what is his commission?

11. **Commission and Purchase Price** Suppose a salesperson gets a commission rate of 12% on the lawnmowers she sells. If the commission on one of the mowers is $24, what is the purchase price of the lawnmower?

12. **Commission and Purchase Price** If an appliance salesperson gets 9% commission on all the appliances she sells, what is the price of a refrigerator if her commission is $67.50?

13. **Commission Rate** If the commission on an $800 washer is $112, what is the commission rate?

14. **Commission Rate** A realtor makes a commission of $3,600 on a $90,000 house he sells. What is his commission rate?

Calculator Problems

The following problems are similar to Problems 1–14. Set them up in the same way, but use a calculator for the calculations.

15. **Sales Tax** The sales tax rate on a certain item is 5.5%. If the purchase price is $216.95, how much is the sales tax? (Round to the nearest cent.)

16. **Purchase Price** If the sales tax rate is 4.75% and the sales tax is $18.95, what is the purchase price? What is the total price? (Both answers should be rounded to the nearest cent.)

17. **Tax Rate** The purchase price for a new suit is $229.50. If the sales tax is $10.33, what is the sales tax rate? (Round to the nearest tenth of a percent.)

Men's Suit
Price $229.50
Tax 10.33
Total Price
$239.83

18. **Commission** If the commission rate for a mobile home salesperson is 11%, what is the commission on the sale of a $15,794 mobile home?

19. **Selling Price** Suppose the commission rate on the sale of used cars is 13%. If the commission on one of the cars is $519.35, what did the car sell for?

20. **Commission Rate** If the commission on the sale of $79.40 worth of clothes is $14.29, what is the commission rate? (Round to the nearest percent.)

Getting Ready for the Next Section

Multiply.

21. 0.05(22,000)

22. 0.176(1,793,000)

23. 0.25(300)

24. 0.12(450)

Divide. Write your answers as decimals.

25. 4 ÷ 25

26. 7 ÷ 35

Subtract.

27. 25 − 21

28. 1,793,000 − 315,568

29. 450 − 54

30. 300 − 75

Add.

31. 396 + 19.8

32. 22,000 + 1,100

Percent Increase or Decrease and Discount

6.5

The table and bar chart below show some statistics compiled by insurance companies regarding stopping distances for automobiles traveling at 20 miles per hour on ice.

	Stopping Distance	Percent Decrease
Regular tires	150 ft	0
Snow tires	151 ft	−1%
Studded snow tires	120 ft	20%
Reinforced tire chains	75 ft	50%

Source: Copyrighted table courtesy of *The Casualty Adjuster's Guide.*

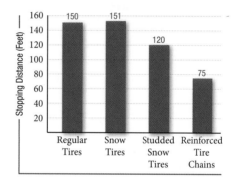

Many times it is more effective to state increases or decreases in terms of percents, rather than the actual number, because with percent we are comparing everything to 100.

EXAMPLE 1   If a person earns $22,000 a year and gets a 5% increase in salary, what is the new salary?

SOLUTION   We can find the dollar amount of the salary increase by finding 5% of $22,000:

$$0.05 \times 22,000 = 1,100$$

The increase in salary is $1,100. The new salary is the old salary plus the raise:

$$
\begin{array}{ll}
\$22,000 & \text{Old salary} \\
+ \quad 1,100 & \text{Raise (5\% of \$22,000)} \\
\hline
\$23,100 & \text{New salary}
\end{array}
$$

EXAMPLE 2   In 1986, there were approximately 1,793,000 arrests for driving under the influence of alcohol or drugs (DUI) in the United States. By 1997, the number of arrests for DUI had decreased 17.6% from the 1986 number. How many people were arrested for DUI in 1997? Round the answer to the nearest thousand.

SOLUTION   The decrease in the number of arrests is 17.6% of 1,793,000, or

$$0.176 \times 1,793,000 = 315,568$$

Subtracting this number from 1,793,000, we have the number of DUI arrests in 1997

$$
\begin{array}{rl}
1,793,000 & \text{\small Number of arrests in 1986} \\
- \quad 315,568 & \text{\small Decrease of 17.6\%} \\
\hline
1,477,432 & \text{\small Number of arrests in 1997}
\end{array}
$$

To the nearest thousand, there were approximately 1,477,000 arrests for DUI in 1997.

EXAMPLE 3   Shoes that usually sell for $25 are on sale for $21. What is the percent decrease in price?

SOLUTION   We must first find the decrease in price. Subtracting the sale price from the original price, we have

$$\$25 - \$21 = \$4$$

The decrease is $4. To find the percent decrease (from the original price), we have

4 *is what percent of* 25?

$$4 = n \cdot 25$$

$$\frac{4}{25} = \frac{n \cdot 25}{25} \qquad \text{\small Divide both sides by 25}$$

$$n = \frac{4}{25} \qquad \text{\small Switch the left and right sides of the equation}$$

$$n = 0.16 \qquad \text{\small Divide}$$

$$n = 16\% \qquad \text{\small Change to a percent}$$

The shoes that sold for $25 have been reduced by 16% to $21. In a problem like this, $25 is the *original* (or *marked*) price, $21 is the *sale price*, $4 is the *discount*, and 16% is the *rate of discount*.

EXAMPLE 4   During a clearance sale, a suit that usually sells for $300 is marked "25% off." What is the discount? What is the sale price?

SOLUTION   To find the discount, we restate the problem as:

What is 25% of 300?

$$n = 0.25 \cdot 300$$
$$n = 75$$

The discount is $75. The sale price is the original price less the discount:

$300   *Original price*
− 75   *Less the discount (25% of $300)*
$225   *Sale price*

EXAMPLE 5   A man buys a washing machine on sale. The machine usually sells for $450, but it is on sale at 12% off. If the sales tax rate is 5%, how much is the total bill for the washer?

SOLUTION   First we have to find the sale price of the washing machine, and we begin by finding the discount:

What is 12% of $450?

$$n = 0.12 \cdot 450$$
$$n = 54$$

The washing machine is marked down $54. The sale price is

$450   *Original price*
− 54   *Discount (12% of $450)*
$396   *Sale price*

Because the sales tax rate is 5%, we find the sales tax as follows:

What is 5% of 396?

$$n = 0.05 \cdot 396$$
$$n = 19.80$$

The sales tax is $19.80. The total price the man pays for the washing machine is

$396.00   *Sale price*
+ 19.80   *Sales tax*
$415.80   *Total price*

GETTING READY FOR CLASS

After reading through the preceding section, respond in your own words and in complete sentences.

1. Suppose the person mentioned in Example 1 was earning $32,000 per year and received the same percent increase in salary. How much more would the raise have been?

2. Suppose the shoes mentioned in Example 3 were on sale for $20, instead of $21. Calculate the new percent decrease in price.

3. Suppose a store owner pays $225 for a suit, and then marks it up $75, to $300. Find the percent increase in price.

4. Compare your answer to Problem 3 above with the problem given in Example 4 of your text. Do you think it is generally true that a 25% discount is equivalent to a $33\frac{1}{3}$% markup?

Problem Set 6.5

Solve each of these problems using the method developed in this section.

1. **Salary Increase** If a person earns $23,000 a year and gets a 7% increase in salary, what is the new salary?

2. **Salary Increase** A computer programmer's yearly income of $57,000 is increased by 8%. What is the dollar amount of the increase, and what is her new salary?

3. **Tuition Increase** The yearly tuition at a college is presently $3,000. Next year it is expected to increase by 17%. What will the tuition at this school be next year?

4. **Price Increase** A supermarket increased the price of cheese that sold for $1.98 per pound by 3%. What is the new price for a pound of this cheese? (Round to the nearest cent.)

5. **Car Value** In one year a new car decreased in value by 20%. If it sold for $16,500 when it was new, what was it worth after 1 year?

6. **Calorie Content** A certain light beer has 20% fewer calories than the regular beer. If the regular beer has 120 calories per bottle, how many calories are in the same-sized bottle of the light beer?

7. **Salary Increase** A person earning $3,500 a month gets a raise of $350 per month. What is the percent increase in salary?

8. **Rate Increase** A student reader is making $6.50 per hour and gets a $0.70 raise. What is the percent increase? (Round to the nearest tenth of a percent.)

9. **Shoe Sale** Shoes that usually sell for $25 are on sale for $20. What is the percent decrease in price?

10. **Enrollment Decrease** The enrollment in a certain elementary school was 410 in 2002. In 2003, the enrollment in the same school was 328. Find the percent decrease in enrollment from 2002 to 2003.

11. **Students to Teachers** The chart shows the student to teacher ratio in the United States from 1975 to 2002. What is the percent decrease in student to teacher ratio from 1975 to 2002? Round to the nearest percent.

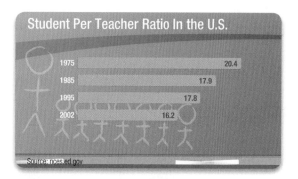

12. **Health Care** The graph shows the rising cost of health care. What is the percent increase in health care costs from 2002 to 2014?

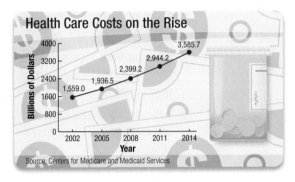

Health Care Costs on the Rise

1,559.0 1,936.5 2,399.2 2,944.2 3,585.7

Billions of Dollars

2002 2005 2008 2011 2014

Year

Source: Centers for Medicare and Medicaid Services

13. **Discount** During a clearance sale, a three-piece suit that usually sells for $300 is marked "15% off." What is the discount? What is the sale price?

14. **Sale Price** On opening day, a new music store offers a 12% discount on all electric guitars. If the regular price on a guitar is $550, what is the sale price?

15. **Total Price** A man buys a washing machine that is on sale. The washing machine usually sells for $450 but is on sale at 20% off. If the sales tax rate in his state is 6%, how much is the total bill for the washer?

16. **Total Price** A bedroom set that normally sells for $1,450 is on sale for 10% off. If the sales tax rate is 5%, what is the total price of the bedroom set if it is bought while on sale?

Calculator Problems

Set up the following problems the same way you set up Problems 1–16. Then use a calculator to do the calculations.

17. **Salary Increase** A teacher making $43,752 per year gets a 6.5% raise. What is the new salary?

18. **Utility Increase** A homeowner had a $15.90 electric bill in December. In January the bill was $17.81. Find the percent increase in the electric bill from December to January. (Round to the nearest whole number.)

19. **Soccer** The rules for soccer state that the playing field must be from 100 to 120 yards long and 55 to 75 yards wide. The 1999 Women's World Cup was played at the Rose Bowl on a playing field 116 yards long and 72 yards wide. The diagram below shows the smallest possible soccer field, the largest possible soccer field, and the soccer field at the Rose Bowl.

Soccer Fields

Smallest   Rose Bowl   Largest

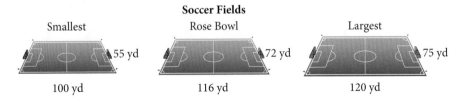

55 yd   72 yd   75 yd

100 yd   116 yd   120 yd

a. **Percent Increase** A team plays on the smallest field, then plays in the Rose Bowl. What is the percent increase in the area of the playing field from the smallest field to the Rose Bowl? Round to the nearest tenth of a percent.

b. **Percent Increase** A team plays a soccer game in the Rose Bowl. The next game is on a field with the largest dimensions. What is the percent increase in the area of the playing field from the Rose Bowl to the largest field? Round to the nearest tenth of a percent.

20. **Football** The diagrams below show the dimensions of playing fields for the National Football League (NFL), the Canadian Football League (CFL), and Arena Football.

Football Fields

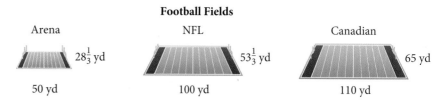

Arena	NFL	Canadian
$28\frac{1}{3}$ yd	$53\frac{1}{3}$ yd	65 yd
50 yd	100 yd	110 yd

a. **Percent Increase** In 1999 Kurt Warner made a successful transition from Arena Football to the NFL, winning the Most Valuable Player award. What was the percent increase in the area of the fields he played on in moving from Arena Football to the NFL? Round to the nearest percent.

b. **Percent Decrease** Doug Flutie played in the Canadian Football League before moving to the NFL. What was the percent decrease in the area of the fields he played on in moving from the CFL to the NFL? Round to the nearest tenth of a percent.

Getting Ready for the Next Section

Multiply. Round to nearest hundredth if necessary.

21. $0.07(2,000)$

22. $0.12(8,000)$

23. $600(0.04)\left(\frac{1}{6}\right)$

24. $900(0.06)\left(\frac{1}{4}\right)$

25. $10,150(0.06)\left(\frac{1}{4}\right)$

26. $10,302.25(0.06)\left(\frac{1}{4}\right)$

Add.

27. $3,210 + 224.7$

28. $900 + 13.50$

29. $10,000 + 150$

30. $10,150 + 152.25$

31. $10,302.25 + 154.53$

32. $10,456.78 + 156.85$

Simplify.

33. $2,000 + 0.07(2,000)$

34. $8,000 + 0.12(8,000)$

35. $3,000 + 0.07(3,000)$

36. $9,000 + 0.12(9,000)$

Interest

Anyone who has borrowed money from a bank or other lending institution, or who has invested money in a savings account, is aware of *interest*. Interest is the amount of money paid for the use of money. If we put $500 in a savings account that pays 6% annually, the interest will be 6% of $500, or 0.06(500) = $30. The amount we invest ($500) is called the *principal*, the percent (6%) is the *interest rate*, and the money earned ($30) is the *interest*.

EXAMPLE 1   A man invests $2,000 in a savings plan that pays 7% per year. How much money will be in the account at the end of 1 year?

SOLUTION   We first find the interest by taking 7% of the principal, $2,000:

$$\text{Interest} = 0.07(\$2,000)$$
$$= \$140$$

The interest earned in 1 year is $140. The total amount of money in the account at the end of a year is the original amount plus the $140 interest:

$2,000	Original investment (principal)
+ 140	Interest (7% of $2,000)
$2,140	Amount after 1 year

The amount in the account after 1 year is $2,140.

EXAMPLE 2   A farmer borrows $8,000 from his local bank at 12%. How much does he pay back to the bank at the end of the year to pay off the loan?

SOLUTION   The interest he pays on the $8,000 is

$$\text{Interest} = 0.12(\$8,000)$$
$$= \$960$$

At the end of the year, he must pay back the original amount he borrowed ($8,000) plus the interest at 12%:

$8,000	Amount borrowed (principal)
+ 960	Interest at 12%
$8,960	Total amount to pay back

The total amount that the farmer pays back is $8,960.

There are many situations in which interest on a loan is figured on other than a yearly basis. Many short-term loans are for only 30 or 60 days. In these cases we can use a formula to calculate the interest that has accumulated. This type of interest is called *simple interest*.

The formula is

$$I = P \cdot R \cdot T$$

where

I = Interest
P = Principal
R = Interest rate (this is the percent)
T = Time (in years, 1 year = 360 days)

We could have used this formula to find the interest in Examples 1 and 2. In those two cases, T is 1. When the length of time is in days rather than years, it is common practice to use 360 days for 1 year, and we write T as a fraction. Examples 3 and 4 illustrate this procedure.

EXAMPLE 3   A student takes out an emergency loan for tuition, books, and supplies. The loan is for $600 at an interest rate of 4%. How much interest does the student pay if the loan is paid back in 60 days?

SOLUTION   The principal P is $600, the rate R is 4% = 0.04, and the time T is $\frac{60}{360}$. Notice that T must be given in years, and 60 days = $\frac{60}{360}$ year. Applying the formula, we have

$$I = P \cdot R \cdot T$$

$$I = 600 \times 0.04 \times \frac{60}{360}$$

$$I = 600 \times 0.04 \times \frac{1}{6} \qquad \frac{60}{360} = \frac{1}{6}$$

$$I = 4 \qquad \text{Multiplication}$$

The interest is $4.

EXAMPLE 4   A woman deposits $900 in an account that pays 6% annually. If she withdraws all the money in the account after 90 days, how much does she withdraw?

SOLUTION   We have P = $900, R = 0.06, and T = 90 days = $\frac{90}{360}$ year. Using these numbers in the formula, we have

$$I = P \cdot R \cdot T$$

$$I = 900 \times 0.06 \times \frac{90}{360}$$

$$I = 900 \times 0.06 \times \frac{1}{4} \qquad \frac{90}{360} = \frac{1}{4}$$

$$I = 13.5 \qquad \text{Multiplication}$$

The interest earned in 90 days is $13.50. If the woman withdraws all the money in her account, she will withdraw

$900.00	Original amount (principal)
+ 13.50	Interest for 90 days
$913.50	Total amount withdrawn

The woman will withdraw $913.50.

A second common kind of interest is *compound interest.* Compound interest includes interest paid on interest. We can use what we know about simple interest to help us solve problems involving compound interest.

EXAMPLE 5   A homemaker puts $3,000 into a savings account that pays 7% compounded annually. How much money is in the account at the end of 2 years?

SOLUTION   Because the account pays 7% annually, the simple interest at the end of 1 year is 7% of $3,000:

$$\text{Interest after 1 year} = 0.07(\$3,000)$$
$$= \$210$$

Because the interest is paid annually, at the end of 1 year the total amount of money in the account is

	$3,000	*Original amount*
+	210	*Interest for 1 year*
	$3,210	*Total in account after 1 year*

The interest paid for the second year is 7% of this new total, or

$$\text{Interest paid the second year} = 0.07(\$3,210)$$
$$= \$224.70$$

At the end of 2 years, the total in the account is

	$3,210.00	*Amount at the beginning of year 2*
+	224.70	*Interest paid for year 2*
	$3,434.70	*Account after 2 years*

At the end of 2 years, the account totals $3,434.70. The total interest earned during this 2-year period is $210 (first year) + $224.70 (second year) = $434.70.

> **Note**
>
> If the interest earned in Example 5 were calculated using the formula for simple interest, $I = P \cdot R \cdot T$, the amount of money in the account at the end of two years would be $3,420.00.

You may have heard of savings and loan companies that offer interest rates that are compounded quarterly. If the interest rate is, say, 6% and it is compounded quarterly, then after every 90 days ($\frac{1}{4}$ of a year) the interest is added to the account. If it is compounded semiannually, then the interest is added to the account every 6 months. Most accounts have interest rates that are compounded daily, which means the simple interest is computed daily and added to the account.

EXAMPLE 6   If $10,000 is invested in a savings account that pays 6% compounded quarterly, how much is in the account at the end of a year?

SOLUTION   The interest for the first quarter ($\frac{1}{4}$ of a year) is calculated using the formula for simple interest:

$$I = P \cdot R \cdot T$$

$$I = \$10,000 \times 0.06 \times \frac{1}{4} \qquad \text{First quarter}$$

$$I = \$150$$

At the end of the first quarter, this interest is added to the original principal. The new principal is $10,000 + $150 = $10,150. Again we apply the formula to calculate the interest for the second quarter:

$$I = \$10,150 \times 0.06 \times \frac{1}{4} \qquad \textit{Second quarter}$$

$$I = \$152.25$$

The principal at the end of the second quarter is $10,150 + $152.25 = $10,302.25. The interest earned during the third quarter is

$$I = \$10,302.25 \times 0.06 \times \frac{1}{4} \qquad \textit{Third quarter}$$

$$I = \$154.53 \qquad \textit{To the nearest cent}$$

The new principal is $10,302.25 + $154.53 = $10,456.78. Interest for the fourth quarter is

$$I = \$10,456.78 \times 0.06 \times \frac{1}{4} \qquad \textit{Fourth quarter}$$

$$I = \$156.85 \qquad \textit{To the nearest cent}$$

The total amount of money in this account at the end of 1 year is

$$\$10,456.78 + \$156.85 = \$10,613.63$$

USING TECHNOLOGY　　*Compound Interest from a Formula*

We can summarize the work above with a formula that allows us to calculate compound interest for any interest rate and any number of compounding periods. If we invest P dollars at an annual interest rate r, compounded n times a year, then the amount of money in the account after t years is given by the formula

$$A = P\left(1 + \frac{r}{n}\right)^{nt}$$

Using numbers from Example 6 to illustrate, we have

P = Principal = $10,000

r = annual interest rate = 0.06

n = number of compounding periods = 4 (interest is compounded quarterly)

t = number of years = 1

Substituting these numbers into the formula above, we have

$$A = 10,000\left(1 + \frac{0.06}{4}\right)^{4 \cdot 1}$$

$$= 10,000(1 + 0.015)^4$$

$$= 10,000(1.015)^4$$

To simplify this last expression on a calculator, we have

Scientific calculator: 10,000 $\boxed{\times}$ 1.015 $\boxed{y^x}$ 4 $\boxed{=}$

Graphing calculator: 10,000 $\boxed{\times}$ 1.015 $\boxed{\wedge}$ 4 $\boxed{\text{ENTER}}$

In either case, the answer is $10,613.63551, which rounds to $10,613.64.

Note

The reason that this answer is different than the result we obtained in Example 6 is that, in Example 6, we rounded each calculation as we did it. The calculator will keep all the digits in all of the intermediate calculations.

GETTING READY FOR CLASS

After reading through the preceding section, respond in your own words and in complete sentences.

1. Suppose the man in Example 1 invested $3,000, instead of $2,000, in the savings plan. How much more interest would he have earned?

2. How much does the student in Example 3 pay back if the loan is paid off after a year, instead of after 60 days?

3. Suppose the homemaker mentioned in Example 5 invests $3,000 in an account that pays only $3\frac{1}{2}\%$ compounded annually. How much is in the account at the end of 2 years?

4. In Example 6, how much money would the account contain at the end of 1 year if it were compounded annually, instead of quarterly?

Problem Set 6.6

These problems are similar to the examples found in this section. They should be set up and solved in the same way. (Problems 1–12 involve simple interest.)

1. **Savings Account** A man invests $2,000 in a savings plan that pays 8% per year. How much money will be in the account at the end of 1 year?

2. **Savings Account** How much simple interest is earned on $5,000 if it is invested for 1 year at 5%?

3. **Savings Account** A savings account pays 7% per year. How much interest will $9,500 invested in such an account earn in a year?

4. **Savings Account** A local bank pays 5.5% annual interest on all savings accounts. If $600 is invested in this type of account, how much will be in the account at the end of a year?

5. **Bank Loan** A farmer borrows $8,000 from his local bank at 7%. How much does he pay back to the bank at the end of the year when he pays off the loan?

6. **Bank Loan** If $400 is borrowed at a rate of 12% for 1 year, how much is the interest?

7. **Bank Loan** A bank lends one of its customers $2,000 at 8% for 1 year. If the customer pays the loan back at the end of the year, how much does he pay the bank?

8. **Bank Loan** If a loan of $2,000 at 20% for 1 year is to be paid back in one payment at the end of the year, how much does the borrower pay the bank?

9. **Student Loan** A student takes out an emergency loan for tuition, books, and supplies. The loan is for $600 with an interest rate of 5%. How much interest does the student pay if the loan is paid back in 60 days?

10. **Short-Term Loan** If a loan of $1,200 at 9% is paid off in 90 days, what is the interest?

11. **Savings Account** A woman deposits $800 in a savings account that pays 5%. If she withdraws all the money in the account after 120 days, how much does she withdraw?

12. **Savings Account** $1,800 is deposited in a savings account that pays 6%. If the money is withdrawn at the end of 30 days, how much interest is earned?

The problems that follow involve compound interest.

13. **Compound Interest** A woman puts $5,000 into a savings account that pays 6% compounded annually. How much money is in the account at the end of 2 years?

14. **Compound Interest** A savings account pays 5% compounded annually. If $10,000 is deposited in the account, how much is in the account at the end of 2 years?

15. **Compound Interest** If $8,000 is invested in a savings account that pays 5% compounded quarterly, how much is in the account at the end of a year?

16. **Compound Interest** Suppose $1,200 is invested in a savings account that pays 6% compounded semiannually. How much is in the account at the end of $1\frac{1}{2}$ years?

Calculator Problems

The following problems should be set up in the same way in which Problems 1–16 have been set up. Then the calculations should be done on a calculator.

17. **Savings Account** A woman invests $917.26 in a savings account that pays 6.25% annually. How much is in the account at the end of a year?

18. **Business Loan** The owner of a clothing store borrows $6,210 for 1 year at 11.5% interest. If he pays the loan back at the end of the year, how much does he pay back?

19. **Compound Interest** Suppose $10,000 is invested in each account below. In each case find the amount of money in the account at the end of 5 years.
    a. Annual interest rate = 6%, compounded quarterly
    b. Annual interest rate = 6%, compounded monthly
    c. Annual interest rate = 5%, compounded quarterly
    d. Annual interest rate = 5%, compounded monthly

20. **Compound Interest** Suppose $5,000 is invested in each account below. In each case find the amount of money in the account at the end of 10 years.
    a. Annual interest rate = 5%, compounded quarterly
    b. Annual interest rate = 6%, compounded quarterly
    c. Annual interest rate = 7%, compounded quarterly
    d. Annual interest rate = 8%, compounded quarterly

Chapter 6 Summary

The Meaning of Percent [6.1]

1. 42% means 42 per hundred

or

$$\frac{42}{100}$$

Percent means "per hundred." It is a way of comparing numbers to the number 100.

Changing Percents to Decimals [6.1]

2. 75% = 0.75

To change a percent to a decimal, drop the percent symbol (%), and move the decimal point two places to the *left*.

Changing Decimals to Percents [6.1]

3. 0.25 = 25%

To change a decimal to a percent, move the decimal point two places to the *right*, and use the % symbol.

Changing Percents to Fractions [6.1]

4. $6\% = \frac{6}{100} = \frac{3}{50}$

To change a percent to a fraction, drop the % symbol, and use a denominator of 100. Reduce the resulting fraction to lowest terms if necessary.

Changing Fractions to Percents [6.1]

5. $\frac{3}{4} = 0.75 = 75\%$

or

$\frac{9}{10} = \frac{90}{100} = 90\%$

To change a fraction to a percent, either write the fraction as a decimal and then change the decimal to a percent, or write the fraction as an equivalent fraction with denominator 100, drop the 100, and use the % symbol.

Basic Word Problems Involving Percents [6.2]

6. Translating to equations, we have:

Type A: $n = 0.14(68)$

Type B: $75n = 25$

Type C: $25 = 0.40n$

There are three basic types of word problems:

Type A: What number is 14% of 68?

Type B: What percent of 75 is 25?

Type C: 25 is 40% of what number?

Applications of Percent [6.3, 6.4, 6.5, 6.6]

To solve them, we write *is* as =, *of* as · (multiply), and *what number* or *what percent* as *n*. We then solve the resulting equation to find the answer to the original question.

There are many different kinds of application problems involving percent. They include problems on income tax, sales tax, commission, discount, percent increase and decrease, and interest. Generally, to solve these problems, we restate them as an equivalent problem of Type A, B, or C above. Problems involving simple interest can be solved using the formula

$$I = P \cdot R \cdot T$$

where I = interest, P = principal, R = interest rate, and T = time (in years). It is standard procedure with simple interest problems to use 360 days = 1 year.

> **COMMON MISTAKES**
>
> 1. A common mistake is forgetting to change a percent to a decimal when working problems that involve percents in the calculations. We always change percents to decimals before doing any calculations.
> 2. Moving the decimal point in the wrong direction when converting percents to decimals or decimals to percents is another common mistake. Remember, *percent* means "per hundred." Rewriting a number expressed as a percent as a decimal will make the numerical part smaller.
>
>   25% = 0.25

Chapter 6 Test Form A

These problems are all taken from examples in your text. Work each problem and check your answers. If you have made a mistake, work the problem again. If you cannot get the correct answer after two tries, look up the correct solution in your text.

1. Write each percent as a decimal.

    a. 37%   **b.** 68%   **c.** 120%   **d.** 0.8%

2. Write each decimal as a percent.

    a. 0.27   **b.** 4.89   **c.** 0.5   **d.** 0.09

3. Change $\frac{3}{8}$ to a percent.

4. What number is 15% of 63?

5. What percent of 78 is 31.9?

6. 34 is 29% of what number?

7. On a 120-question test, a student answered 96 correctly. What percent of the problems did the student work correctly?

8. Suppose the purchase price of a stereo system is $396 and the sales tax is $19.80. What is the tax rate?

9. A real estate agent gets 3% of the price of each house she sells as a commission. If she sells a house for $289,500, how much money did she earn?

10. Shoes that usually sell for $25 are on sale for $21. What is the percent decrease in price?

11. If a person earns $22,000 a year and gets a 5% increase in salary, what is the new salary?

12. A man invests $2,000 in a savings plan that pays 7% per year. How much will be in the account at the end of 1 year?

13. A homemaker puts $3,000 into a savings account that pays 7% compounded annually. How much money is in the account at the end of 2 years?

14. A woman deposits $900 in an account that pays 6% annually. If she withdraws all the money in the account after 90 days, how much does she withdraw?

Chapter Test, Form B

For an alternate, more comprehensive, chapter test, go to MathTV.com and select the test and summary for this chapter of the textbook. Click the worksheet labeled Chapter 6 Test, Form B to download it.

Measurement

The speedometer on your car may give your speed in both miles per hour and kilometers per hour. If you are moving at 60 miles per hour, the speedometer will also read 97 kilometers per hour. The table and chart below show the relationship between these two quantities. The numbers in the second column of the table have been rounded to the nearest whole number.

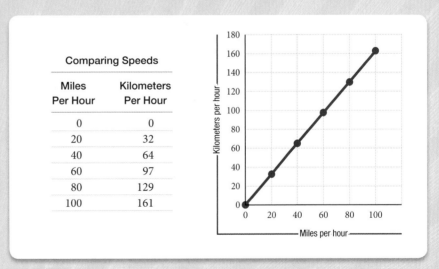

Comparing Speeds

Miles Per Hour	Kilometers Per Hour
0	0
20	32
40	64
60	97
80	129
100	161

In this chapter we look at the process we use to convert from one set of units, such as miles per hour, to another set of units, such as kilometers per hour. You will be interested to know that regardless of the units in question, the method we use is the same in all cases. The method is called *unit analysis* and it is the foundation of this chapter.

Success Skills

iStockphoto.© Tomaz Levstek

If you have made it this far, then you have the study skills necessary to be successful in this course. Success skills are more general in nature and will help you with all your classes and ensure your success in college as well.

Let's start with a question:

Question: What quality is most important for success in any college course?

Answer: Independence. You want to become an independent learner.

We all know people like this. They are generally happy. They don't worry about getting the right instructor, or whether or not things work out every time. They have a confidence that comes from knowing that they are responsible for their success or failure in the goals they set for themselves.

Here are some of the qualities of an independent learner:

- Intends to succeed.
- Doesn't let setbacks deter them.
- Knows their resources.
  - ▷ Instructor's office hours
  - ▷ Math lab
  - ▷ Student Solutions Manual
  - ▷ Group study
  - ▷ Internet
- Doesn't mistake activity for achievement.
- Has a positive attitude.

There are other traits as well. The first step in becoming an independent learner is doing a little self-evaluation and then making of list of traits that you would like to acquire. What skills do you have that align with those of an independent learner? What attributes do you have that keep you from being an independent learner? What qualities would you like to obtain that you don't have now?

Unit Analysis I: Length

Introduction . . .

In this section we will become more familiar with the units used to measure length. We will look at the U.S. system of measurement and the metric system of measurement.

Measuring the length of an object is done by assigning a number to its length. To let other people know what that number represents, we include with it a unit of measure. The most common units used to represent length in the U.S. system are inches, feet, yards, and miles. The basic unit of length is the foot. The other units are defined in terms of feet, as Table 1 shows.

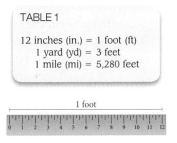

TABLE 1

12 inches (in.) = 1 foot (ft)
1 yard (yd) = 3 feet
1 mile (mi) = 5,280 feet

1 foot

As you can see from the table, the abbreviations for inches, feet, yards, and miles are in., ft, yd, and mi, respectively. What we haven't indicated, even though you may not have realized it, is what 1 foot represents. We have defined all our units associated with length in terms of feet, but we haven't said what a foot is.

There is a long history of the evolution of what is now called a foot. At different times in the past, a foot has represented different arbitrary lengths. Currently, a foot is defined to be exactly 0.3048 meter (the basic measure of length in the metric system), where a meter is 1,650,763.73 wavelengths of the orange-red line in the spectrum of krypton-86 in a vacuum (this doesn't mean much to me either). The reason a foot and a meter are defined this way is that we always want them to measure the same length. Because the wavelength of the orange-red line in the spectrum of krypton-86 will always remain the same, so will the length that a foot represents.

Now that we have said what we mean by 1 foot (even though we may not understand the technical definition), we can go on and look at some examples that involve converting from one kind of unit to another.

EXAMPLE 1 Convert 5 feet to inches.

SOLUTION Because 1 foot = 12 inches, we can multiply 5 by 12 inches to get

$$5 \text{ feet} = 5 \times 12 \text{ inches}$$
$$= 60 \text{ inches}$$

This method of converting from feet to inches probably seems fairly simple. But as we go further in this chapter, the conversions from one kind of unit to another will become more complicated. For these more complicated problems, we need another way to show conversions so that we can be certain to end them with the correct unit of measure. For example, since 1 ft = 12 in., we can say that there are 12 in. per 1 ft or 1 ft per 12 in. That is:

$$\frac{12 \text{ in.}}{1 \text{ ft}} \longleftarrow \text{Per} \quad \text{or} \quad \frac{1 \text{ ft}}{12 \text{ in.}} \longleftarrow \text{Per}$$

We call the expressions $\frac{12 \text{ in.}}{1 \text{ ft}}$ and $\frac{1 \text{ ft}}{12 \text{ in.}}$ *conversion factors*. The fraction bar is read as "per." Both these conversion factors are really just the number 1. That is:

$$\frac{12 \text{ in.}}{1 \text{ ft}} = \frac{12 \text{ in.}}{12 \text{ in.}} = 1$$

We already know that multiplying a number by 1 leaves the number unchanged. So, to convert from one unit to the other, we can multiply by one of the conversion factors without changing value. Both the conversion factors above say the same thing about the units feet and inches. They both indicate that there are 12 inches in every foot. The one we choose to multiply by depends on what units we are starting with and what units we want to end up with. If we start with feet and we want to end up with inches, we multiply by the conversion factor

$$\frac{12 \text{ in.}}{1 \text{ ft}}$$

The units of feet will divide out and leave us with inches.

$$5 \text{ feet} = 5 \cancel{\text{ ft}} \times \frac{12 \text{ in.}}{1 \cancel{\text{ ft}}}$$
$$= 5 \times 12 \text{ in.}$$
$$= 60 \text{ in.}$$

The key to this method of conversion lies in setting the problem up so that the correct units divide out to simplify the expression. We are treating units such as feet in the same way we treated factors when reducing fractions. If a factor is common to the numerator and the denominator, we can divide it out and simplify the fraction. The same idea holds for units such as feet.

We can rewrite Table 1 so that it shows the conversion factors associated with units of length, as shown in Table 2.

Note

We will use this method of converting from one kind of unit to another throughout the rest of this chapter. You should practice using it until you are comfortable with it and can use it correctly. However, it is not the only method of converting units. You may see shortcuts that will allow you to get results more quickly. Use shortcuts if you wish, so long as you can consistently get correct answers and are not using your shortcuts because you don't understand our method of conversion. Use the method of conversion as given here until you are good at it; then use shortcuts if you want to.

TABLE 2

UNITS OF LENGTH IN THE U.S. SYSTEM

The Relationship Between	Is	To Convert From One To The Other, Multiply By
feet and inches	12 in. = 1 ft	$\frac{12 \text{ in.}}{1 \text{ ft}}$ or $\frac{1 \text{ ft}}{12 \text{ in.}}$
feet and yards	1 yd = 3 ft	$\frac{3 \text{ ft}}{1 \text{ yd}}$ or $\frac{1 \text{ yd}}{3 \text{ ft}}$
feet and miles	1 mi = 5,280 ft	$\frac{5,280 \text{ ft}}{1 \text{ mi}}$ or $\frac{1 \text{ mi}}{5,280 \text{ ft}}$

8 ft

 **EXAMPLE 2**   The most common ceiling height in houses is 8 feet. How many yards is this?

SOLUTION   To convert 8 feet to yards, we multiply by the conversion factor $\frac{1 \text{ yd}}{3 \text{ ft}}$ so that feet will divide out and we will be left with yards.

$$8 \text{ ft} = 8 \cancel{\text{ft}} \times \frac{1 \text{ yd}}{3 \cancel{\text{ft}}} \qquad \textit{Multiply by correct conversion factor}$$

$$= \frac{8}{3} \text{ yd} \qquad\qquad 8 \times \frac{1}{3} = \frac{8}{3}$$

$$= 2\frac{2}{3} \text{ yd} \qquad\qquad \textit{Or 2.67 yd to the nearest hundredth}$$

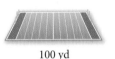 **EXAMPLE 3**   A football field is 100 yards long. How many inches long is a football field?

100 yd

SOLUTION   In this example we must convert yards to feet and then feet to inches. (To make this example more interesting, we are pretending we don't know that there are 36 inches in a yard.) We choose the conversion factors that will allow all the units except inches to divide out.

$$100 \text{ yd} = 100 \cancel{\text{yd}} \times \frac{3 \cancel{\text{ft}}}{1 \cancel{\text{yd}}} \times \frac{12 \text{ in.}}{1 \cancel{\text{ft}}}$$

$$= 100 \times 3 \times 12 \text{ in.}$$
$$= 3,600 \text{ in.}$$

Metric Units of Length

In the metric system the standard unit of length is a meter. A meter is a little longer than a yard (about 3.4 inches longer). The other units of length in the metric system are written in terms of a meter. The metric system uses prefixes to indicate what part of the basic unit of measure is being used. For example, in *milli*meter the prefix *milli* means "one thousandth" of a meter. Table 3 gives the meanings of the most common metric prefixes.

TABLE 3

THE MEANING OF METRIC PREFIXES

Prefix	Meaning
milli	0.001
centi	0.01
deci	0.1
deka	10
hecto	100
kilo	1,000

TABLE 4

METRIC UNITS OF LENGTH

The Relationship Between	Is	To Convert From One To The Other, Multiply By	
millimeters (mm) and meters (m)	1,000 mm = 1 m	$\dfrac{1,000 \text{ mm}}{1 \text{ m}}$ or	$\dfrac{1 \text{ m}}{1,000 \text{ mm}}$
centimeters (cm) and meters	100 cm = 1 m	$\dfrac{100 \text{ cm}}{1 \text{ m}}$ or	$\dfrac{1 \text{ m}}{100 \text{ cm}}$
decimeters (dm) and meters	10 dm = 1 m	$\dfrac{10 \text{ dm}}{1 \text{ m}}$ or	$\dfrac{1 \text{ m}}{10 \text{ dm}}$
dekameters (dam) and meters	1 dam = 10 m	$\dfrac{10 \text{ m}}{1 \text{ dam}}$ or	$\dfrac{1 \text{ dam}}{10 \text{ m}}$
hectometers (hm) and meters	1 hm = 100 m	$\dfrac{100 \text{ m}}{1 \text{ hm}}$ or	$\dfrac{1 \text{ hm}}{100 \text{ m}}$
kilometers (km) and meters	1 km = 1,000 m	$\dfrac{1,000 \text{ m}}{1 \text{ km}}$ or	$\dfrac{1 \text{ km}}{1,000 \text{ m}}$

We can use these prefixes to write the other units of length and conversion factors for the metric system, as given in Table 4.

We use the same method to convert between units in the metric system as we did with the U.S. system. We choose the conversion factor that will allow the units we start with to divide out, leaving the units we want to end up with.

EXAMPLE 4   Convert 25 millimeters to meters.

SOLUTION   To convert from millimeters to meters, we multiply by the conversion factor $\dfrac{1 \text{ m}}{1,000 \text{ mm}}$:

$$25 \text{ mm} = 25 \text{ mm} \times \frac{1 \text{ m}}{1,000 \text{ mm}}$$

$$= \frac{25 \text{ m}}{1,000}$$

$$= 0.025 \text{ m}$$

EXAMPLE 5   Convert 36.5 centimeters to decimeters.

SOLUTION   We convert centimeters to meters and then meters to decimeters:

$$36.5 \text{ cm} = 36.5 \text{ cm} \times \frac{1 \text{ m}}{100 \text{ cm}} \times \frac{10 \text{ dm}}{1 \text{ m}}$$

$$= \frac{36.5 \times 10}{100} \text{ dm}$$

$$= 3.65 \text{ dm}$$

The most common units of length in the metric system are millimeters, centimeters, meters, and kilometers. The other units of length we have listed in our table of metric lengths are not as widely used. The method we have used to convert from one unit of length to another in Examples 1–5 is called *unit analysis*. If you take a chemistry class, you will see it used many times. The same is true of many other science classes as well.

We can summarize the procedure used in unit analysis with the following steps:

Steps Used in Unit Analysis

1. Identify the units you are starting with.
2. Identify the units you want to end with.
3. Find conversion factors that will bridge the starting units and the ending units.
4. Set up the multiplication problem so that all units except the units you want to end with will divide out.

EXAMPLE 6   A sheep rancher is making new lambing pens for the upcoming lambing season. Each pen is a rectangle 6 feet wide and 8 feet long. The fencing material he wants to use sells for $1.36 per foot. If he is planning to build five separate lambing pens (they are separate because he wants a walkway between them), how much will he have to spend for fencing material?

SOLUTION   To find the amount of fencing material he needs for one pen, we find the perimeter of a pen.

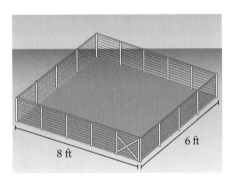

Perimeter = 6 + 6 + 8 + 8 = 28 feet

We set up the solution to the problem using unit analysis. Our starting unit is *pens* and our ending unit is *dollars*. Here are the conversion factors that will form a bridge between pens and dollars:

$$1 \text{ pen} = 28 \text{ feet of fencing}$$
$$1 \text{ foot of fencing} = 1.36 \text{ dollars}$$

Next we write the multiplication problem, using the conversion factors, that will allow all the units except dollars to divide out:

$$5 \text{ pens} = 5 \text{ pens} \times \frac{28 \text{ feet of fencing}}{1 \text{ pen}} \times \frac{1.36 \text{ dollars}}{1 \text{ foot of fencing}}$$

$$= 5 \times 28 \times 1.36 \text{ dollars}$$
$$= \$190.40$$

EXAMPLE 7 In 1993, a ski resort in Vermont advertised their new high-speed chair lift as "the world's fastest chair lift, with a speed of 1,100 feet per second." Show why the speed cannot be correct.

SOLUTION To solve this problem, we can convert feet per second into miles per hour, a unit of measure we are more familiar with on an intuitive level. Here are the conversion factors we will use:

$$1 \text{ mile} = 5,280 \text{ feet}$$

$$1 \text{ hour} = 60 \text{ minutes}$$

$$1 \text{ minute} = 60 \text{ seconds}$$

$$1,100 \text{ ft/second} = \frac{1,100 \text{ feet}}{1 \text{ second}} \times \frac{1 \text{ mile}}{5,280 \text{ feet}} \times \frac{60 \text{ seconds}}{1 \text{ minute}} \times \frac{60 \text{ minutes}}{1 \text{ hour}}$$

$$= \frac{1,100 \times 60 \times 60 \text{ miles}}{5,280 \text{ hours}}$$

$$= 750 \text{ miles/hour}$$

GETTING READY FOR CLASS

After reading through the preceding section, respond in your own words and in complete sentences.

1. Write the relationship between feet and miles. That is, write an equality that shows how many feet are in every mile.

2. Give the metric prefix that means "one hundredth."

3. Give the metric prefix that is equivalent to 1,000.

4. As you know from reading the section in the text, conversion factors are ratios. Write the conversion factor that will allow you to convert from inches to feet. That is, if we wanted to convert 27 inches to feet, what conversion factor would we use?

Problem Set 7.1

Make the following conversions in the U.S. system by multiplying by the appropriate conversion factor. Write your answers as whole numbers or mixed numbers.

1. 5 ft to inches

2. 9 ft to inches

3. 10 ft to inches

4. 20 ft to inches

5. 2 yd to feet

6. 8 yd to feet

7. 4.5 yd to inches

8. 9.5 yd to inches

9. 27 in. to feet

10. 36 in. to feet

11. 2.5 mi to feet

12. 6.75 mi to feet

13. 48 in. to yards

14. 56 in. to yards

Make the following conversions in the metric system by multiplying by the appropriate conversion factor. Write your answers as whole numbers or decimals.

15. 18 m to centimeters

16. 18 m to millimeters

17. 4.8 km to meters

18. 8.9 km to meters

19. 5 dm to centimeters

20. 12 dm to millimeters

21. 248 m to kilometers

22. 969 m to kilometers

23. 67 cm to millimeters

24. 67 mm to centimeters

25. 3,498 cm to meters

26. 4,388 dm to meters

27. 63.4 cm to decimeters

28. 89.5 cm to decimeters

Applying the Concepts

29. Softball If the distance between first and second base in softball is 60 feet, how many yards is it from first to second base?

60 ft

30. Tower Height A transmitting tower is 100 feet tall. How many inches is that?

31. High Jump If a person high jumps 6 feet 8 inches, how many inches is the jump?

32. Desk Width A desk is 48 inches wide. What is the width in yards?

33. Ceiling Height Suppose the ceiling of a home is 2.44 meters above the floor. Express the height of the ceiling in centimeters.

34. **Notebook Width** Standard-sized notebook paper is 21.6 centimeters wide. Express this width in millimeters.

21.6 cm

35. **Dollar Width** A dollar bill is about 6.5 centimeters wide. Express this width in millimeters.

36. **Pencil Length** Most new pencils are 19 centimeters long. Express this length in meters.

37. **Surveying** A unit of measure sometimes used in surveying is the *chain*. There are 80 chains in 1 mile. How many chains are in 37 miles?

38. **Surveying** Another unit of measure used in surveying is a *link;* 1 link is about 8 inches. About how many links are there in 5 feet?

39. **Metric System** A very small unit of measure in the metric system is the *micron* (abbreviated μm). There are 1,000 μm in 1 millimeter. How many microns are in 12 centimeters?

40. **Metric System** Another very small unit of measure in the metric system is the *angstrom* (abbreviated Å). There are 10,000,000 Å in 1 millimeter. How many angstroms are in 15 decimeters?

41. **Horse Racing** In horse racing, 1 *furlong* is 220 yards. How many feet are in 12 furlongs?

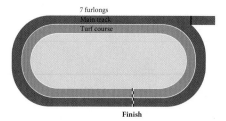

7 furlongs
Main track
Turf course

Finish

42. **Sailing** A *fathom* is 6 feet. How many yards are in 19 fathoms?

43. **Speed Limit** The maximum speed limit on part of Highway 101 in California is 55 miles/hour. Convert 55 miles/hour to feet/second. (Round to the nearest tenth.)

44. **Speed Limit** The maximum speed limit on part of Highway 5 in California is 65 miles/hour. Convert 65 miles/hour to feet/second. (Round to the nearest tenth.)

45. Track and Field A person who runs the 100-yard dash in 10.5 seconds has an average speed of 9.52 yards/second. Convert 9.52 yards/second to miles/hour. (Round to the nearest tenth.)

46. Track and Field A person who runs a mile in 8 minutes has an average speed of 0.125 miles/minute. Convert 0.125 miles/minute to miles/hour.

47. Speed of a Bullet The bullet from a rifle leaves the barrel traveling 1,500 feet/second. Convert 1,500 feet/second to miles/hour. (Round to the nearest whole number.)

48. Speed of a Bullet A bullet from a machine gun on a B-17 Flying Fortress in World War II had a muzzle speed of 1,750 feet/second. Convert 1,750 feet/second to miles/hour. (Round to the nearest whole number.)

49. Farming A farmer is fencing a pasture that is $\frac{1}{2}$ mile wide and 1 mile long. If the fencing material sells for $1.15 per foot, how much will it cost him to fence all four sides of the pasture?

50. Cost of Fencing A family with a swimming pool puts up a chain-link fence around the pool. The fence forms a rectangle 12 yards wide and 24 yards long. If the chain-link fence sells for $2.50 per foot, how much will it cost to fence all four sides of the pool?

51. Farming A 4-H Club group is raising lambs to show at the County Fair. Each lamb eats $\frac{1}{8}$ of a bale of alfalfa a day. If the alfalfa costs $10.50 per bale, how much will it cost to feed one lamb for 120 days?

52. Farming A 4-H Club group is raising pigs to show at the County Fair. Each pig eats 2.4 pounds of grain a day. If the grain costs $5.25 per pound, how much will it cost to feed one pig for 60 days?

Calculator Problems

Set up the following conversions as you have been doing. Then perform the calculations on a calculator.

53. Change 751 miles to feet.

54. Change 639.87 centimeters to meters.

55. Change 4,982 yards to inches.

56. Change 379 millimeters to kilometers.

57. Mount Whitney is the highest point in California. It is 14,494 feet above sea level. Give its height in miles to the nearest tenth.

58. The tallest mountain in the United States is Mount McKinley in Alaska. It is 20,320 feet tall. Give its height in miles to the nearest tenth.

59. California has 3,427 miles of shoreline. How many feet is this?

60. The tip of the television tower at the top of the Empire State Building in New York City is 1,472 feet above the ground. Express this height in miles to the nearest hundredth.

Getting Ready for the Next Section

Perform the indicated operations.

61. 12×12 **62.** 36×24

63. $1 \times 4 \times 2$ **64.** $5 \times 4 \times 2$

65. $10 \times 10 \times 10$ **66.** $100 \times 100 \times 100$

67. $75 \times 43,560$ **68.** $55 \times 43,560$

69. $864 \div 144$ **70.** $1,728 \div 144$

71. $256 \div 640$ **72.** $960 \div 240$

73. $45 \times \dfrac{9}{1}$ **74.** $36 \times \dfrac{9}{1}$

75. $1,800 \times \dfrac{1}{4}$ **76.** $2,000 \times \dfrac{1}{4} \times \dfrac{1}{10}$

77. 1.5×30 **78.** 1.5×45

79. $2.2 \times 1,000$ **80.** $3.5 \times 1,000$

81. 67.5×9 **82.** 43.5×9

Unit Analysis II: Area and Volume

7.2

Figure 1 below gives a summary of the geometric objects we have worked with in previous chapters, along with the formulas for finding the area of each object.

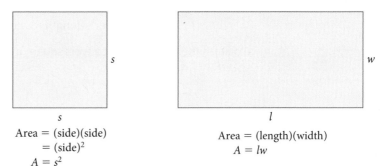

s

w

s

l

Area = (side)(side)
= (side)^2
$A = s^2$

Area = (length)(width)
$A = lw$

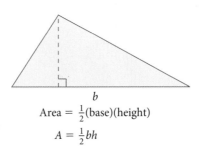

b

Area = $\frac{1}{2}$(base)(height)

$A = \frac{1}{2}bh$

FIGURE 1   Areas of common geometric shapes

EXAMPLE 1   Find the number of square inches in 1 square foot.

SOLUTION   We can think of 1 square foot as 1 ft^2 = 1 ft × ft. To convert from feet to inches, we use the conversion factor 1 foot = 12 inches. Because the unit foot appears twice in 1 ft^2, we multiply by our conversion factor twice.

$$1 \text{ ft}^2 = 1 \text{ ft} \times \text{ft} \times \frac{12 \text{ in.}}{1 \text{ ft}} \times \frac{12 \text{ in.}}{1 \text{ ft}} = 12 \times 12 \text{ in.} \times \text{in.} = 144 \text{ in}^2$$

Now that we know that 1 ft^2 is the same as 144 in^2, we can use this fact as a conversion factor to convert between square feet and square inches. Depending on which units we are converting from, we would use either

$$\frac{144 \text{ in}^2}{1 \text{ ft}^2} \quad \text{or} \quad \frac{1 \text{ ft}^2}{144 \text{ in}^2}$$

EXAMPLE 2   A rectangular poster measures 36 inches by 24 inches. How many square feet of wall space will the poster cover?

SOLUTION   One way to work this problem is to find the number of square inches the poster covers, and then convert square inches to square feet.

$$\text{Area of poster} = \text{length} \times \text{width} = 36 \text{ in.} \times 24 \text{ in.} = 864 \text{ in}^2$$

To finish the problem, we convert square inches to square feet:

$$864 \text{ in}^2 = 864 \cancel{\text{ in}^2} \times \frac{1 \text{ ft}^2}{144 \cancel{\text{ in}^2}}$$

$$= \frac{864}{144} \text{ ft}^2$$

$$= 6 \text{ ft}^2$$

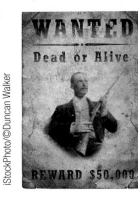

iStockPhoto/©Duncan Walker

Table 1 gives the most common units of area in the U.S. system of measurement, along with the corresponding conversion factors.

TABLE 1

U.S. UNITS OF AREA

The Relationship Between	Is	To Convert From One To The Other, Multiply By	
square inches and square feet	$144 \text{ in}^2 = 1 \text{ ft}^2$	$\dfrac{144 \text{ in}^2}{1 \text{ ft}^2}$ or	$\dfrac{1 \text{ ft}^2}{144 \text{ in}^2}$
square yards and square feet	$9 \text{ ft}^2 = 1 \text{ yd}^2$	$\dfrac{9 \text{ ft}^2}{1 \text{ yd}^2}$ or	$\dfrac{1 \text{ yd}^2}{9 \text{ ft}^2}$
acres and square feet	$1 \text{ acre} = 43{,}560 \text{ ft}^2$	$\dfrac{43{,}560 \text{ ft}^2}{1 \text{ acre}}$ or	$\dfrac{1 \text{ acre}}{43{,}560 \text{ ft}^2}$
acres and square miles	$640 \text{ acres} = 1 \text{ mi}^2$	$\dfrac{640 \text{ acres}}{1 \text{ mi}^2}$ or	$\dfrac{1 \text{ mi}^2}{640 \text{ acres}}$

EXAMPLE 3   A dressmaker orders a bolt of material that is 1.5 yards wide and 30 yards long. How many square feet of material were ordered?

SOLUTION   The area of the material in square yards is

$$A = 1.5 \times 30$$

$$= 45 \text{ yd}^2$$

Converting this to square feet, we have

$$45 \text{ yd}^2 = 45 \cancel{\text{ yd}^2} \times \frac{9 \text{ ft}^2}{1 \cancel{\text{ yd}^2}}$$

$$= 405 \text{ ft}^2$$

EXAMPLE 4   A farmer has 75 acres of land. How many square feet of land does the farmer have?

SOLUTION   Changing acres to square feet, we have

$$75 \text{ acres} = 75 \text{ acres} \times \frac{43,560 \text{ ft}^2}{1 \text{ acre}}$$

$$= 75 \times 43,560 \text{ ft}^2$$
$$= 3,267,000 \text{ ft}^2$$

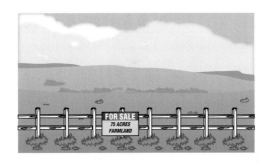

EXAMPLE 5   A new shopping center is to be constructed on 256 acres of land. How many square miles is this?

SOLUTION   Multiplying by the conversion factor that will allow acres to divide out, we have

$$256 \text{ acres} = 256 \text{ acres} \times \frac{1 \text{ mi}^2}{640 \text{ acres}}$$

$$= \frac{256}{640} \text{ mi}^2$$

$$= 0.4 \text{ mi}^2$$

Units of area in the metric system are considerably simpler than those in the U.S. system because metric units are given in terms of powers of 10. Table 2 lists the conversion factors that are most commonly used.

TABLE 2

METRIC UNITS OF AREA

The Relationship Between	Is	To Convert From One To The Other, Multiply By	
square millimeters and square centimeters	$1 \text{ cm}^2 = 100 \text{ mm}^2$	$\dfrac{100 \text{ mm}^2}{1 \text{ cm}^2}$ or	$\dfrac{1 \text{ cm}^2}{100 \text{ mm}^2}$
square centimeters and square decimeters	$1 \text{ dm}^2 = 100 \text{ cm}^2$	$\dfrac{100 \text{ cm}^2}{1 \text{ dm}^2}$ or	$\dfrac{1 \text{ dm}^2}{100 \text{ cm}^2}$
square decimeters and square meters	$1 \text{ m}^2 = 100 \text{ dm}^2$	$\dfrac{100 \text{ dm}^2}{1 \text{ m}^2}$ or	$\dfrac{1 \text{ m}^2}{100 \text{ dm}^2}$
square meters and ares (a)	$1 \text{ a} = 100 \text{ m}^2$	$\dfrac{100 \text{ m}^2}{1 \text{ a}}$ or	$\dfrac{1 \text{ a}}{100 \text{ m}^2}$
ares and hectares (ha)	$1 \text{ ha} = 100 \text{ a}$	$\dfrac{100 \text{ a}}{1 \text{ ha}}$ or	$\dfrac{1 \text{ ha}}{100 \text{ a}}$

EXAMPLE 6   How many square millimeters are in 1 square meter?

SOLUTION   We start with 1 m² and end up with square millimeters:

$$1 \text{ m}^2 = 1 \, \cancel{\text{m}^2} \times \frac{100 \, \cancel{\text{dm}^2}}{1 \, \cancel{\text{m}^2}} \times \frac{100 \, \cancel{\text{cm}^2}}{1 \, \cancel{\text{dm}^2}} \times \frac{100 \text{ mm}^2}{1 \, \cancel{\text{cm}^2}}$$

$$= 100 \times 100 \times 100 \text{ mm}^2$$

$$= 1{,}000{,}000 \text{ mm}^2$$

Units of Measure for Volume

Table 3 lists the units of volume in the U.S. system and their conversion factors.

TABLE 3

UNITS OF VOLUME IN THE U.S. SYSTEM

The Relationship Between	Is	To Convert From One To The Other, Multiply By
cubic inches (in³) and cubic feet (ft³)	1 ft³ = 1,728 in³	$\frac{1{,}728 \text{ in}^3}{1 \text{ ft}^3}$ or $\frac{1 \text{ ft}^3}{1{,}728 \text{ in}^3}$
cubic feet and cubic yards (yd³)	1 yd³ = 27 ft³	$\frac{27 \text{ ft}^3}{1 \text{ yd}^3}$ or $\frac{1 \text{ yd}^3}{27 \text{ ft}^3}$
fluid ounces (fl oz) and pints (pt)	1 pt = 16 fl oz	$\frac{16 \text{ fl oz}}{1 \text{ pt}}$ or $\frac{1 \text{ pt}}{16 \text{ fl oz}}$
pints and quarts (qt)	1 qt = 2 pt	$\frac{2 \text{ pt}}{1 \text{ qt}}$ or $\frac{1 \text{ qt}}{2 \text{ pt}}$
quarts and gallons (gal)	1 gal = 4 qt	$\frac{4 \text{ qt}}{1 \text{ gal}}$ or $\frac{1 \text{ gal}}{4 \text{ qt}}$

EXAMPLE 7   What is the capacity (volume) in pints of a 1-gallon container of milk?

SOLUTION   We change from gallons to quarts and then quarts to pints by multiplying by the appropriate conversion factors as given in Table 3.

$$1 \text{ gal} = 1 \, \cancel{\text{gal}} \times \frac{4 \, \cancel{\text{qt}}}{1 \, \cancel{\text{gal}}} \times \frac{2 \text{ pt}}{1 \, \cancel{\text{qt}}}$$

$$= 1 \times 4 \times 2 \text{ pt}$$

$$= 8 \text{ pt}$$

A 1-gallon container has the same capacity as 8 1-pint containers.

EXAMPLE 8   A dairy herd produces 1,800 quarts of milk each day. How many gallons is this equivalent to?

SOLUTION   Converting 1,800 quarts to gallons, we have

$$1{,}800 \text{ qt} = 1{,}800 \text{ qt} \times \frac{1 \text{ gal}}{4 \text{ qt}}$$

$$= \frac{1{,}800}{4} \text{ gal}$$

$$= 450 \text{ gal}$$

We see that 1,800 quarts is equivalent to 450 gallons.

In the metric system the basic unit of measure for volume is the liter. A liter is the volume enclosed by a cube that is 10 cm on each edge, as shown in Figure 2. We can see that a liter is equivalent to 1,000 cm³.

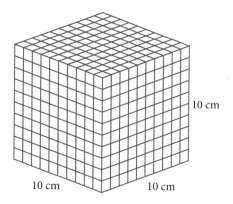

$$1 \text{ liter} = 10 \text{ cm} \times 10 \text{ cm} \times 10 \text{ cm}$$
$$= 1{,}000 \text{ cm}^3$$

FIGURE 2

The other units of volume in the metric system use the same prefixes we encountered previously. The units with prefixes centi, deci, and deka are not as common as the others, so in Table 4 we include only liters, milliliters, hectoliters, and kiloliters.

Note

As you can see from the table and the discussion above, a cubic centimeter (cm³) and a milliliter (mL) are equal. Both are one thousandth of a liter. It is also common in some fields (like medicine) to abbreviate the term cubic centimeter as cc. Although we will use the notation mL when discussing volume in the metric system, you should be aware that 1 mL = 1 cm³ = 1 cc.

TABLE 4

METRIC UNITS OF VOLUME

The Relationship Between	Is	To Convert From One To The Other, Multiply By	
milliliters (mL) and liters	1 liter (L) = 1,000 mL	$\frac{1{,}000 \text{ mL}}{1 \text{ liter}}$ or	$\frac{1 \text{ liter}}{1{,}000 \text{ mL}}$
hectoliters (hL) and liters	100 liters = 1 hL	$\frac{100 \text{ liters}}{1 \text{ hL}}$ or	$\frac{1 \text{ hL}}{100 \text{ liters}}$
kiloliters (kL) and liters	1,000 liters (L) = 1 kL	$\frac{1{,}000 \text{ liters}}{1 \text{ kL}}$ or	$\frac{1 \text{ kL}}{1{,}000 \text{ liters}}$

Here is an example of conversion from one unit of volume to another in the metric system.

EXAMPLE 9   A sports car has a 2.2-liter engine. What is the displacement (volume) of the engine in milliliters?

SOLUTION   Using the appropriate conversion factor from Table 4, we have

$$2.2 \text{ liters} = 2.2 \text{ } \cancel{\text{liters}} \times \frac{1,000 \text{ mL}}{1 \text{ } \cancel{\text{liter}}}$$

$$= 2.2 \times 1,000 \text{ mL}$$
$$= 2,200 \text{ mL}$$

GETTING READY FOR CLASS

After reading through the preceding section, respond in your own words and in complete sentences.

1. Write the formula for the area of each of the following:

  a. a square of side s.

  b. a rectangle with length l and width w.

2. What is the relationship between square feet and square inches?

3. Fill in the numerators below so that each conversion factor is equal to 1.

  a. $\dfrac{\text{qt}}{1 \text{ gal}}$ **b.** $\dfrac{\text{mL}}{1 \text{ liter}}$ **c.** $\dfrac{\text{acres}}{1 \text{ mi}^2}$

4. Write the conversion factor that will allow us to convert from square yards to square feet.

Problem Set 7.2

Use the tables given in this section to make the following conversions. Be sure to show the conversion factor used in each case.

1. 3 ft² to square inches
2. 5 ft² to square inches
3. 288 in² to square feet
4. 720 in² to square feet
5. 30 acres to square feet
6. 92 acres to square feet
7. 2 mi² to acres
8. 7 mi² to acres
9. 1,920 acres to square miles
10. 3,200 acres to square miles
11. 12 yd² to square feet
12. 20 yd² to square feet
13. 17 cm² to square millimeters
14. 150 mm² to square centimeters
15. 2.8 m² to square centimeters
16. 10 dm² to square millimeters
17. 1,200 mm² to square meters
18. 19.79 cm² to square meters
19. 5 a to square meters
20. 12 a to square centimeters
21. 7 ha to ares
22. 3.6 ha to ares
23. 342 a to hectares
24. 986 a to hectares

Applying the Concepts

25. **Sports** The diagrams below show the dimensions of playing fields for the National Football League (NFL), the Canadian Football League, and Arena Football. Find the area of each field and then convert each area to acres. Round answers to the nearest hundredth.

Football Fields

Arena   NFL   Canadian

Arena: $28\frac{1}{3}$ yd, 50 yd
NFL: $53\frac{1}{3}$ yd, 100 yd
Canadian: 65 yd, 110 yd

26. **Soccer** The rules for soccer state that the playing field must be from 100 to 120 yards long and 55 to 75 yards wide. The 1999 Women's World Cup was played at the Rose Bowl on a playing field 116 yards long and 72 yards wide. The diagram below shows the smallest possible soccer field, the largest possible soccer field, and the soccer field at the Rose Bowl. Find the area of each one and then convert the area of each to acres. Round to the nearest tenth, if necessary.

Soccer Fields

Smallest   Rose Bowl   Largest

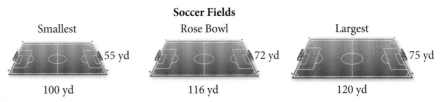

Smallest: 55 yd, 100 yd
Rose Bowl: 72 yd, 116 yd
Largest: 75 yd, 120 yd

355

27. **Swimming Pool** A public swimming pool measures 100 meters by 30 meters and is rectangular. What is the area of the pool in ares?

28. **Construction** A family decides to put tiles in the entryway of their home. The entryway has an area of 6 square meters. If each tile is 5 centimeters by 5 centimeters, how many tiles will it take to cover the entryway?

29. **Landscaping** A landscaper is putting in a brick patio. The area of the patio is 110 square meters. If the bricks measure 10 centimeters by 20 centimeters, how many bricks will it take to make the patio? Assume no space between bricks.

30. **Sewing** A dressmaker is using a pattern that requires 2 square yards of material. If the material is on a bolt that is 54 inches wide, how long must a piece of material be cut from the bolt to be sure there is enough material for the pattern?

Make the following conversions using the conversion factors given in Tables 3 and 4.

31. 5 yd³ to cubic feet	**32.** 3.8 yd³ to cubic feet	**33.** 3 pt to fluid ounces
34. 8 pt to fluid ounces	**35.** 2 gal to quarts	**36.** 12 gal to quarts
37. 2.5 gal to pints	**38.** 7 gal to pints	**39.** 15 qt to fluid ounces
40. 5.9 qt to fluid ounces	**41.** 64 pt to gallons	**42.** 256 pt to gallons
43. 12 pt to quarts	**44.** 18 pt to quarts	**45.** 243 ft³ to cubic yards
46. 864 ft³ to cubic yards	**47.** 5 L to milliliters	**48.** 9.6 L to milliliters
49. 127 mL to liters	**50.** 93.8 mL to liters	**51.** 4 kL to milliliters
52. 3 kL to milliliters	**53.** 14.92 kL to liters	**54.** 4.71 kL to liters

Applying the Concepts

55. **Filling Coffee Cups** If a regular-size coffee cup holds about $\frac{1}{2}$ pint, about how many cups can be filled from a 1-gallon coffee maker?

56. **Filling Glasses** If a regular-size drinking glass holds about 0.25 liter of liquid, how many glasses can be filled from a 750-milliliter container?

57. **Capacity of a Refrigerator** A refrigerator has a capacity of 20 cubic feet. What is the capacity of the refrigerator in cubic inches?

58. **Volume of a Tank** The gasoline tank on a car holds 18 gallons of gas. What is the volume of the tank in quarts?

59. **Filling Glasses** How many 8-fluid-ounce glasses of water will it take to fill a 3-gallon aquarium?

60. **Filling a Container** How many 5-milliliter test tubes filled with water will it take to fill a 1-liter container?

Calculator Problems

Set up the following problems as you have been doing. Then use a calculator to perform the actual calculations. Round all answers to two decimal places where appropriate.

61. **Geography** Lake Superior is the largest of the Great Lakes. It covers 31,700 square miles of area. What is the area of Lake Superior in acres?

62. **Geography** The state of California consists of 156,360 square miles of land and 2,330 square miles of water. Write the total area (both land and water) in acres.

63. **Geography** Death Valley National Monument contains 2,067,795 acres of land. How many square miles is this?

64. **Geography** The Badlands National Monument in South Dakota was established in 1929. It covers 243,302 acres of land. What is the area in square miles?

65. Convert 93.4 qt to gallons.

66. Convert 7,362 fl oz to gallons.

67. How many cubic feet are contained in 796 cubic yards?

68. The engine of a car has a displacement of 440 cubic inches. What is the displacement in cubic feet?

69. **Volume of Water** The Grand Coulee Dam holds 10,585,000 cubic yards of water. What is the volume of water in cubic feet?

70. **Volume of Water** Hoover Dam was built in 1936 on the Colorado River in Nevada. It holds a volume of 4,400,000 cubic yards of water. What is this volume in cubic feet?

Getting Ready for the Next Section

Perform the indicated operations.

71. 12×16

72. 15×16

73. $3 \times 2,000$

74. $5 \times 2,000$

75. $3 \times 1,000 \times 100$

76. $5 \times 1,000 \times 100$

77. 50×250

78. 75×200

79. $12,500 \times \dfrac{1}{1,000}$

80. $15,000 \times \dfrac{1}{1,000}$

Unit Analysis III: Weight

The most common units of weight in the U.S. system are ounces, pounds, and tons. The relationships among these units are given in Table 1.

TABLE 1

UNITS OF WEIGHT IN THE U.S. SYSTEM

The Relationship Between	Is	To Convert From One To The Other, Multiply By
ounces (oz) and pounds (lb)	1 lb = 16 oz	$\dfrac{16\ oz}{1\ lb}$ or $\dfrac{1\ lb}{16\ oz}$
pounds and tons (T)	1 T = 2,000 lb	$\dfrac{2,000\ lb}{1\ T}$ or $\dfrac{1\ T}{2,000\ lb}$

EXAMPLE 1   Convert 12 pounds to ounces.

SOLUTION   Using the conversion factor from the table, and applying the method we have been using, we have

$$12\ lb = 12\ \cancel{lb} \times \frac{16\ oz}{1\ \cancel{lb}}$$
$$= 12 \times 16\ oz$$
$$= 192\ oz$$

12 pounds is equivalent to 192 ounces.

EXAMPLE 2   Convert 3 tons to pounds.

SOLUTION   We use the conversion factor from the table. We have

$$3\ T = 3\ \cancel{T} \times \frac{2,000\ lb}{1\ \cancel{T}}$$
$$= 6,000\ lb$$

6,000 pounds is the equivalent of 3 tons.

In the metric system the basic unit of weight is a gram. We use the same prefixes we have already used to write the other units of weight in terms of grams. Table 2 lists the most common metric units of weight and their conversion factors.

TABLE 2

METRIC UNITS OF WEIGHT

The Relationship Between	Is	To Convert From One To The Other, Multiply By
milligrams (mg) and grams (g)	1 g = 1,000 mg	$\dfrac{1,000\ mg}{1\ g}$ or $\dfrac{1\ g}{1,000\ mg}$
centigrams (cg) and grams (g)	1 g = 100 cg	$\dfrac{100\ cg}{1\ g}$ or $\dfrac{1\ g}{100\ cg}$
kilograms (kg) and grams	1,000 g = 1 kg	$\dfrac{1,000\ g}{1\ kg}$ or $\dfrac{1\ kg}{1,000\ g}$
metric tons (t) and kilograms	1,000 kg = 1 t	$\dfrac{1,000\ kg}{1\ t}$ or $\dfrac{1\ t}{1,000\ kg}$

EXAMPLE 3   Convert 3 kilograms to centigrams.

SOLUTION   We convert kilograms to grams and then grams to centigrams:

$$3 \text{ kg} = 3 \text{ kg} \times \frac{1{,}000 \text{ g}}{1 \text{ kg}} \times \frac{100 \text{ cg}}{1 \text{ g}}$$

$$= 3 \times 1{,}000 \times 100 \text{ cg}$$

$$= 300{,}000 \text{ cg}$$

EXAMPLE 4   A bottle of vitamin C contains 50 tablets. Each tablet contains 250 milligrams of vitamin C. What is the total number of grams of vitamin C in the bottle?

SOLUTION   We begin by finding the total number of milligrams of vitamin C in the bottle. Since there are 50 tablets, and each contains 250 mg of vitamin C, we can multiply 50 by 250 to get the total number of milligrams of vitamin C:

$$\text{Milligrams of vitamin C} = 50 \times 250 \text{ mg}$$

$$= 12{,}500 \text{ mg}$$

Next we convert 12,500 mg to grams:

$$12{,}500 \text{ mg} = 12{,}500 \text{ mg} \times \frac{1 \text{ g}}{1{,}000 \text{ mg}}$$

$$= \frac{12{,}500}{1{,}000} \text{ g}$$

$$= 12.5 \text{ g}$$

The bottle contains 12.5 g of vitamin C.

GETTING READY FOR CLASS

After reading through the preceding section, respond in your own words and in complete sentences.

1. What is the relationship between pounds and ounces?
2. Write the conversion factor used to convert from pounds to ounces.
3. Write the conversion factor used to convert from milligrams to grams.
4. What is the relationship between grams and kilograms?

Problem Set 7.3

Use the conversion factors in Tables 1 and 2 to make the following conversions.

1. 8 lb to ounces

2. 5 lb to ounces

3. 2 T to pounds

4. 5 T to pounds

5. 192 oz to pounds

6. 176 oz to pounds

7. 1,800 lb to tons

8. 10,200 lb to tons

9. 1 T to ounces

10. 3 T to ounces

11. $3\frac{1}{2}$ lb to ounces

12. $5\frac{1}{4}$ lb to ounces

13. $6\frac{1}{2}$ T to pounds

14. $4\frac{1}{5}$ T to pounds

15. 2 kg to grams

16. 5 kg to grams

17. 4 cg to milligrams

18. 3 cg to milligrams

19. 2 kg to centigrams

20. 5 kg to centigrams

21. 5.08 g to centigrams

22. 7.14 g to centigrams

23. 450 cg to grams

24. 979 cg to grams

25. 478.95 mg to centigrams

26. 659.43 mg to centigrams

27. 1,578 mg to grams

28. 1,979 mg to grams

29. 42,000 cg to kilograms

30. 97,000 cg to kilograms

Applying the Concepts

31. Fish Oil A bottle of fish oil contains 60 soft gels, each containing 800 mg of the omega-3 fatty acid. How many total grams of the omega-3 fatty acid are in this bottle?

32. Fish Oil A bottle of fish oil contains 50 soft gels, each containing 300 mg of the omega-6 fatty acid. How many total grams of the omega-6 fatty acid are in this bottle?

33. B-Complex A certain B-complex vitamin supplement contains 50 mg of riboflavin, or vitamin B_2. A bottle contains 80 vitamins. How many total grams of riboflavin are in this bottle?

34. B-Complex A certain B-complex vitamin supplement contains 30 mg of thiamine, or vitamin B_1. A bottle contains 80 vitamins. How many total grams of thiamine are in this bottle?

35. Aspirin A bottle of low-strength aspirin contains 120 tablets. Each tablet contains 81 mg of aspirin. How many total grams of aspirin are in this bottle?

36. Aspirin A bottle of maximum-strength aspirin contains 90 tablets. Each tablet contains 500 mg of aspirin. How many total grams of aspirin are in this bottle?

37. **Vitamin C** A certain brand of vitamin C contains 500 mg per tablet. A bottle contains 240 vitamins. How many total grams of vitamin C are in this bottle?

38. **Vitamin C** A certain brand of vitamin C contains 600 mg per tablet. A bottle contains 150 vitamins. How many total grams of vitamin C are in this bottle?

Coca Cola Bottles The soft drink Coke is sold throughout the world. Although the size of the bottle varies between different countries, a "six-pack" is sold everywhere. For each of the problems below, find the number of liters in a "6-pack" from the given bottle size.

Write each fraction or mixed number as a decimal.

Country	Bottle size	Liters in a 6-pack
39. Estonia	500 mL	
40. Israel	350 mL	
41. Jordan	250 mL	
42. Kenya	300 mL	

Getting Ready for the Next Section

Perform the indicated operations.

43. 8×2.54 **44.** 9×3.28 **45.** $3 \times 1.06 \times 2$ **46.** $3 \times 5 \times 3.79$

47. $80.5 \div 1.61$ **48.** $96.6 \div 1.61$ **49.** $125 \div 2.20$ **50.** $165 \div 2.20$

51. $2,000 \div 16.39$ (Round your answer to the nearest whole number.)

52. $2,200 \div 16.39$ (Round your answer to the nearest whole number.)

53. $\frac{9}{5}(120) + 32$ **54.** $\frac{9}{5}(40) + 32$ **55.** $\frac{5(102 - 32)}{9}$ **56.** $\frac{5(101.6 - 32)}{9}$

Converting Between the Two Systems and Temperature

7.4

Because most of us have always used the U.S. system of measurement in our everyday lives, we are much more familiar with it on an intuitive level than we are with the metric system. We have an intuitive idea of how long feet and inches are, how much a pound weighs, and what a square yard of material looks like. The metric system is actually much easier to use than the U.S. system. The reason some of us have such a hard time with the metric system is that we don't have the feel for it that we do for the U.S. system. We have trouble visualizing how long a meter is or how much a gram weighs. The following list is intended to give you something to associate with each basic unit of measurement in the metric system:

1. A meter is just a little longer than a yard.
2. The length of the edge of a sugar cube is about 1 centimeter.
3. A liter is just a little larger than a quart.
4. A sugar cube has a volume of approximately 1 milliliter.
5. A paper clip weighs about 1 gram.
6. A 2-pound can of coffee weighs about 1 kilogram.

TABLE 1

ACTUAL CONVERSION FACTORS BETWEEN THE METRIC AND U.S. SYSTEMS OF MEASUREMENT

The Relationship Between	Is	To Convert From One To The Other, Multiply By
Length		
inches and centimeters	2.54 cm = 1 in.	$\dfrac{2.54\ \text{cm}}{1\ \text{in.}}$ or $\dfrac{1\ \text{in.}}{2.54\ \text{cm}}$
feet and meters	1 m = 3.28 ft	$\dfrac{3.28\ \text{ft}}{1\ \text{m}}$ or $\dfrac{1\ \text{m}}{3.28\ \text{ft}}$
miles and kilometers	1.61 km = 1 mi	$\dfrac{1.61\ \text{km}}{1\ \text{mi}}$ or $\dfrac{1\ \text{mi}}{1.61\ \text{km}}$
Area		
square inches and square centimeters	6.45 cm² = 1 in²	$\dfrac{6.45\ \text{cm}^2}{1\ \text{in}^2}$ or $\dfrac{1\ \text{in}^2}{6.45\ \text{cm}^2}$
square meters and square yards	1.196 yd² = 1 m²	$\dfrac{1.196\ \text{yd}^2}{1\ \text{m}^2}$ or $\dfrac{1\ \text{m}^2}{1.196\ \text{yd}^2}$
acres and hectares	1 ha = 2.47 acres	$\dfrac{2.47\ \text{acres}}{1\ \text{ha}}$ or $\dfrac{1\ \text{ha}}{2.47\ \text{acres}}$
Volume		
cubic inches and milliliters	16.39 mL = 1 in³	$\dfrac{16.39\ \text{mL}}{1\ \text{in}^3}$ or $\dfrac{1\ \text{in}^3}{16.39\ \text{mL}}$
liters and quarts	1.06 qt = 1 liter	$\dfrac{1.06\ \text{qt}}{1\ \text{liter}}$ or $\dfrac{1\ \text{liter}}{1.06\ \text{qt}}$
gallons and liters	3.79 liters = 1 gal	$\dfrac{3.79\ \text{liters}}{1\ \text{gal}}$ or $\dfrac{1\ \text{gal}}{3.79\ \text{liters}}$
Weight		
ounces and grams	28.3 g = 1 oz	$\dfrac{28.3\ \text{g}}{1\ \text{oz}}$ or $\dfrac{1\ \text{oz}}{28.3\ \text{g}}$
kilograms and pounds	2.20 lb = 1 kg	$\dfrac{2.20\ \text{lb}}{1\ \text{kg}}$ or $\dfrac{1\ \text{kg}}{2.20\ \text{lb}}$

There are many other conversion factors that we could have included in Table 1. We have listed only the most common ones. Almost all of them are approximations. That is, most of the conversion factors are decimals that have been rounded to the nearest hundredth. If we want more accuracy, we obtain a table that has more digits in the conversion factors.

EXAMPLE 1   Convert 8 inches to centimeters.

SOLUTION   Choosing the appropriate conversion factor from Table 1, we have

$$8 \text{ in.} = 8 \text{ in.} \times \frac{2.54 \text{ cm}}{1 \text{ in.}}$$

$$= 8 \times 2.54 \text{ cm} \qquad \text{\textit{This calculation should be done on a calculator}}$$

$$= 20.32 \text{ cm}$$

EXAMPLE 2   Convert 80.5 kilometers to miles.

SOLUTION   Using the conversion factor that takes us from kilometers to miles, we have

$$80.5 \text{ km} = 80.5 \text{ km} \times \frac{1 \text{ mi}}{1.61 \text{ km}}$$

$$= \frac{80.5}{1.61} \text{ mi} \qquad \text{\textit{This calculation should be done on a calculator}}$$

$$= 50 \text{ mi}$$

So 50 miles is equivalent to 80.5 kilometers. If we travel at 50 miles per hour in a car, we are moving at the rate of 80.5 kilometers per hour.

EXAMPLE 3   Convert 3 liters to pints.

SOLUTION   Because Table 1 doesn't list a conversion factor that will take us directly from liters to pints, we first convert liters to quarts, and then convert quarts to pints.

$$3 \text{ liters} = 3 \text{ liters} \times \frac{1.06 \text{ qt}}{1 \text{ liter}} \times \frac{2 \text{ pt}}{1 \text{ qt}}$$

$$= 3 \times 1.06 \times 2 \text{ pt} \qquad \text{\textit{This calculation should be done on a calculator}}$$

$$= 6.36 \text{ pt}$$

EXAMPLE 4   The engine in a car has a 2-liter displacement. What is the displacement in cubic inches?

SOLUTION   We convert liters to milliliters and then milliliters to cubic inches:

$$2 \text{ liters} = 2 \text{ liters} \times \frac{1,000 \text{ mL}}{1 \text{ liter}} \times \frac{1 \text{ in}^3}{16.39 \text{ mL}}$$

$$= \frac{2 \times 1,000}{16.39} \text{ in}^3 \qquad \text{\textit{This calculation should be done on a calculator}}$$

$$= 122 \text{ in}^3 \qquad \text{\textit{To the nearest cubic inch}}$$

EXAMPLE 5   If a person weighs 125 pounds, what is her weight in kilograms?

SOLUTION   Converting from pounds to kilograms, we have

$$125 \text{ lb} = 125 \text{ lb} \times \frac{1 \text{ kg}}{2.20 \text{ lb}}$$

$$= \frac{125}{2.20} \text{kg}$$

$$= 56.8 \text{ kg} \qquad \textit{To the nearest tenth}$$

Temperature

We end this section with a discussion of temperature in both systems of measurement. In the U.S. system we measure temperature on the Fahrenheit scale. On this scale, water boils at 212 degrees and freezes at 32 degrees. When we write 32 degrees measured on the Fahrenheit scale, we use the notation

32°F   (read, "32 degrees Fahrenheit")

In the metric system the scale we use to measure temperature is the Celsius scale (formerly called the centigrade scale). On this scale, water boils at 100 degrees and freezes at 0 degrees. When we write 100 degrees measured on the Celsius scale, we use the notation

100°C   (read, "100 degrees Celsius")

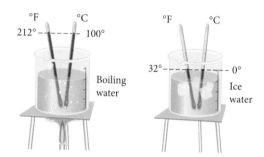

Table 2 is intended to give you a feel for the relationship between the two temperature scales. Table 3 gives the formulas, in both symbols and words, that are used to convert between the two scales.

TABLE 2

Situation	Temperature Fahrenheit	Temperature Celsius
Water freezes	32°F	0°C
Room temperature	68°F	20°C
Normal body temperature	98.6°F	37°C
Water boils	212°F	100°C
Bake cookies	350°F	176.7°C
Broil meat	554°F	290°C

TABLE 3

To Convert From	Formula In Symbols	Formula In Words
Fahrenheit to Celsius	$C = \dfrac{5(F - 32)}{9}$	Subtract 32, multiply by 5, and then divide by 9.
Celsius to Fahrenheit	$F = \dfrac{9}{5}C + 32$	Multiply by $\dfrac{9}{5}$, and then add 32.

The following examples show how we use the formulas given in Table 3.

EXAMPLE 6   Convert 120°C to degrees Fahrenheit.

SOLUTION   We use the formula

$$F = \frac{9}{5}C + 32$$

and replace C with 120:

When $C = 120$

the formula $F = \frac{9}{5}C + 32$

becomes $F = \frac{9}{5}(120) + 32$

$F = 216 + 32$

$F = 248$

We see that 120°C is equivalent to 248°F; they both mean the same temperature.

EXAMPLE 7   A man with the flu has a temperature of 102°F. What is his temperature on the Celsius scale?

SOLUTION   When $F = 102$

the formula $C = \dfrac{5(F - 32)}{9}$

becomes $C = \dfrac{5(102 - 32)}{9}$

$C = \dfrac{5(70)}{9}$

$C = 38.9$ *Rounded to the nearest tenth*

The man's temperature, rounded to the nearest tenth, is 38.9°C on the Celsius scale.

GETTING READY FOR CLASS

After reading through the preceding section, respond in your own words and in complete sentences.

1. Write the equality that gives the relationship between centimeters and inches.

2. Write the equality that gives the relationship between grams and ounces.

3. Fill in the numerators below so that each conversion factor is equal to 1.

    a. $\dfrac{\text{ft}}{1 \text{ meter}}$ b. $\dfrac{\text{qt}}{1 \text{ liter}}$ c. $\dfrac{\text{lb}}{1 \text{ kg}}$

4. Is it a hot day if the temperature outside is 37°C?

Problem Set 7.4

Use Tables 1 and 3 to make the following conversions.

1. 6 in. to centimeters
2. 1 ft to centimeters
3. 4 m to feet
4. 2 km to feet
5. 6 m to yards
6. 15 mi to kilometers
7. 20 mi to meters (round to the nearest hundred meters)
8. 600 m to yards
9. 5 m² to square yards
10. 2 in² to square centimeters
11. 10 ha to acres
12. 50 a to acres
13. 500 in³ to milliliters
14. 400 in³ to liters
15. 2 L to quarts
16. 15 L to quarts
17. 20 gal to liters
18. 15 gal to liters
19. 12 oz to grams
20. 1 lb to grams (round to the nearest 10 grams)
21. 15 kg to pounds
22. 10 kg to ounces
23. 185°C to degrees Fahrenheit
24. 20°C to degrees Fahrenheit
25. 86°F to degrees Celsius
26. 122°F to degrees Celsius

Calculator Problems

Set up the following problems as we have set up the examples in this section. Then use a calculator for the calculations and round your answers to the nearest hundredth.

27. 10 cm to inches
28. 100 mi to kilometers
29. 25 ft to meters
30. 400 mL to cubic inches
31. 49 qt to liters
32. 65 L to gallons
33. 500 g to ounces
34. 100 lb to kilograms

35. **Weight** Give your weight in kilograms.

36. **Height** Give your height in meters and centimeters.

37. **Sports** The 100-yard dash is a popular race in track. How far is 100 yards in meters?

38. **Engine Displacement** A 351-cubic-inch engine has a displacement of how many liters?

39. **Sewing** 25 square yards of material is how many square meters?

40. **Weight** How many grams does a 5 lb 4 oz roast weigh?

41. **Speed** 55 miles per hour is equivalent to how many kilometers per hour?

42. **Capacity** A 1-quart container holds how many liters?

43. **Sports** A high jumper jumps 6 ft 8 in. How many meters is this?

44. **Farming** A farmer owns 57 acres of land. How many hectares is that?

45. **Body Temperature** A person has a temperature of 101°F. What is the person's temperature, to the nearest tenth, on the Celsius scale?

46. **Air Temperature** If the temperature outside is 30°C, is it a better day for water skiing or for snow skiing?

Getting Ready for the Next Section

Perform the indicated operations.

47. $15 + 60$

48. $25 + 60$

49. $\begin{array}{r} 37 \\ + 45 \\ \hline \end{array}$

50. $\begin{array}{r} 27 \\ + 46 \\ \hline \end{array}$

51. $3 + 0.25$

52. $2 + 0.75$

53. $82 - 60$

54. $73 - 60$

55. $\begin{array}{r} 75 \\ - 34 \\ \hline \end{array}$

56. $\begin{array}{r} 85 \\ - 42 \\ \hline \end{array}$

57. 12×4

58. 8×4

59. $3 \times 60 + 15$

60. $2 \times 65 + 45$

61. $3 + 16 \times \dfrac{1}{60}$

62. $2 + 45 \times \dfrac{1}{60}$

63. If fish costs $6.00 per pound, find the cost of 15 pounds.

64. If fish costs $5.00 per pound, find the cost of 14 pounds.

Operations with Time and Mixed Units

Many occupations require the use of a time card. A time card records the number of hours and minutes at work. At the end of a work week the hours and minutes are totaled separately, and then the minutes are converted to hours.

In this section we will perform operations with mixed units of measure. Mixed units are used when we use 2 hours 30 minutes, rather than 2 and a half hours, or 5 feet 9 inches, rather than five and three-quarter feet. As you will see, many of these types of problems arise in everyday life.

The relationship between	is	To convert from one to the other, multiply by
minutes and seconds	1 min = 60 sec	$\frac{1\ min}{60\ sec}$ or $\frac{60\ sec}{1\ min}$
hours and minutes	1 hr = 60 min	$\frac{1\ hr}{60\ min}$ or $\frac{60\ min}{1\ hr}$

EXAMPLE 1   Convert 3 hours 15 minutes to
a. Minutes   **b.** Hours

SOLUTION   **a.** To convert to minutes, we multiply the hours by the conversion factor then add minutes:

$$3 \text{ hr } 15 \text{ min} = 3 \text{ hr} \times \frac{60 \text{ min}}{1 \text{ hr}} + 15 \text{ min}$$
$$= 180 \text{ min} + 15 \text{ min}$$
$$= 195 \text{ min}$$

b. To convert to hours, we multiply the minutes by the conversion factor then add hours:

$$3 \text{ hr } 15 \text{ min} = 3 \text{ hr} + 15 \text{ min} \times \frac{1 \text{ hr}}{60 \text{ min}}$$
$$= 3 \text{ hr} + 0.25 \text{ hr}$$
$$= 3.25 \text{ hr}$$

EXAMPLE 2   Add 5 minutes 37 seconds and 7 minutes 45 seconds.

SOLUTION   First, we align the units properly

$$
\begin{array}{rr}
5 \text{ min} & 37 \text{ sec} \\
+\ 7 \text{ min} & 45 \text{ sec} \\
\hline
12 \text{ min} & 82 \text{ sec}
\end{array}
$$

Since there are 60 seconds in every minute, we write 82 seconds as 1 minute 22 seconds. We have

$$12 \text{ min } 82 \text{ sec} = 12 \text{ min} + 1 \text{ min } 22 \text{ sec}$$
$$= 13 \text{ min } 22 \text{ sec}$$

The idea of adding the units separately is similar to adding mixed fractions. That is, we align the whole numbers with the whole numbers and the fractions with the fractions.

Similarly, when we subtract units of time, we "borrow" 60 seconds from the minutes column, or 60 minutes from the hours column.

EXAMPLE 3   Subtract 34 minutes from 8 hour 15 minutes.

SOLUTION   Again, we first line up the numbers in the hours column, and then the numbers in the minutes column:

$$
\begin{array}{ll}
8\text{ hr} & 15\text{ min} \\
- & 34\text{ min} \\
\hline
\end{array}
\qquad
\begin{array}{ll}
7\text{ hr} & 75\text{ min} \\
- & 34\text{ min} \\
\hline
7\text{ hr} & 41\text{ min}
\end{array}
$$

Next we see how to multiply and divide using units of measure.

EXAMPLE 4   Jake purchases 4 halibut. The fish cost $6.00 per pound, and each weighs 3 lb 12 oz. What is the cost of the fish?

SOLUTION   First, we multiply each unit by 4:

$$
\begin{array}{ll}
3\text{ lb} & 12\text{ oz} \\
\times & 4 \\
\hline
12\text{ lb} & 48\text{ oz}
\end{array}
$$

To convert the 48 ounces to pounds, we multiply the ounces by the conversion factor

$$
\begin{aligned}
12\text{ lb }48\text{ oz} &= 12\text{ lb} + 48\text{ oz} \times \frac{1\text{ lb}}{16\text{ oz}} \\
&= 12\text{ lb} + 3\text{ lb} \\
&= 15\text{ lb}
\end{aligned}
$$

Finally, we multiply the 15 lb and $6.00/lb for a total price of $90.00.

GETTING READY FOR CLASS

After reading through the preceding section, respond in your own words and in complete sentences.

1. Explain the difference between saying *2 and a half hours* and saying *2 hours and 50 minutes*.
2. How are operations with mixed units of measure similar to operations with mixed numbers?
3. Why do we borrow a 60 from the minutes column for the seconds column when subtracting in Example 3?
4. Give an example of when you may have to use multiplication with mixed units of measure.

Problem Set 7.5

Use the tables of conversion factors given in this section and other sections in this chapter to make the following conversions. (Round your answers to the nearest hundredth.)

1. 4 hours 30 minutes to
  a. Minutes
  b. Hours

2. 2 hours 45 minutes to
  a. Minutes
  b. Hours

3. 5 hours 20 minutes to
  a. Minutes
  b. Hours

4. 4 hours 40 minutes to
  a. Minutes
  b. Hours

5. 6 minutes 30 seconds to
  a. Seconds
  b. Minutes

6. 8 minutes 45 seconds to
  a. Seconds
  b. Minutes

7. 5 minutes 20 seconds to
  a. Seconds
  b. Minutes

8. 4 minutes 40 seconds to
  a. Seconds
  b. Minutes

9. 2 pounds 8 ounces to
  a. Ounces
  b. Pounds

10. 3 pounds 4 ounces to
  a. Ounces
  b. Pounds

11. 4 pounds 12 ounces to
  a. Ounces
  b. Pounds

12. 5 pounds 16 ounces to
  a. Ounces
  b. Pounds

13. 4 feet 6 inches to
  a. Inches
  b. Feet

14. 3 feet 3 inches to
  a. Inches
  b. Feet

15. 5 feet 9 inches to
  a. Inches
  b. Feet

16. 3 feet 4 inches to
  a. Inches
  b. Feet

17. 2 gallons 1 quart
  a. Quarts
  b. Gallons

18. 3 gallons 2 quarts
  a. Quarts
  b. Gallons

Perform the indicated operation. Again, remember to use the appropriate conversion factor.

19. Add 4 hours 47 minutes and 6 hours 13 minutes.

20. Add 5 hours 39 minutes and 2 hours 21 minutes.

21. Add 8 feet 10 inches and 13 feet 6 inches.

22. Add 16 feet 7 inches and 7 feet 9 inches.

23. Add 4 pounds 12 ounces and 6 pounds 4 ounces.

24. Add 11 pounds 9 ounces and 3 pounds 7 ounces.

25. Subtract 2 hours 35 minutes from 8 hours 15 minutes.

26. Subtract 3 hours 47 minutes from 5 hours 33 minutes.

27. Subtract 3 hours 43 minutes from 7 hours 30 minutes.

28. Subtract 1 hour 44 minutes from 6 hours 22 minutes.

29. Subtract 4 hours 17 minutes from 5 hours 9 minutes.

30. Subtract 2 hours 54 minutes from 3 hours 7 minutes.

Applying the Concepts

Triathlon The Ironman Triathlon World Championship, held each October in Kona on the island of Hawaii, consists of three parts: a 2.4-mile ocean swim, a 112-mile bike race, and a 26.2-mile marathon. The table shows the results from the 2003 event.

Triathlete	Swim Time (Hr:Min:Sec)	Bike Time (Hr:Min:Sec)	Run Time (Hr:Min:Sec)	Total Time (Hr:Min:Sec)
Peter Reid	0:50:36	4:40:04	2:47:38	
Lori Bowden	0:56:51	5:09:00	3:02:10	

31. Fill in the total time column.

32. How much faster was Peter's total time than Lori's?

33. How much faster was Peter than Lori in the swim?

34. How much faster was Peter than Lori in the run?

35. Cost of Fish Fredrick is purchasing four whole salmon. The fish cost $4.00 per pound, and each weighs 6 lb 8 oz. What is the cost of the fish?

36. Cost of Steak Mike is purchasing eight top sirloin steaks. The meat costs $4.00 per pound, and each steak weighs 1 lb 4 oz. What is the total cost of the steaks?

37. Stationary Bike Maggie rides a stationary bike for 1 hour and 15 minutes, 4 days a week. After 2 weeks, how many hours has she spent riding the stationary bike?

38. Gardening Scott works in his garden for 1 hour and 5 minutes, 3 days a week. After 4 weeks, how many hours has Scott spent gardening?

39. Cost of Fabric Allison is making a quilt. She buys 3 yards and 1 foot each of six different fabrics. The fabrics cost $7.50 a yard. How much will Allison spend?

40. Cost of Lumber Trish is building a fence. She buys six fence posts at the lumberyard, each measuring 5 ft 4 in. The lumber costs $3 per foot. How much will Trish spend?

41. Cost of Avocados Jacqueline is buying six avocados. Each avocado weighs 8 oz. How much will they cost her if avocados cost $2.00 a pound?

42. Cost of Apples Mary is purchasing 12 apples. Each apple weighs 4 oz. If the cost of the apples is $1.50 a pound, how much will Mary pay?

Chapter 7 Summary

Conversion Factors [7.1, 7.2, 7.3, 7.4, 7.5]

EXAMPLES

1. Convert 5 feet to inches.
$$5 \text{ ft} = 5 \text{ ft} \times \frac{12 \text{ in.}}{1 \text{ ft}}$$
$$= 5 \times 12 \text{ in.}$$
$$= 60 \text{ in.}$$

To convert from one kind of unit to another, we choose an appropriate conversion factor from one of the tables given in this chapter. For example, if we want to convert 5 feet to inches, we look for conversion factors that give the relationship between feet and inches. There are two conversion factors for feet and inches:

$$\frac{12 \text{ in.}}{1 \text{ ft}} \quad \text{and} \quad \frac{1 \text{ ft}}{12 \text{ in.}}$$

Length [7.1]

2. Convert 8 feet to yards.
$$8 \text{ ft} = 8 \text{ ft} \times \frac{1 \text{ yd}}{3 \text{ ft}}$$
$$= \frac{8}{3} \text{ yd}$$
$$= 2\frac{2}{3} \text{ yd}$$

U.S. SYSTEM

The Relationship Between	Is	To Convert From One To The Other, Multiply By
feet and inches	12 in. = 1 ft	$\frac{12 \text{ in.}}{1 \text{ ft}}$ or $\frac{1 \text{ ft}}{12 \text{ in.}}$
feet and yards	1 yd = 3 ft	$\frac{3 \text{ ft}}{1 \text{ yd}}$ or $\frac{1 \text{ yd}}{3 \text{ ft}}$
feet and miles	1 mi = 5,280 ft	$\frac{5,280 \text{ ft}}{1 \text{ mi}}$ or $\frac{1 \text{ mi}}{5,280 \text{ ft}}$

3. Convert 25 millimeters to meters.
$$25 \text{ mm}$$
$$= 25 \text{ mm} \times \frac{1 \text{ m}}{1,000 \text{ mm}}$$
$$= \frac{25 \text{ m}}{1,000}$$
$$= 0.025 \text{ m}$$

METRIC SYSTEM

The Relationship Between	Is	To Convert From One To The Other, Multiply By
millimeters (mm) and meters (m)	1,000 mm = 1 m	$\frac{1,000 \text{ mm}}{1 \text{ m}}$ or $\frac{1 \text{ m}}{1,000 \text{ mm}}$
centimeters (cm) and meters	100 cm = 1 m	$\frac{100 \text{ cm}}{1 \text{ m}}$ or $\frac{1 \text{ m}}{100 \text{ cm}}$
decimeters (dm) and meters	10 dm = 1 m	$\frac{10 \text{ dm}}{1 \text{ m}}$ or $\frac{1 \text{ m}}{10 \text{ dm}}$
dekameters (dam) and meters	1 dam = 10 m	$\frac{10 \text{ m}}{1 \text{ dam}}$ or $\frac{1 \text{ dam}}{10 \text{ m}}$
hectometers (hm) and meters	1 hm = 100 m	$\frac{100 \text{ m}}{1 \text{ hm}}$ or $\frac{1 \text{ hm}}{100 \text{ m}}$
kilometers (km) and meters	1 km = 1,000 m	$\frac{1,000 \text{ m}}{1 \text{ km}}$ or $\frac{1 \text{ km}}{1,000 \text{ m}}$

Area [7.2]

4. Convert 256 acres to square miles.

256 acres

$= 256 \text{ acres} \times \dfrac{1 \text{ mi}^2}{640 \text{ acres}}$

$= \dfrac{256}{640} \text{ mi}^2$

$= 0.4 \text{ mi}^2$

U.S. SYSTEM

The Relationship Between	Is	To Convert From One To The Other, Multiply By	
square inches and square feet	144 in² = 1 ft²	$\dfrac{144 \text{ in}^2}{1 \text{ ft}^2}$ or	$\dfrac{1 \text{ ft}^2}{144 \text{ in}^2}$
square yards and square feet	9 ft² = 1 yd²	$\dfrac{9 \text{ ft}^2}{1 \text{ yd}^2}$ or	$\dfrac{1 \text{ yd}^2}{9 \text{ ft}^2}$
acres and square feet	1 acre = 43,560 ft²	$\dfrac{43,560 \text{ ft}^2}{1 \text{ acre}}$ or	$\dfrac{1 \text{ acre}}{43,560 \text{ ft}^2}$
acres and square miles	640 acres = 1 mi²	$\dfrac{640 \text{ acres}}{1 \text{ mi}^2}$ or	$\dfrac{1 \text{ mi}^2}{640 \text{ acres}}$

METRIC SYSTEM

The Relationship Between	Is	To Convert From One To The Other, Multiply By	
square millimeters and square centimeters	1 cm² = 100 mm²	$\dfrac{100 \text{ mm}^2}{1 \text{ cm}^2}$ or	$\dfrac{1 \text{ cm}^2}{100 \text{ mm}^2}$
square centimeters and square decimeters	1 dm² = 100 cm²	$\dfrac{100 \text{ cm}^2}{1 \text{ dm}^2}$ or	$\dfrac{1 \text{ dm}^2}{100 \text{ cm}^2}$
square decimeters and square meters	1 m² = 100 dm²	$\dfrac{100 \text{ dm}^2}{1 \text{ m}^2}$ or	$\dfrac{1 \text{ m}^2}{100 \text{ dm}^2}$
square meters and ares (a)	1 a = 100 m²	$\dfrac{100 \text{ m}^2}{1 \text{ a}}$ or	$\dfrac{1 \text{ a}}{100 \text{ m}^2}$
ares and hectares (ha)	1 ha = 100 a	$\dfrac{100 \text{ a}}{1 \text{ ha}}$ or	$\dfrac{1 \text{ ha}}{100 \text{ a}}$

Volume [7.2]

U.S. SYSTEM

The Relationship Between	Is	To Convert From One To The Other, Multiply By	
cubic inches (in³) and cubic feet (ft³)	1 ft³ = 1,728 in³	$\dfrac{1,728 \text{ in}^3}{1 \text{ ft}^3}$ or	$\dfrac{1 \text{ ft}^3}{1,728 \text{ in}^3}$
cubic feet and cubic yards (yd³)	1 yd³ = 27 ft³	$\dfrac{27 \text{ ft}^3}{1 \text{ yd}^3}$ or	$\dfrac{1 \text{ yd}^3}{27 \text{ ft}^3}$
fluid ounces (fl oz) and pints (pt)	1 pt = 16 fl oz	$\dfrac{16 \text{ fl oz}}{1 \text{ pt}}$ or	$\dfrac{1 \text{ pt}}{16 \text{ fl oz}}$
pints and quarts (qt)	1 qt = 2 pt	$\dfrac{2 \text{ pt}}{1 \text{ qt}}$ or	$\dfrac{1 \text{ qt}}{2 \text{ pt}}$
quarts and gallons (gal)	1 gal = 4 qt	$\dfrac{4 \text{ qt}}{1 \text{ gal}}$ or	$\dfrac{1 \text{ gal}}{4 \text{ qt}}$

5. Convert 2.2 liters to milliliters.

2.2 liters
$= 2.2 \text{ liters} \times \dfrac{1{,}000 \text{ mL}}{1 \text{ liter}}$
$= 2.2 \times 1{,}000 \text{ mL}$
$= 2{,}200 \text{ mL}$

METRIC SYSTEM		
The Relationship Between	Is	To Convert From One To The Other, Multiply By
milliliters (mL) and liters	1 liter (L) = 1,000 mL	$\dfrac{1{,}000 \text{ mL}}{1 \text{ liter}}$ or $\dfrac{1 \text{ liter}}{1{,}000 \text{ mL}}$
hectoliters (hL) and liters	100 liters = 1 hL	$\dfrac{100 \text{ liters}}{1 \text{ hL}}$ or $\dfrac{1 \text{ hL}}{100 \text{ liters}}$
kiloliters (kL) and liters	1,000 liters (L) = 1 kL	$\dfrac{1{,}000 \text{ liters}}{1 \text{ kL}}$ or $\dfrac{1 \text{ kL}}{1{,}000 \text{ liters}}$

Weight [7.3]

6. Convert 12 pounds to ounces.

$12 \text{ lb} = 12 \text{ lb} \times \dfrac{16 \text{ oz}}{1 \text{ lb}}$
$= 12 \times 16 \text{ oz}$
$= 192 \text{ oz}$

U.S. SYSTEM		
The Relationship Between	Is	To Convert From One To The Other, Multiply By
ounces (oz) and pounds (lb)	1 lb = 16 oz	$\dfrac{16 \text{ oz}}{1 \text{ lb}}$ or $\dfrac{1 \text{ lb}}{16 \text{ oz}}$
pounds and tons (T)	1 T = 2,000 lb	$\dfrac{2{,}000 \text{ lb}}{1 \text{ T}}$ or $\dfrac{1 \text{ T}}{2{,}000 \text{ lb}}$

7. Convert 3 kilograms to centigrams.

3 kg
$= 3 \text{ kg} \times \dfrac{1{,}000 \text{ g}}{1 \text{ kg}} \times \dfrac{100 \text{ cg}}{1 \text{ g}}$
$= 3 \times 1{,}000 \times 100 \text{ cg}$
$= 300{,}000 \text{ cg}$

METRIC SYSTEM		
The Relationship Between	Is	To Convert From One To The Other, Multiply By
milligrams (mg) and grams (g)	1 g = 1,000 mg	$\dfrac{1{,}000 \text{ mg}}{1 \text{ g}}$ or $\dfrac{1 \text{ g}}{1{,}000 \text{ mg}}$
centigrams (cg) and grams	1 g = 100 cg	$\dfrac{100 \text{ cg}}{1 \text{ g}}$ or $\dfrac{1 \text{ g}}{100 \text{ cg}}$
kilograms (kg) and grams	1,000 g = 1 kg	$\dfrac{1{,}000 \text{ g}}{1 \text{ kg}}$ or $\dfrac{1 \text{ kg}}{1{,}000 \text{ g}}$
metric tons (t) and kilograms	1,000 kg = 1 t	$\dfrac{1{,}000 \text{ kg}}{1 \text{ t}}$ or $\dfrac{1 \text{ t}}{1{,}000 \text{ kg}}$

Converting Between the Systems [7.4]

8. Convert 8 inches to centimeters.

$8 \text{ in.} = 8 \text{ in.} \times \dfrac{2.54 \text{ cm}}{1 \text{ in.}}$

$= 8 \times 2.54 \text{ cm}$

$= 20.32 \text{ cm}$

CONVERSION FACTORS

The Relationship Between	Is	To Convert From One To The Other, Multiply By
Length		
inches and centimeters	2.54 cm = 1 in.	$\dfrac{2.54 \text{ cm}}{1 \text{ in.}}$ or $\dfrac{1 \text{ in.}}{2.54 \text{ cm}}$
feet and meters	1 m = 3.28 ft	$\dfrac{3.28 \text{ ft}}{1 \text{ m}}$ or $\dfrac{1 \text{ m}}{3.28 \text{ ft}}$
miles and kilometers	1.61 km = 1 mi	$\dfrac{1.61 \text{ km}}{1 \text{ mi}}$ or $\dfrac{1 \text{ mi}}{1.61 \text{ km}}$
Area		
square inches and square centimeters	6.45 cm² = 1 in²	$\dfrac{6.45 \text{ cm}^2}{1 \text{ in}^2}$ or $\dfrac{1 \text{ in}^2}{6.45 \text{ cm}^2}$
square meters and square yards	1.196 yd² = 1 m²	$\dfrac{1.196 \text{ yd}^2}{1 \text{ m}^2}$ or $\dfrac{1 \text{ m}^2}{1.196 \text{ yd}^2}$
acres and hectares	1 ha = 2.47 acres	$\dfrac{2.47 \text{ acres}}{1 \text{ ha}}$ or $\dfrac{1 \text{ ha}}{2.47 \text{ acres}}$
Volume		
cubic inches and milliliters	16.39 mL = 1 in³	$\dfrac{16.39 \text{ mL}}{1 \text{ in}^3}$ or $\dfrac{1 \text{ in}^3}{16.39 \text{ mL}}$
liters and quarts	1.06 qt = 1 liter	$\dfrac{1.06 \text{ qt}}{1 \text{ liter}}$ or $\dfrac{1 \text{ liter}}{1.06 \text{ qt}}$
gallons and liters	3.79 liters = 1 gal	$\dfrac{3.79 \text{ liters}}{1 \text{ gal}}$ or $\dfrac{1 \text{ gal}}{3.79 \text{ liters}}$
Weight		
ounces and grams	28.3 g = 1 oz	$\dfrac{28.3 \text{ g}}{1 \text{ oz}}$ or $\dfrac{1 \text{ oz}}{28.3 \text{ g}}$
kilograms and pounds	2.20 lb = 1 kg	$\dfrac{2.20 \text{ lb}}{1 \text{ kg}}$ or $\dfrac{1 \text{ kg}}{2.20 \text{ lb}}$

Temperature [7.4]

9. Convert 120°C to degrees Fahrenheit.

$F = \dfrac{9}{5}C + 32$

$F = \dfrac{9}{5}(120) + 32$

$F = 216 + 32$

$F = 248$

To Convert From	Formula In Symbols	Formula In Words
Fahrenheit to Celsius	$C = \dfrac{5(F - 32)}{9}$	Subtract 32, multiply by 5, and then divide by 9.
Celsius to Fahrenheit	$F = \dfrac{9}{5}C + 32$	Multiply by $\dfrac{9}{5}$, and then add 32.

Time [7.5]

10. Convert 3 hours 45 minutes to minutes.

$= 3 \text{ hr} \times \dfrac{60 \text{ min}}{1 \text{ hr}} + 45 \text{ min}$

$= 180 \text{ min} + 45 \text{ min}$

$= 225 \text{ min}$

The relationship between	is	To convert from one to the other, multiply by
minutes and seconds	1 min = 60 sec	$\dfrac{1 \text{ min}}{60 \text{ sec}}$ or $\dfrac{60 \text{ sec}}{1 \text{ min}}$
hours and minutes	1 hr = 60 min	$\dfrac{1 \text{ hr}}{60 \text{ min}}$ or $\dfrac{60 \text{ min}}{1 \text{ hr}}$

Chapter 7 Test Form A

These problems are all taken from examples in your text. Work each problem and check your answers. If you have made a mistake, work the problem again. If you cannot get the correct answer after two tries, look up the correct solution in your text.

1. A football field is 100 yards long. Convert 100 yards to inches.

2. Convert 25 millimeters to meters.

3. Convert 36.5 centimeters to decimeters.

4. Find the number of square inches in 1 square foot.

5. How many square millimeters are in 1 square meter?

6. A dairy herd produces 1,800 quarts of milk each day. How many gallons is this?

7. Convert 12 pounds to ounces.

8. Convert 3 kilograms to centigrams.

9. Convert 80.5 kilometers to miles.

10. The engine in a car has a 2-liter displacement. What is the displacement in cubic inches?

11. A man with the flu has a temperature of 102 (degrees) F. What is his temperature in degrees Celsius?

12. Convert 3 hours 15 minutes to:

    a. Minutes   b. Hours

13. Subtract 34 minutes from 8 hours 15 minutes

Chapter Test, Form B

For an alternate, more comprehensive, chapter test, go to MathTV.com and select the test and summary for this chapter of the textbook. Click the worksheet labeled Chapter 7 Test, Form B to download it.

Geometry

Chapter Outline

The Getty Museum in Los Angeles is an example of the partnership of geometry and design. Whether inside the museum at the exhibits, or outside by the buildings and gardens, you will see how simple geometric shapes have enhanced the design of the site. In this chapter we investigate some of the more common characteristics we use to describe these shapes. We will focus on perimeter, circumference, area, and volume. From time to time we will use items from the Getty Museum to assist us in our investigations.

Year	Visitors (Millions)
1997	1.7
1998	1.5
1999	1.3
2000	1.3
2001	1.3

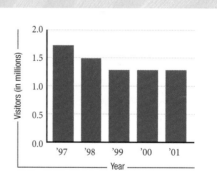

Success Skills

Never mistake activity for achievement.

— John Wooden, legendary UCLA basketball coach

You may think that the John Wooden quote above has to do with being productive and efficient, or using your time wisely, but it is really about being honest with yourself. I have had students come to me after failing a test saying, "I can't understand why I got such a low grade after I put so much time in studying." One student even had help from a tutor and felt she understood everything that we covered. After asking her a few questions, it became clear that she spent all her time studying with a tutor and the tutor was doing most of the work. The tutor can work all the homework problems, but the student cannot. She has mistaken activity for achievement.

Can you think of situations in your life when you are mistaking activity for achievement?

How would you describe someone who is mistaking activity for achievement in the way they study for their math class?

Which of the following best describes the idea behind the John Wooden quote?

▸ Always be efficient.
▸ Don't kid yourself.
▸ Take responsibility for your own success.
▸ Study with purpose.

The world around us is filled with things that can be described in mathematical terms using circumference. Here are a few of the items we will use to explore perimeter and circumference.

Perimeter

We begin this section by reviewing the definition of a polygon, and the definition of perimeter.

> **DEFINITION**
>
> A **polygon** is a closed geometric figure, with at least three sides, in which each side is a straight line segment.

The most common polygons are squares, rectangles, and triangles.

> **DEFINITION**
>
> The **perimeter** of any polygon is the sum of the lengths of the sides, and it is denoted with the letter P.

To find the perimeter of a polygon we add all the lengths of the sides together.
Here are the most common polygons, along with the formula for the perimeter of each.

 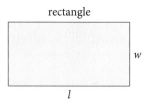 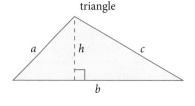

square

rectangle

triangle

$P = 4s$

$P = 2l + 2w$

$P = a + b + c$

We can justify our formulas as follows. If each side of a square is s units long, then the perimeter is found by adding all four sides together:

$$\text{Perimeter} = P = s + s + s + s = 4s$$

Likewise, if a rectangle has a length of l and a width of w, then to find the perimeter we add all four sides together:

$$\text{Perimeter} = P = l + l + w + w = 2l + 2w$$

EXAMPLE 1   Find the perimeter of the given rectangle.

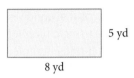

5 yd

8 yd

SOLUTION   The given rectangle has a width of 5 yards and a length of 8 yards. We can use the formula for $P = 2l + 2w$ to find the perimeter.

We have: $P = 2(8) + 2(5)$
$$= 16 + 10$$
$$= 26 \text{ yards}$$

EXAMPLE 2   Find the perimeter of each of the following stamps. Write your answer as a decimal, rounded to the nearest tenth, if necessary.

a. Each side is 35.0 millimeters

b. Base = $2\dfrac{5}{8}$ inches
Other two sides = $1\dfrac{7}{8}$ inches

c. Length = 1.56 inches
Width = 0.99 inches

SOLUTION   We can add all the sides together, or we can apply our formulas. Let's apply the formulas.

   a. $P = 4s = 4 \cdot 35 = 140$ mm

   b. $P = a + b + c = 2\dfrac{5}{8} + 1\dfrac{7}{8} + 1\dfrac{7}{8} = 6\dfrac{3}{8} \approx 6.4$ in.

   c. $P = 2l + 2w = 2(1.56) + 2(0.99) = 5.1$ in.

Circumference

The *circumference* of a circle is the distance around the outside, just as the perimeter of a polygon is the distance around the outside. The circumference of a circle can be found by measuring its radius or diameter and then using the appropriate formula. The *radius* of a circle is the distance from the center of the circle to the circle itself. The radius is denoted by the letter r. The *diameter* of a circle is the distance from one side to the other, through the center. The diameter is denoted by the letter d. In the figure below we can see that the diameter is twice the radius, or

$$d = 2r$$

The relationship between the circumference and the diameter or radius is not as obvious. As a matter of fact, it takes some fairly complicated mathematics to show just what the relationship between the circumference and the diameter is.

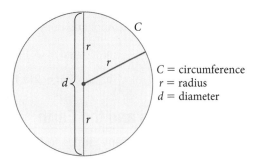

$C = \text{circumference}$
$r = \text{radius}$
$d = \text{diameter}$

If you took a string and actually measured the circumference of a circle by wrapping the string around the circle and then measured the diameter of the same circle, you would find that the ratio of the circumference to the diameter, $\frac{C}{d}$, would be approximately equal to 3.14. The actual ratio of C to d in any circle is an irrational number. It can't be written in decimal form. We use the symbol π (Greek *pi*) to represent this ratio. In symbols, the relationship between the circumference and the diameter in any circle is

$$\frac{C}{d} = \pi$$

Knowing what we do about the relationship between division and multiplication, we can rewrite this formula as

$$C = \pi d$$

This is the formula for the circumference of a circle. When we do the actual calculations, we will use the approximation 3.14 for π.

Because $d = 2r$, the same formula written in terms of the radius is

$$C = 2\pi r$$

EXAMPLE 3   Find the circumference of each coin.

a. 1 Euro coin (Round to the nearest whole number.)

Diameter = 23.25 millimeters

b. Susan B. Anthony dollar (Round to the nearest hundredth.)

Radius = 0.52 inch

SOLUTION   Applying our formulas for circumference we have:

a. $C = \pi d \approx (3.14)(23.25) \approx 73$ mm

b. $C = 2\pi r \approx 2(3.14)(0.52) \approx 3.27$ in.

Circles and the Earth

There are many circles found on the surface of the earth. The most familiar are the latitude and longitude lines. Of these circles, the ones with the largest circumference are called *great circles.* All of the longitude lines are great circles. Of the latitude lines, only the equator is a great circle.

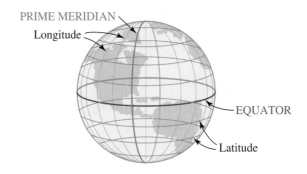

EXAMPLE 4   If the circumference of the earth is approximately 24,900 miles at the equator, what is the diameter of the earth to the nearest 10 miles?

SOLUTION   We substitute 24,900 for C in the formula $C = \pi d$, and then we solve for d.

$$24,900 = \pi d$$

$$24,900 \approx 3.14d \qquad \text{Substitute 3.14 for } \pi$$

$$d \approx \frac{24,900}{3.14} \qquad \text{Divide each side by 3.14}$$

$$d \approx 7,930 \text{ miles}$$

GETTING READY FOR CLASS

After reading through the preceding section, respond in your own words and in complete sentences.

1. What is the perimeter of a polygon?

2. How are perimeter and circumference related?

3. How do you find the perimeter of a square if each side is 15 inches long?

4. A rectangle has a width of 24 feet and a length of 37 feet. How do you find the perimeter?

Problem Set 8.1

Find the perimeter of each figure. The first two figures are squares.

1.

8 in.

2.

9 cm

3.

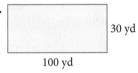

30 yd

100 yd

4.

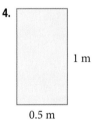

1 m

0.5 m

5.

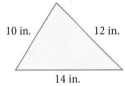

10 in.   12 in.

14 in.

6.

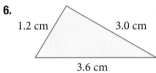

1.2 cm   3.0 cm

3.6 cm

7.

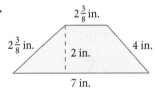

$2\frac{3}{8}$ in.

$2\frac{3}{8}$ in.   4 in.

2 in.

7 in.

8.

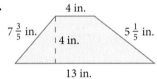

4 in.

$7\frac{3}{5}$ in. $5\frac{1}{5}$ in.

4 in.

13 in.

9.

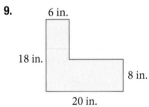

6 in.

18 in.

8 in.

20 in.

10.

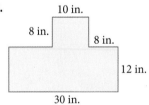

10 in.

8 in.   8 in.

12 in.

30 in.

11.

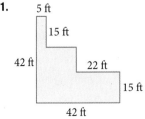

5 ft

15 ft

42 ft   22 ft

15 ft

42 ft

12.

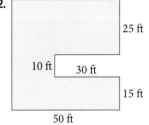

25 ft

10 ft   30 ft

15 ft

50 ft

386

Find the perimeter of each figure. Use 3.14 for π.

13.

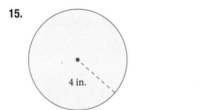

Half circle

4 in.

4 in.

14.

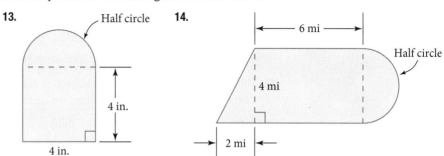

6 mi

4 mi

2 mi

Half circle

Find the circumference of each circle. Use 3.14 for π.

15.

4 in.

16.

2 in.

Applying the Concepts

17. Art at the Getty. "The Musicians' Brawl" by Georges de La Tour is on display at the Getty Museum. The painting is oil on canvas and was created sometime between 1625 and 1630. Find the perimeter.

$33\frac{3}{4}$ in.

$55\frac{1}{2}$ in.

18. Art at the Getty. "Still Life with Blue Pot" by Paul Cézanne has been displayed at the Getty Museum. Painted about 1900, it is watercolor over graphite. Find the perimeter.

$18\frac{7}{8}$ in.

$24\frac{7}{8}$ in.

19. Circumference A dinner plate has a radius of 6 inches. Find the circumference.

20. Circumference A salad plate has a radius of 3 inches. Find the circumference.

21. Circumference The radius of the earth is approximately 3,900 miles. Find the circumference of the earth at the equator. (The equator is a circle around the earth that divides the earth into two equal halves.)

22. Circumference The radius of the moon is approximately 1,100 miles. Find the circumference of the moon around its equator.

23. Perimeter of a Banknote A 10-euro banknote has a width of 67 millimeters and a length of 127 millimeters. Find the perimeter.

24. Perimeter of a Dollar A $10 bill has a width of 2.56 inches and a length of 6.14 inches. Find the perimeter.

25. Circumference of a Coin The $1 coin here depicts Sacagawea and her infant son. The diameter of the coin is 26.5 millimeters. Find the circumference.

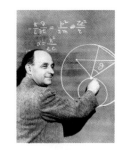

26. Circumference of a Stamp The stamp shown here was issued in Germany in 2000 to commemorate the 100th anniversary of soccer. Find the circumference of the circle if the radius is 14.5 millimeters.

27. Perimeter of a Stamp A U.S. stamp of the Mexican artist Frida Kahlo was issued in 2001. It is the first U.S. stamp to honor an Hispanic woman. The image area of the stamp has a width of 0.84 inch and a length of 1.41 inches. Find the perimeter of the image.

28. Perimeter of a Stamp A U.S. stamp issued in 2001 to honor the Italian scientist Enrico Fermi caused some discussion. Some of the mathematics in the upper left corner of the stamp is incorrect. The image area of the stamp has a width of 21.4 millimeters and a length of 35.8 millimeters. Find the perimeter of the image.

29. Perimeter of the Sierpinski Triangle The diagram shows one stage of what is known as the Sierpinski triangle. Each triangle in the diagram has three equal sides. The large triangle is made up of 4 smaller triangles. If each side of the large triangle is 2 inches, and each side of the smaller triangles is 1 inch, what is the perimeter of the shaded region?

30. Perimeter of the Sierpinski Triangle The diagram here shows another stage of the Sierpinski triangle. Each triangle in the diagram has three equal sides. The largest triangle is made up of a number of smaller triangles. If each side of the large triangle is 2 inches, and each side of the smallest triangles is 0.5 inches, what is the perimeter of the shaded region?

31. Geometry Suppose a rectangle has a perimeter of 12 inches. If the length and the width are whole numbers, give all the possible values for the width. Assume the width is shorter side and the length is the longer side.

32. Geometry Suppose a rectangle has a perimeter of 10 inches. If the length and the width are whole numbers, give all the possible values for the width. Assume the width is the shorter side and the length is the longer side.

33. Geometry If a rectangle has a perimeter of 20 feet, is it possible for the rectangle to be a square? Explain your answer.

34. Geometry If a rectangle has a perimeter of 10 feet, is it possible for the rectangle to be a square? Explain your answer.

35. Geometry A rectangle has a perimeter of 9.5 inches. If the length is 2.75 inches, find the width.

36. Geometry A rectangle has a perimeter of 11 inches. If the width is 2.5 inches, find the length.

Getting Ready for the Next Section

Simplify each expression. Round your answers to the nearest hundredth.

37. $\frac{1}{2} \cdot 6 \cdot 3$

38. $\frac{1}{2} \cdot 4 \cdot 2$

39. $\frac{1}{2}\left(2\frac{5}{8}\right)\left(1\frac{1}{4}\right)$

40. $\frac{1}{2}(6.6)(3.3)$

41. $3.14(14.5)^2$

42. $3.14(5)^2$

43. $144 - 36(3.14)$

44. $100 - 25(3.14)$

Recall that the area of a flat object is a measure of the amount of surface the object has. The area of the rectangle below is 8 square centimeters, because it takes 8 square centimeters to cover it.

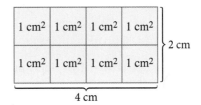

Area

As we have noted previously, the area of this rectangle can also be found by multiplying the length and the width.

$$\begin{aligned} \text{Area} &= (\text{length}) \cdot (\text{width}) \\ &= (4 \text{ centimeters}) \cdot (2 \text{ centimeters}) \\ &= (4 \cdot 2) \cdot (\text{centimeters} \cdot \text{centimeters}) \\ &= 8 \text{ square centimeters} \end{aligned}$$

From this example, and others, we conclude that the area of any rectangle is the product of the length and width.

Here are the most common geometric figures along with the formula for the area of each one. The only formulas that are new to us are the ones that accompany the parallelogram and the circle.

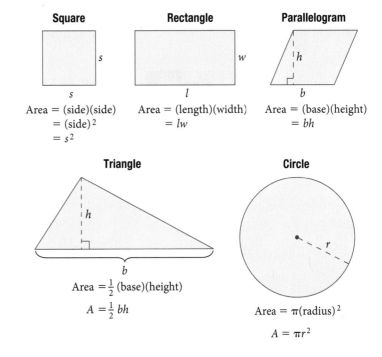

EXAMPLE 1   The parallelogram below has a base of 5 centimeters and a height of 2 centimeters. Find the area.

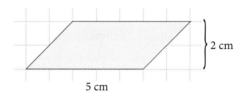

5 cm

2 cm

SOLUTION   If we apply our formula we have

$$\text{Area} = (\text{base})(\text{height})$$
$$A = bh$$
$$= 5 \cdot 2$$
$$= 10 \text{ cm}^2$$

Or, we could simply count the number of square centimeters it takes to cover the object. There are 8 complete squares and 4 half-squares, giving a total of 10 squares for an area of 10 square centimeters. Counting the squares in this manner helps us see why the formula for the area of a parallelogram is the product of the base and the height.

To justify our formula in general, we simply rearrange the parts to form a rectangle.

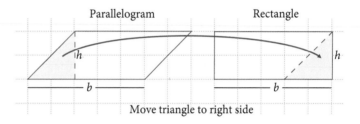

Parallelogram   Rectangle

Move triangle to right side

EXAMPLE 2   The triangle below has a base of 6 centimeters and a height of 3 centimeters. Find the area.

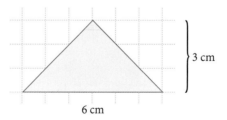

6 cm

3 cm

SOLUTION   If we apply our formula we have

$$\text{Area} = \frac{1}{2}(\text{base})(\text{height})$$

$$A = \frac{1}{2}bh$$

$$= \frac{1}{2} \cdot 6 \cdot 3$$

$$= 9 \text{ cm}^2$$

As was the case in Example 1, we can also count the number of square centimeters it takes to cover the triangle. There are 6 complete squares and 6 half-squares, giving a total of 9 squares for an area of 9 square centimeters.

EXAMPLE 3   Find the area of each of the following stamps.

a. Each side is 35.0 millimeters

b. Write your answer as a decimal, rounded to the nearest hundredth.

Base = $2\frac{5}{8}$ inches

Height = $1\frac{1}{4}$ inches

c. Round to the nearest hundredth.

Length = 1.56 inches
Width = 0.99 inches

SOLUTION   Applying our formulas for area we have

a. $A = s^2 = (35 \text{ mm})^2 = 1{,}225 \text{ mm}^2$

b. $A = \frac{1}{2}bh = \frac{1}{2}(2\frac{5}{8} \text{ in.})(1\frac{1}{4} \text{ in.}) = \frac{1}{2} \cdot \frac{21}{8} \cdot \frac{5}{4} \text{ in}^2 \approx 1.64 \text{ in}^2$

c. $A = lw = (1.56 \text{ in.})(0.99 \text{ in.}) \approx 1.54 \text{ in}^2$

EXAMPLE 4   The circle shown in the stamp in Example 3a has a radius of 14.5 millimeters. Find the area of the circle to the nearest whole number.

SOLUTION   Using our formula for the area of a circle, and using 3.14 for π, we have

$$A = \pi r^2 \qquad \text{Formula for area}$$
$$\approx 3.14(14.5)^2 \qquad \text{Substitute in values}$$
$$\approx 3.14(210.25) \qquad \text{Square 14.5}$$
$$\approx 660.185 \text{ mm}^2 \qquad \text{Multiply}$$
$$\approx 660 \text{ mm}^2 \qquad \text{Round to the nearest whole number}$$

EXAMPLE 5   Find the area of the shaded portion of this figure.

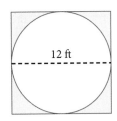

12 ft

SOLUTION   We have a circle inscribed in a square. We notice the diameter of the circle is the same length as one side of the square. To find the area of the shaded region, we subtract the area of the circle from the area of the square as follows:

$$A = 12^2 - 6^2\pi$$
$$= 144 - 36\pi$$
$$\approx 30.96 \text{ ft}^2$$

GETTING READY FOR CLASS

After reading through the preceding section, respond in your own words and in complete sentences.

1. Suppose a rectangle is 8 inches long and 3 inches wide. How many square inches will it take to cover the rectangle?
2. A rectangle measures 4 feet by 6 feet. What units will you assign to the perimeter and to the area?
3. What does π represent?
4. How do you find the area of a circle?

Problem Set 8.2

Find the area enclosed by each figure.

1.

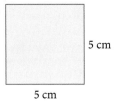

5 cm

5 cm

2.

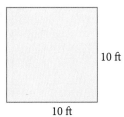

10 ft

10 ft

3.

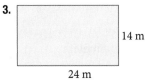

14 m

24 m

4.

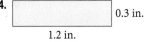

0.3 in.

1.2 in.

5.

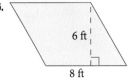

6 ft

10 ft

6.

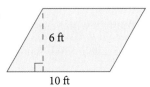

6 ft

8 ft

7.

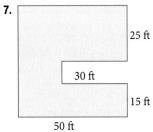

25 ft

30 ft

15 ft

50 ft

8.

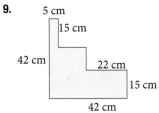

5 ft   5 ft

4 ft

8 ft

15 ft

9.

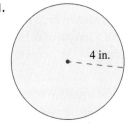

5 cm

15 cm

42 cm

22 cm

15 cm

42 cm

10.

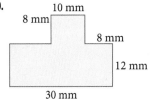

10 mm

8 mm

8 mm

12 mm

30 mm

11.

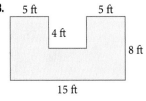

4 in.

12.

2 in.

13.

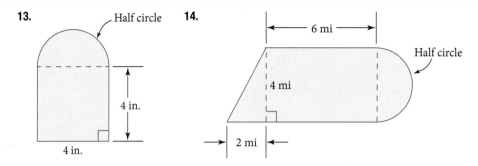

Half circle

4 in.

4 in.

14.

6 mi

Half circle

4 mi

2 mi

15. Find the area of each object.

    a. A square with side 10 inches   **b.** A circle with radius 10 inches

16. Find the area of each object.

    a. A square with side 6 centimeters

    b. A triangle with a base and height of 6 centimeters

    c. A circle with radius 6 centimeters (Round to the nearest whole number.)

17. Find the area of the triangle with base 19 inches and height 14 inches.

18. Find the area of the triangle with base 13 inches and height 8 inches.

19. The base of a triangle is $\frac{4}{3}$ feet and the height is $\frac{2}{3}$ feet. Find the area.

20. The base of a triangle is $\frac{8}{7}$ feet and the height is $\frac{14}{5}$ feet. Find the area.

Applying the Concepts

21. Area A swimming pool is 20 feet wide and 40 feet long. If it is surrounded by square tiles, each of which is 1 foot by 1 foot, how many tiles are there surrounding the pool?

22. Area A garden is rectangular with a width of 8 feet and a length of 12 feet. If it is surrounded by a walkway 2 feet wide, how many square feet of area does the walkway cover?

23. Area of a Stamp A stamp of the Mexican artist Frida Kahlo was issued in 2001. The image area of the stamp has a width of 0.84 inches and a length of 1.41 inches. Find the area of the image. Round to the nearest hundredth.

24. Area of a Stamp A stamp of the Italian scientist Enrico Fermi was issued in 2001. The image area of the stamp has a width of 21.4 millimeters and a length of 35.8 millimeters. Find the area of the image. Round to the nearest whole number.

25. Area of a Euro A 10-euro banknote has a width of 67 millimeters and a length of 127 millimeters. Find the area.

26. Area of a Dollar The $10 bill shown here has a width of 2.56 inches and a length of 6.14 inches. Find the area. Round to the nearest hundredth.

27. Comparing Areas The side of a square is 5 feet long. If all four sides are increased by 2 feet, by how much is the area increased?

28. **Comparing Areas** The length of a side in a square is 20 inches. If all four sides are decreased by 4 inches, by how much is the area decreased?

29. **Area of a Coin** The diameter of this $1 coin is 26.5 mm. Find the area of one side of the coin. Round to the nearest hundredth.

30. **Area of a Coin** The Susan B. Anthony dollar shown at right has a radius of 0.52 inches. Find the area of one side of the coin to the nearest hundredth.

31. **a.** Each side of the red square in the corner is 1 centimeter, and all squares are the same size. On the grid below, draw three more squares. Each side of the first one will be 2 centimeters, each side of the second square will be 3 centimeters, and each side of the third square will be 4 centimeters.

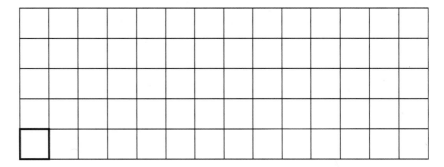

　b. Use the squares you have drawn above to complete each of the following tables.

PERIMETERS OF SQUARES

Length of each Side (in Centimeters)	Perimeter (in Centimeters)
1	
2	
3	
4	

AREAS OF SQUARES

Length of Each Side (in Centimeters)	Area (in Square Centimeters)
1	
2	
3	
4	

32. a. The lengths of the sides of the squares in the grid below are all 1 centimeter. The red square has a perimeter of 12 centimeters. On the grid below, draw two different rectangles, each with a perimeter of 12 centimeters.

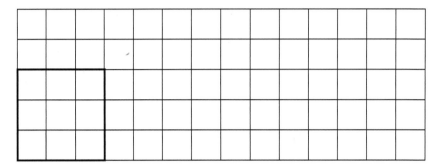

b. Find the area of each of the three figures in part a.

33. The circle here is said to be *inscribed* in the square. If the area of the circle is 64π square centimeters, find the length of one of the diagonals of the square (the distance from *A* to *D*). Round to the nearest tenth.

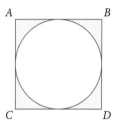

34. The painting below is "An Allegory of Passion" by Hans Holbein the Younger and is on display at the Getty Museum. The painting is oil on panel and was created sometime between 1532-1536. Find the diameter, circumference, and area of the circle that is inside of the square.

$17\frac{7}{8}$ in.

The J. Paul Getty Museum, Los Angeles

Getting Ready for the Next Section

Simplify.

35. $2 \cdot 3 \cdot 4$

36. $2(12)(15)$

37. $12 + 20 + 40$

38. $54 + 40 + 46$

39. $2(3.14)(0.125)(6)$

40. $314 \div 2$

41. $78.5 + 311.5 + 157$

42. $\frac{1}{2}[4(3.14)(100)]$

Surface Area

You have probably heard that 70% of the Earth's surface is water and only 30% land. In this section we will learn how to compute the surface area of any sphere, such as a planet, given its radius. We will also find the surface area of three dimensional shapes, such as cubes and cylinders.

NASA

Surface Area of a Rectangular Solid

The figure below shows a closed box with length l, width w, and height h. The surfaces of the box are labeled as sides, top, bottom, front, and back. A box like this is called a *rectangular solid.* In general, a rectangular solid is a closed figure in which all sides are rectangular that meet at right angles.

A box with dimensions l, w, and h

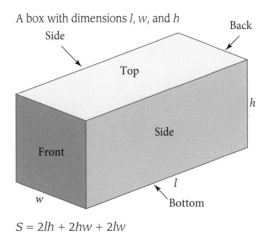

$$S = 2lh + 2hw + 2lw$$

To find the *surface area* of the box, we add the areas of each of the six surfaces that are labeled in the Figure.

$$\text{Surface area} = \text{side} + \text{side} + \text{front} + \text{back} + \text{top} + \text{bottom}$$
$$S = l \cdot h + l \cdot h + h \cdot w + h \cdot w + l \cdot w + l \cdot w$$
$$= 2lh + 2hw + 2lw$$

EXAMPLE 1   Find the surface area of the box shown here.

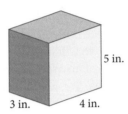

5 in.

3 in.   4 in.

SOLUTION   To find the surface area we find the area of each surface individually, and then we add them together:

$$\text{Surface area} = 2(3 \text{ in.})(4 \text{ in.}) + 2(3 \text{ in.})(5 \text{ in.}) + 2(4 \text{ in.})(5 \text{ in.})$$
$$= 24 \text{ in}^2 + 30 \text{ in}^2 + 40 \text{ in}^2$$
$$= 94 \text{ in}^2$$

The total surface area is 94 square inches.

Surface Area of a Cylinder

Here are the formulas for the surface area of some right circular cylinders. A right cylinder is a cylinder whose base is a circle and whose sides are perpendicular to the base.

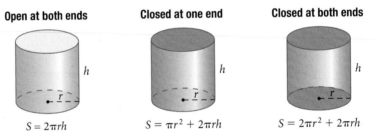

Open at both ends   **Closed at one end**   **Closed at both ends**

h h h

r r r

$S = 2\pi rh$ $S = \pi r^2 + 2\pi rh$ $S = 2\pi r^2 + 2\pi rh$

EXAMPLE 2   The drinking straw shown below has a radius of 0.125 inch and a length of 6 inches. How much material was used to make the straw?

SOLUTION   Since a straw is a cylinder that is open at both ends, we find the amount of material needed to make the straw by calculating the surface area.

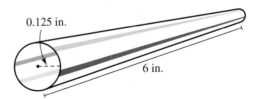

0.125 in.

6 in.

$$S = 2\pi rh$$
$$\approx 2(3.14)(0.125)(6)$$
$$= 4.71 \text{ in}^2$$

It takes 4.71 square inches of material to make the straw.

Surface Area of a Sphere

The figure below shows a sphere and the formula for its surface area.

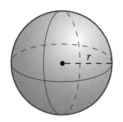

Surface Area = $4\pi(\text{radius})^2$

$S = 4\pi r^2$

EXAMPLE 3   The figure below is composed of a right circular cylinder with half a sphere on top. (A half-sphere is called a hemisphere.) Find the surface area of the figure assuming it is closed on the bottom.

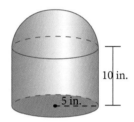

10 in.

5 in.

SOLUTION   The total surface area is found by adding the surface area of the cylinder to the surface area of the hemisphere.

S = surface area of bottom of cylinder + surface area of side of cylinder + surface area of hemisphere

$$= \pi r^2 + 2\pi rh + \frac{1}{2}(4\pi r^2)$$

$$\approx (3.14)(5)^2 + 2(3.14)(5)(10) + \frac{1}{2}[4(3.14)(5^2)]$$

$$= 78.5 + 314 + 157$$

$$= 549.5 \text{ in}^2$$

The total surface area is 549.5 square inches.

GETTING READY FOR CLASS

After reading through the preceding section, respond in your own words and in complete sentences.

1. What is a rectangular solid?
2. How do the formulas for a cylinder open at both ends and a cylinder closed at both ends differ?
3. What is a hemisphere?
4. How are a circle and a sphere related?

Find the surface area of each figure.

1.

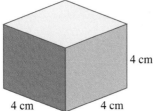

4 cm

4 cm   4 cm

2.

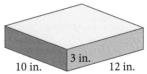

10 in.   3 in.   12 in.

3.

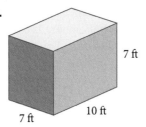

7 ft

7 ft   10 ft

4.

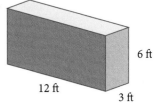

6 ft

12 ft   3 ft

5.

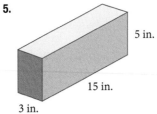

5 in.

15 in.

3 in.

6.

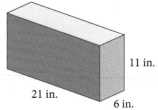

11 in.

21 in.   6 in.

Assume all cylinders are closed at both ends. Round to the nearest hundredth, if necessary.

7.

8 ft

2 ft

8.

8 ft

4 ft

9.

4 ft

2 ft

10.

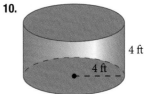

4 ft

4 ft

11.

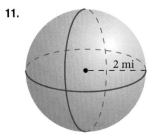

2 mi

12.

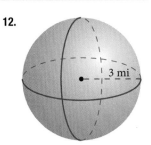

3 mi

13.

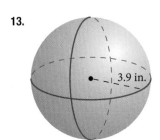

3.9 in.

14.

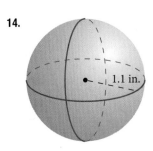

1.1 in.

15.

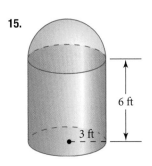

6 ft

3 ft

16.

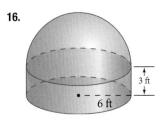

3 ft

6 ft

Applying the Concepts

17. Surface Area of a Coin The $1 coin depicts Sacagawea and her infant son. The diameter of the coin is 26.5 mm, and the thickness is 2.00 mm. Find the surface area of the coin to the nearest hundredth.

18. Surface Area of a Coin A Susan B. Anthony dollar has a radius of 0.52 inch and a thickness of 0.0079 inch. Find the surface area of the coin. Round to the nearest ten thousandth.

19. Travertine at the Getty. Over 1.2 million square feet of travertine stone from Italy was used to construct the Getty Museum. It was mined in large slabs that measured 6 meters high, 12 meters wide, and 2 meters deep. Find the surface area of one of these slabs.

20. Travertine at the Getty. After large slabs of travertine were mined in Italy, they were cut into smaller blocks that were used to construct the Getty Museum. According to the website **www.getty.edu**, "on average, each block at the Getty Center is 76 × 76 centimeters and weighs 115 kilograms, with a typical thickness of 8 centimeters." Find the surface area of one of these stones.

Making a Cylinder. Make an $8\frac{1}{2}$ by 11 inch piece of notebook paper into a cylinder as shown. Use the diagrams to help with Problems 21 and 22.

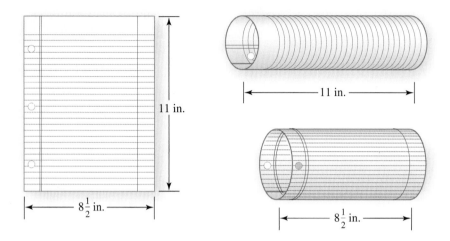

21. **a.** If the length of the cylinder is 11 inches, what is the largest possible radius? Round to the nearest thousandth.
   b. Using the radius from Part a, and the formula for the surface area of a cylinder, find the surface area of the rolled piece of notebook paper. Round to the nearest tenth.
   c. Find the area of the original piece of notebook paper by multiplying length by width. Does this area match the area you found in Part b?

22. **a.** If the length of the cylinder is $8\frac{1}{2}$ inches, what is the largest possible radius? Round to the nearest thousandth.
   b. Using the radius from Part a, and the formula for the surface area of a cylinder, find the surface area of the rolled piece of notebook paper. Round to the nearest tenth.
   c. Find the area of the original piece of notebook paper by multiplying length by width. Does this area match the area you found in Part b?

23. A living room is 10 feet long and 8 feet wide. If the ceiling is 8 feet high, what is the total surface area of the four walls?

24. A family room is 12 feet wide and 14 feet long. If the ceiling is 8 feet high, what is the total surface area of the four walls?

Extending the Concepts

25. The surface of the earth is 70% water. If the diameter of the earth is 8,000 miles, how many square miles of the earth are covered by land.

26. The surface of the earth is 70% water. How many square kilometers of the earth are covered by land if the diameter is 12,874 kilometers?

Surface Area of a Cone. The surface area of a cone is given by the formula $S = \pi r^2 + \pi r l$ where l is called the slant height. We can use the Pythagorean Theorem to find the slant height (see right figure). Using the formula, find the slant height and surface area of the following cones.

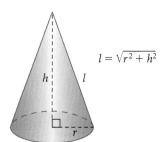

$l = \sqrt{r^2 + h^2}$

27.

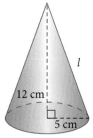

28.

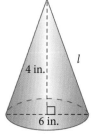

Getting Ready for the Next Section

Simplify each expression.

29. 3^3

30. $15 \cdot 3 \cdot 5$

31. $\dfrac{1}{2} \cdot \dfrac{4}{3}$

32. 0.125^2

Simplify each expression. Round to the nearest thousandth.

33. $3.14(0.125)^2(6)$

34. $\dfrac{2}{3}(392.6)$

35. $3.14(25)(10)$

36. $785 \div 3$

Volume

Next, we move up one dimension and consider what is called volume. Volume is the measure of the space enclosed by a solid. For instance, if each edge of a cube is 3 feet long, then we can think of the cube as being made up of a number of smaller cubes, each of which is 1 foot long, 1 foot wide, and 1 foot high. Each of these smaller cubes is called a cubic foot. To count the number of them in the larger cube, think of the large cube as having three layers. You can see that the top layer contains 9 cubic feet. Because there are three layers, the total number of cubic feet in the large cube is $9 \cdot 3 = 27$.

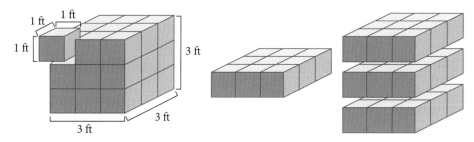

Volume of a Rectangular Solid

On the other hand, if we multiply the length, the width, and the height of the cube, we have the same result:

$$\text{Volume} = (3 \text{ feet})(3 \text{ feet})(3 \text{ feet})$$
$$= (3 \cdot 3 \cdot 3)(\text{feet} \cdot \text{feet} \cdot \text{feet})$$
$$= 27 \text{ cubic feet}$$

For the present we will confine our discussion of volume to volumes of *rectangular solids*. Rectangular solids are the three-dimensional equivalents of rectangles: Opposite sides are parallel, and any two sides that meet, meet at right angles. A rectangular solid is shown below, along with the formula used to calculate its volume.

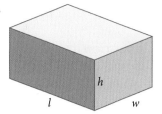

$$\text{Volume} = (\text{length})(\text{width})(\text{height})$$
$$V = lwh$$

EXAMPLE 1   Find the volume of a rectangular solid with length 15 inches, width 3 inches, and height 5 inches.

SOLUTION   To find the volume we apply the formula:

$$V = l \cdot w \cdot h$$
$$= (15 \text{ in.})(3 \text{ in.})(5 \text{ in.})$$
$$= 225 \text{ in}^3$$

Volume of a Cylinder

Here is the formula for the volume of a right circular cylinder, a cylinder whose base is a circle and whose sides are perpendicular to the base.

Volume = $\pi(\text{radius})^2(\text{height})$
$$V = \pi r^2 h$$

EXAMPLE 2   The drinking straw shown below has a radius of 0.125 inch and a length of 6 inches. To the nearest thousandth, find the volume of liquid that it will hold.

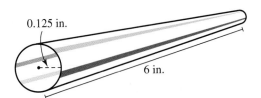

0.125 in.

6 in.

SOLUTION   The total volume is found from the formula for the volume of a right circular cylinder. In this case, the radius is $r = 0.125$, and the height is $h = 6$. We approximate π with 3.14.

$$V = \pi r^2 h$$
$$\approx (3.14)(0.125)^2(6)$$
$$\approx (3.14)(0.015625)(6)$$
$$\approx 0.294 \text{ in}^3 \text{ to the nearest thousandth}$$

Volume of a Cone

Next, we have the formula for the volume of a cone. As you can see, the relevant dimensions of a cone are the radius of its circular base and its height. You will also notice that this formula involves both a fraction, the number $\frac{1}{3}$, and a decimal, the number π, for which we have been using 3.14.

Volume $= \frac{1}{3}\pi(\text{radius})^2(\text{height})$

$V = \frac{1}{3}\pi r^2 h$

EXAMPLE 3   Find the volume of the given cone.

5 cm

3 cm

SOLUTION   The volume is found by using the formula for the volume of a cone. In this case, the radius = 3 cm, and the height = 5 cm. Again, we use 3.14 for π.

$$V = \frac{1}{3}(3.14)(3^2)(5)$$

$$\approx \frac{1}{3}(3.14)(9)(5)$$

$$\approx (3.14)(3)(5)$$

$$\approx (3.14)(15)$$

$$\approx 47.1 \text{ cm}^3$$

Volume of a Sphere

Next we have a sphere and the formula for its volume. Once again, the formula contains both the fraction $\frac{4}{3}$ and the number π.

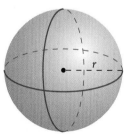

Volume $= \frac{4}{3}\pi(\text{radius})^3$

$= \frac{4}{3}\pi r^3$

EXAMPLE 4   The figure here is composed of a right circular cylinder with half a sphere on top. (A half-sphere is called a hemisphere.) To the nearest tenth, find the total volume enclosed by the figure.

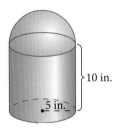

10 in.

5 in.

SOLUTION   The total volume is found by adding the volume of the cylinder to the volume of the hemisphere.

V = volume of cylinder + volume of hemisphere

$$= \pi r^2 h + \frac{1}{2} \cdot \frac{4}{3} \pi r^3$$

$$\approx (3.14)(5)^2(10) + \frac{1}{2} \cdot \frac{4}{3}(3.14)(5)^3$$

$$= (3.14)(25)(10) + \frac{1}{2} \cdot \frac{4}{3}(3.14)(125)$$

$$= 785 + \frac{2}{3}(392.5) \qquad \text{Multiply: } \frac{1}{2} \cdot \frac{4}{3} = \frac{4}{6} = \frac{2}{3}$$

$$= 785 + \frac{785}{3} \qquad\qquad \text{Multiply: } 2(392.5) = 785$$

$$= 785 + 261.7 \qquad\quad \text{Divide 785 by 3, and round to the nearest tenth}$$

$$= 1{,}046.7 \text{ in}^3$$

GETTING READY FOR CLASS

After reading through the preceding section, respond in your own words and in complete sentences.

1. If the dimensions of a rectangular solid are given in inches, what units will be associated with the volume?
2. What is the relationship between area and volume?
3. What formulas from this section involve both a fraction and a decimal?
4. State the volume formula for a cube, a cylinder, a cone, and a sphere.

Problem Set 8.4

Find the volume of each figure. Round to the nearest hundredth, if necessary.

1.

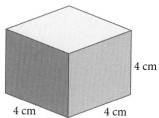

4 cm

4 cm   4 cm

2.

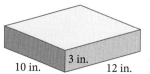

3 in.

10 in.   12 in.

3.

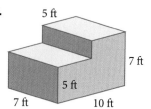

5 ft

7 ft

5 ft

7 ft   10 ft

4.

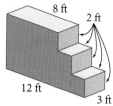

8 ft   2 ft

12 ft

3 ft

5.

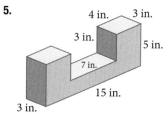

4 in.   3 in.

3 in.

5 in.

7 in.

15 in.

3 in.

6.

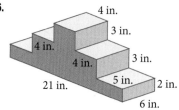

4 in.

3 in.

4 in.

4 in.   3 in.

21 in.   5 in.   2 in.

6 in.

7.

8 ft

2 ft

8.

8 ft

4 ft

9.

4 ft

2 ft

10.

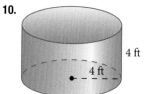

4 ft

4 ft

11.

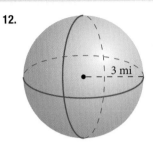

2 mi

12.

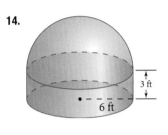

3 mi

13.

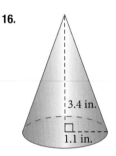

6 ft

3 ft

14.

3 ft

6 ft

15.

7.1 in.

3.9 in.

16.

3.4 in.

1.1 in.

Applying the Concepts

17. Volume of a Coin The $1 coin depicts Sacagawea and her infant son. The diameter of the coin is 26.5 mm, and the thickness is 2.00 mm. Find the volume of the coin to the nearest hundredth.

18. Volume of a Coin The Susan B. Anthony dollar shown here has a radius of 0.52 inch and a thickness of 0.0079 inch. Find the volume of the coin. Round to the nearest ten thousandth.

19. Ice Cream An ice-cream cone has a radius of 2.3 cm and a height of 6.2 cm. The cone is filled with ice-cream and one additional "scoop" in the shape of a sphere with the same radius as the cone is placed on top. What is the amount of ice cream in cubic cm? Round to the nearest hundredth.

20. Ice Cream An ice-cream cone has a radius of 1.7 inches and a height of 4.6 inches. The cone is filled with ice-cream and one and a half additional "scoops" in the shape of a sphere with the same radius as the cone are placed on top. What is the amount of ice cream in cubic inches? Round to the nearest hundredth.

Engine Size The size of an engine is a measure of volume. To calculate the size of an engine is to find the volume of a cylinder and multiply by the number of cylinders. The size of a cylinder is given as a bore size, or diameter, and a stroke size, or height, of the cylinder.

21. Find the size of an 8-cylinder Chevy big-block engine with a bore of 4.25 inches and a stroke of 3.76 inches. Round to the nearest hundredth.

22. Find the size of an 8-cylinder Chevy big-block engine with a bore of 4.47 inches and a stroke of 4.00 inches. Round to the nearest hundredth.

Triangular Pyramid. The formula for finding the volume of a triangular pyramid is one-third the area of the base times the height, or $V = \frac{1}{3} Bh$, where B = the area of the base.

Find the volume of each pyramid. Round to the nearest hundredth.

23.

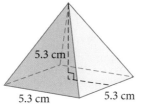

5.3 cm

5.3 cm   5.3 cm

24.

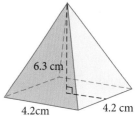

6.3 cm

4.2cm   4.2 cm

25.

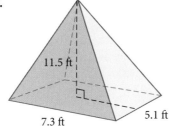

11.5 ft

7.3 ft   5.1 ft

26.

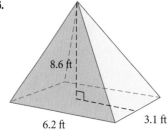

8.6 ft

6.2 ft   3.1 ft

27. **Travertine at the Getty.** After the large slabs of travertine were mined in Italy, they were cut into smaller blocks that were used to construct the Getty Museum. According to the website **www.getty.edu**, "on the average each block at the Getty Center is 76×76 centimeters and weighs 115 kilograms, with a typical thickness of 8 centimeters." Find the volume of an average block of travertine.

28. **Travertine at the Getty.** Find the volume of a large slab of travertine if the slab is 6 meters high, 12 meters wide, and 2 meters deep.

Chapter 8 Summary

EXAMPLES

1. a. Find the perimeter and the area.

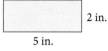

2 in.

5 in.

$P = 2 \cdot 5 + 2 \cdot 2$
$\quad = 14$ in.
$A = 5 \cdot 2$
$\quad = 10$ in²

Formulas for Perimeter and Area of Polygons [8.1, 8.2]

Square

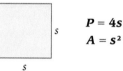

$P = 4s$
$A = s^2$

Rectangle

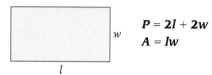

$P = 2l + 2w$
$A = lw$

b. Find the area.

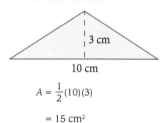

3 cm

10 cm

$A = \dfrac{1}{2}(10)(3)$
$\quad = 15$ cm²

Triangle

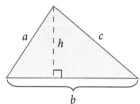

$P = a + b + c$
$A = \dfrac{1}{2} bh$

Parallelogram

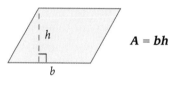

$A = bh$

Formulas for Diameter and Radius of a Circle [8.1, 8.2]

2. If the radius of a circle is 5.7 feet, find the diameter.
$d = 2(5.7)$
$\quad = 11.4$ ft

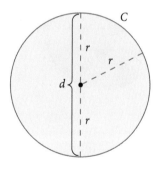

C = circumference
r = radius
d = diameter
$d = 2r$
$r = \dfrac{d}{2}$

Formulas for Circumference and Area of a Circle [8.1, 8.2]

3. Find the circumference and the area.

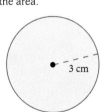

3 cm

$C = 2(3.14)(3)$
$\quad = 18.84$ cm
$A = 3.14(3)^2$
$\quad = 28.26$ cm²

r

$C = 2\pi r$
$A = \pi r^2$

Formulas for Surface Area and Volume [8.3, 8.4]

4. Find the volume and the surface area.

a.

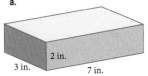

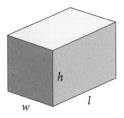

$V = 7 \cdot 3 \cdot 2$
$\quad = 42 \text{ in}^3$
$S = 2(3)(2) + 2(3)(7) + 2(2)(7)$
$\quad = 12 + 42 + 28$
$\quad = 82 \text{ in}^2$

Volume = _lwh_
Surface Area = 2_lh_ + 2_hw_ + 2_lw_

b.

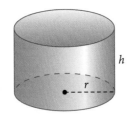

$V = (3.14)(2)^2 \cdot 4$
$\quad = 50.24 \text{ mm}^3$
$S = 2(3.14)(2)(2) + 2(3.14)(2)(4)$
$\quad = 75.4 \text{ mm}^2$

Volume = _πr²h_
Surface Area = 2_πr²_ + 2_πrh_

c.

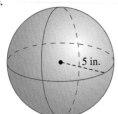

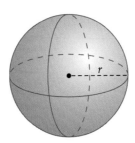

$V = \frac{4}{3}(3.14)5^3$
$\quad = 523 \text{ in}^3$, to the nearest whole number
$S = 4(3.14)5^2$
$\quad = 314 \text{ in}^2$

Volume = $\frac{4}{3}πr^3$
Surface Area = 4_πr²_

d. (Volume only.)

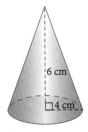

$V = \frac{1}{3}(3.14) \cdot 4^2(6)$

$\quad = 100.5 \text{ cm}^2$

Volume = $\frac{1}{3}πr^2h$

Chapter 8 Test Form A

These problems are all taken from examples in your text. Work each problem and check your answers. If you have made a mistake, work the problem again. If you cannot get the correct answer after two tries, look up the correct solution in your text.

1. Find the perimeter of a rectangle with a width of 5 yards and a length of 8 yards.

2. Find the circumference of each coin.

   a. Diameter = 23.25 millimeters   b. Radius = 0.52 inch

3. A parallelogram has a base of 5 centimeters and a height of 2 centimeters. Find the area.

4. A circular stamp has a radius of 14.5 millimeters. Find its area to the nearest whole number.

5. Find the surface area of a box with a width of 3 inches, a length of 4 inches and a height of 5 inches.

6. A drinking straw has a radius of 0.125 inch and a length of 6 inches. How much material was used to make the straw?

7. Find the volume of a rectangular solid with length 15 inches, width 3 inches, and height 5 inches.

8. Find the volume of a cone with a radius 3 centimeters and height 5 centimeters.

Chapter Test, Form B

For an alternate, more comprehensive, chapter test, go to MathTV.com and select the test and summary for this chapter of the textbook. Click the worksheet labeled Chapter 8 Test, Form B to download it.

Introduction to Algebra

9

Chapter Outline

iStockphoto.com © mabe123

The table below gives the record low temperature, in degrees Fahrenheit, for each month of the year in Jackson Hole, Wyoming. The accompanying line graph is drawn from the information in the table. Line graphs displaying negative numbers are one of the items that we will study in this chapter.

Month	Temperature
January	−50°F
February	−44°F
March	−32°F
April	−5°F
May	12°F
June	19°F
July	24°F
August	18°F
September	14°F
October	2°F
November	−27°F
December	−49°F

You can see that temperatures below zero are given as negative numbers. Most of this chapter deals with negative numbers. Here is the type of question we will ask in this chapter: Can you find the difference in record low temperatures between September and November from the table or the graph above?

The table gives the record low temperature for September as 14°F and the record low for November as −27°F. Because the word *difference* is associated with subtraction, our question is answered with the following problem:

$$\left(\begin{array}{c}\text{Difference in September} \\ \text{and November temperatures}\end{array}\right) = \left(\begin{array}{c}\text{September low} \\ \text{temperature}\end{array}\right) - \left(\begin{array}{c}\text{November low} \\ \text{temperature}\end{array}\right) = 14 - (-27)$$

In order to work these problems, we need rules for addition, subtraction, multiplication, and division with negative numbers. That is our main goal for this chapter.

Success Skills

iStockphoto © Silvia Boratti

Don't complain about anything, ever.

Do you complain to your classmates about your teacher? If you do, it could be getting in the way of your success in the class.

I have students that tell me that they like the way I teach and that they are enjoying my class. I have other students, in the same class, that complain to each other about me. They say I don't explain things well enough. Are the complaining students giving themselves a reason for not doing well in the class? I think so. They are shifting the responsibility for their success from themselves to me. It's not their fault they are not doing well, it's mine. When these students are alone, trying to do homework, they start thinking about how unfair everything is and they lose their motivation to study. Without intending to, they have set themselves up to fail by making their complaints more important than their progress in the class.

What happens when you stop complaining? You put yourself back in charge of your success. When there is no one to blame if things don't go well, you are more likely to do well. I have had students tell me that, once they stopped complaining about a class, the teacher became a better teacher and they started to actually enjoy going to class.

If you find yourself complaining to your friends about a class or a teacher, make a decision to stop. When other people start complaining to each other about the class or the teacher, walk away; don't participate in the complaining session. Try it for a day, or a week, or for the rest of the term. It may be difficult to do at first, but I'm sure you will like the results, and if you don't, you can always go back to complaining.

Positive and Negative Numbers

Suppose you have a balance of $20 in your checkbook and then write a check for $30. You are now overdrawn by $10. How will you write this new balance? One way is with a negative number. You could write the balance as −$10, which is a negative number.

RECORD ALL CHARGES OR CREDITS THAT AFFECT YOUR ACCOUNT					
NUMBER	DATE	DESCRIPTION OF TRANSACTION	PAYMENT/DEBIT (-)	DEPOSIT/CREDIT (+)	BALANCE
					$20 00
1501	9/15	Campus Bookstore	$30 00		-$10 00

Negative numbers can be used to describe other situations as well—for instance, temperature below zero and distance below sea level.

To see the relationship between negative and positive numbers, we can extend the number line as shown in Figure 1. We first draw a straight line and label a convenient point with 0. This is called the *origin*, and it is usually in the middle of the line. We then label positive numbers to the right (as we have done previously), and negative numbers to the left.

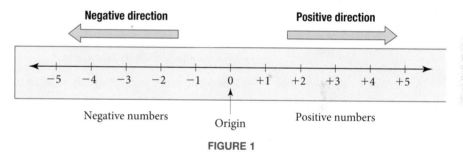

FIGURE 1

Note

A number, other than 0, with no sign (+ or −) in front of it is assumed to be positive. That is, 5 = +5.

The numbers increase going from left to right. If we move to the right, we are moving in the positive direction. If we move to the left, we are moving in the negative direction. *Any number to the left of another number is considered to be smaller than the number to its right.*

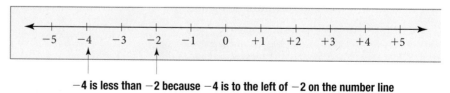

−4 is less than −2 because −4 is to the left of −2 on the number line

FIGURE 2

We see from the line that every negative number is less than every positive number.

> **Notation**
>
> If a and b are any two numbers on the number line, then
>
> $a < b$ is read "a is less than b"
>
> $a > b$ is read "a is greater than b"

In algebra we can use inequality symbols when comparing numbers.

As you can see, the inequality symbols always point to the smaller of the two numbers being compared. Here are some examples that illustrate how we use the inequality symbols.

 **EXAMPLE 1**   Explain the meaning of each expression.

a. $3 < 5$ **b.** $0 > 100$ **c.** $-3 < 5$ **d.** $-5 < -2$

SOLUTION

a. $3 < 5$ is read "3 is less than 5." Note that it would also be correct to write $5 > 3$. Both statements, "3 is less than 5" and "5 is greater than 3," have the same meaning. The inequality symbols always point to the smaller number.

b. $0 > 100$ is a false statement, because 0 is less than 100, not greater than 100. To write a true inequality statement using the numbers 0 and 100, we would have to write either $0 < 100$ or $100 > 0$.

c. $-3 < 5$ is a true statement, because -3 is to the left of 5 on the number line, and, therefore, it must be less than 5. Another statement that means the same thing is $5 > -3$.

d. $-5 < -2$ is a true statement, because -5 is to the left of -2 on the number line, meaning that -5 is less than -2. Both statements $-5 < -2$ and $-2 > -5$ have the same meaning; they both say that -5 is a smaller number than -2.

It is sometimes convenient to talk about only the numerical part of a number and disregard the sign (+ or −) in front of it. The following definition gives us a way of doing this.

> **DEFINITION**
>
> The **absolute value** of a number is its distance from 0 on the number line. We denote the absolute value of a number with vertical lines. For example, the absolute value of -3 is written $|-3|$.

The absolute value of a number is never negative because it is a distance, and a distance is always measured in positive units (unless it happens to be 0).

EXAMPLE 2   Simplify each expression.

a. $|5|$ **b.** $|-3|$ **c.** $|-7|$

SOLUTION

a. $|5| = 5$ The number 5 is 5 units from 0.
b. $|-3| = 3$ The number -3 is 3 units from 0.
c. $|-7| = 7$ The number -7 is 7 units from 0.

DEFINITION

Two numbers that are the same distance from 0 but in opposite directions from 0 are called *opposites*.* The notation for the opposite of a is $-a$.

EXAMPLE 3   Give the opposite of each of the following numbers:

$$5, 7, 1, -5, -8$$

SOLUTION   The opposite of 5 is -5.
The opposite of 7 is -7.
The opposite of 1 is -1.
The opposite of -5 is $-(-5)$, or 5.
The opposite of -8 is $-(-8)$, or 8.

We see from this example that the opposite of every positive number is a negative number, and, likewise, the opposite of every negative number is a positive number. The last two parts of Example 3 illustrate the following property:

Property

If a represents any positive number, then it is always true that
$$-(-a) = a$$

In other words, this property states that the opposite of a negative number is a positive number.

It should be evident now that the symbols $+$ and $-$ can be used to indicate several different ideas in mathematics. In the past we have used them to indicate addition and subtraction. They can also be used to indicate the direction a number is from 0 on the number line. For instance, the number $+3$ (read "positive 3") is the number that is 3 units from zero in the positive direction. On the other hand, the number -3 (read "negative 3") is the number that is 3 units from 0 in the negative direction. The symbol $-$ can also be used to indicate the opposite of a number, as in $-(-2) = 2$. The interpretation of the symbols $+$ and $-$ depends on the situation in which they are used. For example:

$3 + 5$ The $+$ sign indicates addition.
$7 - 2$ The $-$ sign indicates subtraction.
-7 The $-$ sign is read "negative" 7.
$-(-5)$ The first $-$ sign is read "the opposite of." The second $-$ sign is read "negative" 5.

*In some books opposites are called *additive inverses*.

This may seem confusing at first, but as you work through the problems in this chapter you will get used to the different interpretations of the symbols + and −.

We should mention here that the set of whole numbers along with their opposites forms the set of *integers*. That is:

$$\text{Integers} = \{\ldots, -3, -2, -1, 0, 1, 2, 3, \ldots\}$$

GETTING READY FOR CLASS

After reading through the preceding section, respond in your own words and in complete sentences.

1. Write the statement "3 is less than 5" in symbols.
2. What is the absolute value of a number?
3. Describe what we mean by numbers that are "opposites" of each other.
4. If you locate two different numbers on the real number line, which one will be the smaller number?

Problem Set 9.1

Write each of the following in words.

1. $4 < 7$ **2.** $0 < 10$ **3.** $5 > -2$ **4.** $8 > -8$

5. $-10 < -3.$ **6.** $-20 < -5$ **7.** $0 > -4$ **8.** $0 > -100$

Write each of the following in symbols.

9. 30 is greater than -30.   **10.** -30 is less than 30.

11. -10 is less than 0.   **12.** 0 is greater than -10.

13. -3 is greater than -15.   **14.** -15 is less than -3.

Place either $<$ or $>$ between each of the following pairs of numbers so that the resulting statement is true.

15. 3   7   **16.** 17   0   **17.** 7 -5 **18.** 2 -13

19. -6 0   **20.** -14 0   **21.** -12 -2 **22.** -20 -1

23. $-\dfrac{1}{2}$ $-\dfrac{3}{4}$ **24.** $-\dfrac{6}{7}$ $\dfrac{5}{6}$ **25.** -0.75 0.25 **26.** -1 -3.5

27. -0.1 -0.01 **28.** -0.04 -0.4 **29.** -3 $|6|$ **30.** $|8|$ -2

31. 15 $|-4|$ **32.** 20 $|-6|$ **33.** $|-2|$ $|-7|$ **34.** $|-3|$ $|-1|$

Find each of the following absolute values.

35. $|2|$ **36.** $|7|$ **37.** $|100|$ **38.** $|10,000|$

39. $|-8|$ **40.** $|-9|$ **41.** $|-231|$ **42.** $|-457|$

43. $\left|-\dfrac{3}{4}\right|$ **44.** $\left|-\dfrac{1}{10}\right|$ **45.** $|-200|$ **46.** $|-350|$

47. $|8|$ **48.** $|9|$ **49.** $|231|$ **50.** $|457|$

Give the opposite of each of the following numbers.

51. 3   **52.** -5 **53.** -2 **54.** 15   **55.** 75   **56.** -32

57. 0   **58.** 1   **59.** -0.123 **60.** -3.45 **61.** $\dfrac{7}{8}$ **62.** $\dfrac{1}{100}$

Simplify each of the following.

63. $-(-2)$ **64.** $-(-5)$ **65.** $-(-8)$ **66.** $-(-3)$

67. $-|-2|$ **68.** $-|-5|$ **69.** $-|-8|$ **70.** $-|-3|$

71. What number is its own opposite?

72. Is $|a| = a$ always a true statement?

73. If n is a negative number, is $-n$ positive or negative?

74. If n is a positive number, is $-n$ positive or negative?

Estimating

Work Problems 75–80 mentally, without pencil and paper or a calculator.

75. Is −60 closer to 0 or −100?

76. Is −20 closer to 0 or −30?

77. Is −10 closer to −20 or 20?

78. Is −20 closer to −40 or 10?

79. Is −362 closer to −360 or −370?

80. Is −368 closer to −360 or −370?

Applying the Concepts

81. Temperature and Altitude Yamina is flying from Phoenix to San Francisco on a Boeing 737 jet. When the plane reaches an altitude of 33,000 feet, the temperature outside the plane is 61 degrees below zero Fahrenheit. Represent this temperature with a negative number. If the temperature outside the plane gets warmer by 10 degrees, what will the new temperature be?

82. Temperature Change At 11:00 in the morning in Superior, Wisconsin, Jim notices the temperature is 15 degrees below zero Fahrenheit. Write this temperature as a negative number. At noon it has warmed up by 8 degrees. What is the temperature at noon?

83. Temperature Change At 10:00 in the morning in White Bear Lake, Wisconsin, Zach notices the temperature is 5 degrees below zero Fahrenheit. Write this temperature as a negative number. By noon the temperature has dropped another 10 degrees. What is the temperature at noon?

84. Snorkeling Steve is snorkeling in the ocean near his home in Maui. At one point he is 6 feet below the surface. Represent this situation with a negative number. If he descends another 6 feet, what negative number will represent his new position?

Table 2 lists various wind chill temperatures. The top row gives air temperature, while the first column gives wind speed, in miles per hour. The numbers within the table indicate how cold the weather will feel. For example, if the thermometer reads 30°F and the wind is blowing at 15 miles per hour, the wind chill temperature is 9°F.

TABLE 2

WIND CHILL TEMPERATURES

Air Temperatures (°F)

Wind Speed	30°	25°	20°	15°	10°	5°	0°	−5°
10 mph	16°	10°	3°	−3°	−9°	−15°	−22°	−27°
15 mph	9°	2°	−5°	−11°	−18°	−25°	−31°	−38°
20 mph	4°	−3°	−10°	−17°	−24°	−31°	−39°	−46°
25 mph	1°	−7°	−15°	−22°	−29°	−36°	−44°	−51°
30 mph	−2°	−10°	−18°	−25°	−33°	−41°	−49°	−56°

85. Wind Chill Find the wind chill temperature if the thermometer reads 25°F and the wind is blowing at 25 miles per hour.

86. Wind Chill Find the wind chill temperature if the thermometer reads 10°F and the wind is blowing at 25 miles per hour.

87. Wind Chill Which will feel colder: a day with an air temperature of 10°F and a 25-mph wind, or a day with an air temperature of −5°F and a 10-mph wind?

88. Wind Chill Which will feel colder: a day with an air temperature of 15°F and a 20-mph wind, or a day with an air temperature of 5°F and a 10-mph wind?

Table 3 lists the record low temperatures for each month of the year for Lake Placid, New York. Table 4 lists the record high temperatures for the same city.

TABLE 3

**RECORD LOW TEMPERATURES
FOR LAKE PLACID, NEW YORK**

Month	Temperature
January	−36°F
February	−30°F
March	−14°F
April	−2°F
May	19°F
June	22°F
July	35°F
August	30°F
September	19°F
October	15°F
November	−11°F
December	−26°F

TABLE 4

**RECORD HIGH TEMPERATURES
FOR LAKE PLACID, NEW YORK**

Month	Temperature
January	54°F
February	59°F
March	69°F
April	82°F
May	90°F
June	93°F
July	97°F
August	93°F
September	90°F
October	87°F
November	67°F
December	60°F

89. Temperature Figure 5 is a bar chart of the information in Table 3. Construct a scatter diagram of the same information. Then connect the dots in the scatter diagram to obtain a line graph of that same information. (Notice that we have used the numbers 1 through 12 to represent the months January through December.)

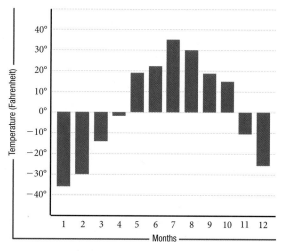

FIGURE 5 A bar chart of Table 3

90. Temperature Figure 6 is a bar chart of the information in Table 4. Construct a scatter diagram of the same information. Then connect the dots in the scatter diagram to obtain a line graph of that same information. (Again, we have used the numbers 1 through 12 to represent the months January through December.)

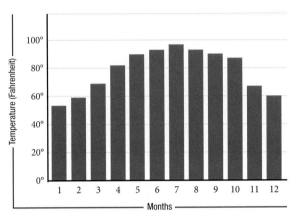

FIGURE 6 A bar chart of Table 4

Getting Ready for the Next Section

Add or subtract.

91. $10 + 15$

92. $12 + 15$

93. $15 - 10$

94. $15 - 12$

95. $10 - 5 - 3 + 4$

96. $12 - 3 - 7 + 5$

97. $[3 + 10] + [8 - 2]$

98. $[2 + 12] + [7 - 5]$

99. $\dfrac{3}{8} - \dfrac{1}{8}$

100. $\dfrac{5}{6} - \dfrac{2}{6}$

101. $\dfrac{1}{10} + \dfrac{4}{5} - \dfrac{3}{20}$

102. $\dfrac{1}{2} + \dfrac{3}{4} - \dfrac{5}{8}$

103. $4.75 + 2.25$

104. $5.76 + 3.24$

105. $6.89 - 3.42$

106. $8.55 - 6.88$

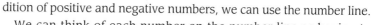

Addition with Negative Numbers

9.2

Suppose you are in Las Vegas playing blackjack and you lose $3 on the first hand and then you lose $5 on the next hand. If you represent winning with positive numbers and losing with negative numbers, how will you represent the results from your first two hands? Since you lost $3 and $5 for a total of $8, one way to represent the situation is with addition of negative numbers:

$$(-\$3) + (-\$5) = -\$8$$

From this example we see that the sum of two negative numbers is a negative number. To generalize addition of positive and negative numbers, we can use the number line.

We can think of each number on the number line as having two characteristics: (1) a *distance* from 0 (absolute value) and (2) a *direction* from 0 (positive or negative). The distance from 0 is represented by the numerical part of the number (like the 5 in the number −5), and its direction is represented by the + or − sign in front of the number.

We can visualize addition of numbers on the number line by thinking in terms of distance and direction from 0. Let's begin with a simple problem we know the answer to. We interpret the sum 3 + 5 on the number line as follows:

1. The first number is 3, which tells us "start at the origin, and move 3 units in the positive direction."

2. The + sign is read "and then move."

3. The 5 means "5 units in the positive direction."

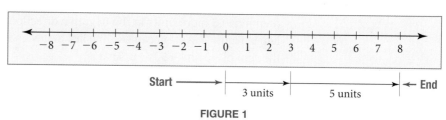

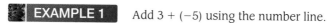

FIGURE 1

Figure 1 shows these steps. To summarize, 3 + 5 means to start at the origin (0), move 3 units in the *positive* direction, and then move 5 units in the *positive* direction. We end up at 8, which is the sum we are looking for: 3 + 5 = 8.

 EXAMPLE 1   Add 3 + (−5) using the number line.

SOLUTION   We start at the origin, move 3 units in the positive direction, and then move 5 units in the negative direction, as shown in Figure 2. The last arrow ends at −2, which must be the sum of 3 and −5. That is:

$$3 + (-5) = -2$$

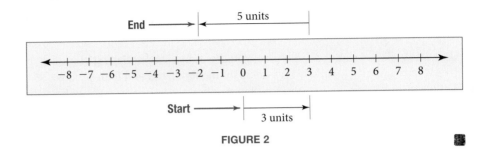

FIGURE 2

EXAMPLE 2   Add $-3 + 5$ using the number line.

SOLUTION   We start at the origin, move 3 units in the negative direction, and then move 5 units in the positive direction, as shown in Figure 3. We end up at 2, which is the sum of -3 and 5. That is:

$$-3 + 5 = 2$$

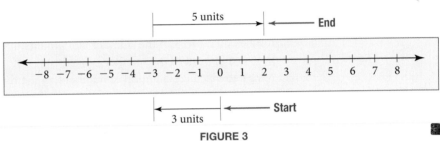

FIGURE 3

EXAMPLE 3   Add $-3 + (-5)$ using the number line.

SOLUTION   We start at the origin, move 3 units in the negative direction, and then move 5 more units in the negative direction. This is shown on the number line in Figure 4. As you can see, the last arrow ends at -8. We must conclude that the sum of -3 and -5 is -8. That is:

$$-3 + (-5) = -8$$

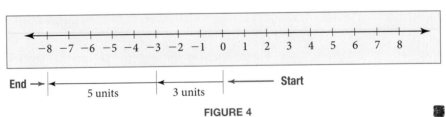

FIGURE 4

Adding numbers on the number line as we have done in these first three examples gives us a way of visualizing addition of positive and negative numbers. We want to be able to write a rule for addition of positive and negative numbers that doesn't involve the number line. The number line is a way of justifying the rule we will write. Here is a summary of the results we have so far:

$$3 + 5 = 8 \qquad -3 + 5 = 2$$
$$3 + (-5) = -2 \qquad -3 + (-5) = -8$$

Looking over these results, we write the following rule for adding any two numbers:

RULE

1. To add two numbers with the same sign: Simply add their absolute values, and use the common sign. If both numbers are positive, the answer is positive. If both numbers are negative, the answer is negative.
2. To add two numbers with different signs: Subtract the smaller absolute value from the larger absolute value. The answer will have the sign of the number with the larger absolute value.

The following examples show how the rule is used. You will find that the rule for addition is consistent with all the results obtained using the number line.

EXAMPLE 4   Add all combinations of positive and negative 10 and 15.

SOLUTION

$$10 + 15 = 25$$
$$10 + (-15) = -5$$
$$-10 + 15 = 5$$
$$-10 + (-15) = -25$$

Notice that when we add two numbers with the same sign, the answer also has that sign. When the signs are not the same, the answer has the sign of the number with the larger absolute value.

Once you have become familiar with the rule for adding positive and negative numbers, you can apply it to more complicated sums.

EXAMPLE 5   Simplify: $10 + (-5) + (-3) + 4$

SOLUTION   Adding left to right, we have:

$$10 + (-5) + (-3) + 4 = 5 + (-3) + 4 \qquad 10 + (-5) = 5$$
$$= 2 + 4 \qquad\qquad 5 + (-3) = 2$$
$$= 6$$

EXAMPLE 6   Simplify: $[-3 + (-10)] + [8 + (-2)]$

SOLUTION   We begin by adding the numbers inside the brackets.

$$[-3 + (-10)] + [8 + (-2)] = [-13] + [6]$$
$$= -7$$

EXAMPLE 7   Add: $-4.75 + (-2.25)$

SOLUTION   Because both signs are negative, we add absolute values. The answer will be negative.

$$-4.75 + (-2.25) = -7.00$$

EXAMPLE 8   Add: $3.42 + (-6.89)$

SOLUTION   The signs are different, so we subtract the smaller absolute value from the larger absolute value. The answer will be negative, because 6.89 is larger than 3.42 and the sign in front of 6.89 is $-$.

$$3.42 + (-6.89) = -3.47$$

EXAMPLE 9   Add: $\dfrac{3}{8} + \left(-\dfrac{1}{8}\right)$

SOLUTION   We subtract absolute values. The answer will be positive, because $\frac{3}{8}$ is positive.

$$\dfrac{3}{8} + \left(-\dfrac{1}{8}\right) = \dfrac{2}{8}$$

$$= \dfrac{1}{4} \qquad \text{Reduce to lowest terms}$$

EXAMPLE 10   Add: $\dfrac{1}{10} + \left(-\dfrac{4}{5}\right) + \left(-\dfrac{3}{20}\right)$

SOLUTION   To begin, change each fraction to an equivalent fraction with an LCD of 20.

$$\dfrac{1}{10} + \left(-\dfrac{4}{5}\right) + \left(-\dfrac{3}{20}\right) = \dfrac{1 \cdot 2}{10 \cdot 2} + \left(-\dfrac{4 \cdot 4}{5 \cdot 4}\right) + \left(-\dfrac{3}{20}\right)$$

$$= \dfrac{2}{20} + \left(-\dfrac{16}{20}\right) + \left(-\dfrac{3}{20}\right)$$

$$= -\dfrac{14}{20} + \left(-\dfrac{3}{20}\right)$$

$$= -\dfrac{17}{20}$$

GETTING READY FOR CLASS

After reading through the preceding section, respond in your own words and in complete sentences.

1. Explain how you would use the number line to add 3 and 5.
2. If two numbers are negative, such as -3 and -5, what sign will their sum have?
3. If you add two numbers with different signs, how do you determine the sign of the answer?
4. With respect to addition with positive and negative numbers, does the phrase "two negatives make a positive" make any sense?

Problem Set 9.2

Draw a number line from -10 to $+10$ and use it to add the following numbers.

1. $2 + 3$ **2.** $2 + (-3)$ **3.** $-2 + 3$ **4.** $-2 + (-3)$

5. $5 + (-7)$ **6.** $-5 + 7$ **7.** $-4 + (-2)$ **8.** $-8 + (-2)$

9. $10 + (-6)$ **10.** $-9 + 3$ **11.** $7 + (-3)$ **12.** $-7 + 3$

13. $-4 + (-5)$ **14.** $-2 + (-7)$

Combine the following by using the rule for addition of positive and negative numbers. (Your goal is to be fast and accurate at addition, with the latter being more important.)

15. $7 + 8$ **16.** $9 + 12$ **17.** $5 + (-8)$ **18.** $4 + (-11)$

19. $-6 + (-5)$ **20.** $-7 + (-2)$ **21.** $-10 + 3$ **22.** $-14 + 7$

23. $-1 + (-2)$ **24.** $-5 + (-4)$ **25.** $-11 + (-5)$ **26.** $-16 + (-10)$

27. $4 + (-12)$ **28.** $9 + (-1)$ **29.** $-85 + (-42)$ **30.** $-96 + (-31)$

31. $-121 + 170$ **32.** $-130 + 158$ **33.** $-375 + 409$ **34.** $-765 + 213$

Complete the following tables.

35.

FIRST NUMBER a	SECOND NUMBER b	THEIR SUM a + b
5	-3	
5	-4	
5	-5	
5	-6	
5	-7	

36.

FIRST NUMBER a	SECOND NUMBER b	THEIR SUM a + b
-5	3	
-5	4	
-5	5	
-5	6	
-5	7	

37.

FIRST NUMBER x	SECOND NUMBER y	THEIR SUM x + y
-5	-3	
-5	-4	
-5	-5	
-5	-6	
-5	-7	

38.

FIRST NUMBER x	SECOND NUMBER y	THEIR SUM x + y
30	-20	
-30	20	
-30	-20	
30	20	
-30	0	

Add the following numbers left to right.

39. $10 + (-18) + 4$ **40.** $-2 + 4 + (-6)$

41. $24 + (-6) + (-8)$ **42.** $35 + (-5) + (-30)$

43. $-201 + (-143) + (-101)$ **44.** $-27 + (-56) + (-89)$

45. $-321 + 752 + (-324)$ **46.** $-571 + 437 + (-502)$

47. $-8 + 3 + (-5) + 9$ **48.** $-9 + 2 + (-10) + 3$

49. $-2 + (-5) + (-6) + (-7)$ **50.** $-8 + (-3) + (-4) + (-7)$

51. $15 + (-30) + 18 + (-20)$

52. $20 + (-15) + 30 + (-18)$

53. $-78 + (-42) + 57 + 13$

54. $-89 + (-51) + 65 + 17$

Use the rule for order of operations to simplify each of the following.

55. $(-8 + 5) + (-6 + 2)$

56. $(-3 + 1) + (-9 + 4)$

57. $(-10 + 4) + (-3 + 12)$

58. $(-11 + 5) + (-3 + 2)$

59. $20 + (-30 + 50) + 10$

60. $30 + (-40 + 20) + 50$

61. $108 + (-456 + 275)$

62. $106 + (-512 + 318)$

63. $[5 + (-8)] + [3 + (-11)]$

64. $[8 + (-2)] + [5 + (-7)]$

65. $[57 + (-35)] + [19 + (-24)]$

66. $[63 + (-27)] + [18 + (-24)]$

Use the rule for addition of numbers to add the following fractions and decimals.

67. $-1.3 + (-2.5)$

68. $-9.1 + (-4.5)$

69. $24.8 + (-10.4)$

70. $29.5 + (-21.3)$

71. $-5.35 + 2.35 + (-6.89)$

72. $-9.48 + 5.48 + (-4.28)$

73. $-\dfrac{5}{6} + \left(-\dfrac{1}{6}\right)$

74. $-\dfrac{7}{9} + \left(-\dfrac{2}{9}\right)$

75. $\dfrac{3}{7} + \left(-\dfrac{5}{7}\right)$

76. $\dfrac{11}{13} + \left(-\dfrac{12}{13}\right)$

77. $-\dfrac{2}{5} + \dfrac{3}{5} + \left(-\dfrac{4}{5}\right)$

78. $-\dfrac{6}{7} + \dfrac{4}{7} + \left(-\dfrac{1}{7}\right)$

79. $-3.8 + 2.54 + 0.4$

80. $-9.6 + 5.15 + 0.8$

81. $-2.89 + (-1.4) + 0.09$

82. $-3.99 + (-1.42) + 0.06$

83. $\dfrac{1}{2} + \left(-\dfrac{3}{4}\right)$

84. $\dfrac{3}{5} + \left(-\dfrac{7}{10}\right)$

85. $-\dfrac{2}{3} + \left(-\dfrac{3}{5}\right) + \left(-\dfrac{1}{15}\right)$

86. $-\dfrac{3}{4} + \left(-\dfrac{1}{3}\right) + \left(-\dfrac{5}{12}\right)$

87. Find the sum of -8, -10, and -3.

88. Find the sum of -4, 17, and -6.

89. What number do you add to 8 to get 3?

90. What number do you add to 10 to get 4?

91. What number do you add to -3 to get -7?

92. What number do you add to -5 to get -8?

93. What number do you add to -4 to get 3?

94. What number do you add to -7 to get 2?

95. If the sum of -3 and 5 is increased by 8, what number results?

96. If the sum of -9 and -2 is increased by 10, what number results?

Estimating

Work Problems 97–104 mentally, without pencil and paper or a calculator.

97. The answer to the problem 251 + 249 is closest to which of the following numbers?
   a. 500   **b.** 0   **c.** −500

98. The answer to the problem 251 + (−249) is closest to which of the following numbers?
   a. 500   **b.** 0   **c.** −500

99. The answer to the problem −251 + 249 is closest to which of the following numbers?
   a. 500   **b.** 0   **c.** −500

100. The answer to the problem −251 + (−249) is closest to which of the following numbers?
   a. 500   **b.** 0   **c.** −500

101. The sum of 77 and 22 is closest to which of the following numbers?
   a. −100   **b.** −60 **c.** 60   **d.** 100

102. The sum of −77 and 22 is closest to which of the following numbers?
   a. −100   **b.** −60 **c.** 60   **d.** 100

103. The sum of 77 and −22 is closest to which of the following numbers?
   a. −100   **b.** −60 **c.** 60   **d.** 100

104. The sum of −77 and −22 is closest to which of the following numbers?
   a. −100   **b.** −60 **c.** 60   **d.** 100

Applying the Concepts

105. Checkbook Balance Ethan has a balance of −$40 in his checkbook. If he deposits $100 and then writes a check for $50, what is the new balance in his checkbook?

		RECORD ALL CHARGES OR CREDITS THAT AFFECT YOUR ACCOUNT				
						BALANCE
NUMBER	DATE	DESCRIPTION OF TRANSACTION	PAYMENT/DEBIT (-)	DEPOSIT/CREDIT (+)		-$40 00
	9/20	Deposit		$100 00		
1502	9/21	Vons Market	$50 00			

106. Checkbook Balance Kendra has a balance of −$20 in her checkbook. If she deposits $45 and then writes a check for $15, what is the new balance in her checkbook?

107. Gambling While gambling in Las Vegas, a person wins $74 playing blackjack and then loses $141 on roulette. Use positive and negative numbers to write this situation in symbols. Then give the person's net loss or gain.

108. **Gambling** While playing blackjack, a person loses $17 on his first hand, then wins $14, and then loses $21. Write this situation using positive and negative numbers and addition; then simplify.

109. **Stock Gain/Loss** Suppose a certain stock gains 3 points on the stock exchange on Monday and then loses 5 points on Tuesday. Express the situation using positive and negative numbers, and then give the net gain or loss of the stock for this 2-day period.

110. **Stock Gain/Loss** A stock gains 2 points on Wednesday, then loses 1 on Thursday, and gains 3 on Friday. Use positive and negative numbers and addition to write this situation in symbols, and then simplify.

111. **Distance** The distance between two numbers on the number line is 10. If one of the numbers is 3, what are the two possibilities for the other number?

112. **Distance** The distance between two numbers is 8. If one of the numbers is −5, what are the two possibilities for the other number?

Getting Ready for the Next Section

Give the opposite of each number.

113. 2   114. 3   115. −4   116. −5   117. $\dfrac{2}{5}$ 118. $\dfrac{3}{8}$

119. −30   120. −15   121. 60.3   122. 70.4

123. Subtract 3 from 5.   124. Subtract 2 from 8.

125. Find the difference of 7 and 4.   126. Find the difference of 8 and 6.

Subtraction with Negative Numbers   9.3

Earlier in this chapter we asked how we would represent the final balance in a checkbook if the original balance was $20 and we wrote a check for $30. We decided that the final balance would be −$10. We can summarize the whole situation with subtraction:

$$\$20 - \$30 = -\$10$$

		RECORD ALL CHARGES OR CREDITS THAT AFFECT YOUR ACCOUNT			BALANCE	
NUMBER	DATE	DESCRIPTION OF TRANSACTION	PAYMENT/DEBIT (-)	DEPOSIT/CREDIT (+)	$20	00
1501	9/15	Campus Bookstore	$30 00		-$10	00

From this we see that subtracting 30 from 20 gives us −10. Another example that gives the same answer but involves addition is this:

$$20 + (-30) = -10$$

From the two examples above, we find that subtracting 30 gives the same result as adding −30. We use this kind of reasoning to give a definition for subtraction that will allow us to use the rules we developed for addition to do our subtraction problems. Here is that definition:

> ## DEFINITION
>
> **Subtraction** If a and b represent any two numbers, then it is always true that
>
> $$a - b = a + (-b)$$
>
> To subtract b Add its opposite, $-b$
>
> In words: Subtracting a number is equivalent to adding its opposite.

Let's see if this definition conflicts with what we already know to be true about subtraction.

From previous experience we know that

$$5 - 2 = 3$$

We can get the same answer by using the definition we just gave for subtraction. Instead of subtracting 2, we can add its opposite, −2. Here is how it looks:

$$5 - 2 = 5 + (-2) \qquad \text{Change subtraction to addition of the opposite}$$

$$= 3 \qquad \text{Apply the rule for addition of positive and negative numbers}$$

The result is the same whether we use our previous knowledge of subtraction or the new definition. The new definition is essential when the problems begin to get more complicated.

Note

A real-life analogy to Example 1 would be: "If the temperature were 7° below 0 and then it dropped another 2°, what would the temperature be then?"

EXAMPLE 1   Subtract: $-7 - 2$

SOLUTION   We have never subtracted a positive number from a negative number before. We must apply our definition of subtraction:

$$-7 - 2 = -7 + (-2) \qquad \text{Instead of subtracting 2,}$$
$$\text{we add its opposite, } -2$$
$$= -9 \qquad \text{Apply the rule for addition}$$

EXAMPLE 2   Subtract: $12 - (-6)$

SOLUTION   The first $-$ sign is read "subtract," and the second one is read "negative." The problem in words is "12 subtract negative 6." We can use the definition of subtraction to change this to the addition of positive 6:

$$12 - (-6) = 12 + 6 \qquad \text{Subtracting } -6 \text{ is equivalent}$$
$$\text{to adding } +6$$
$$= 18 \qquad \text{Addition}$$

EXAMPLE 3   The following table shows the relationship between subtraction and addition:

Subtraction	Addition of the opposite	Answer
$7 - 9$	$7 + (-9)$	-2
$-7 - 9$	$-7 + (-9)$	-16
$7 - (-9)$	$7 + 9$	16
$-7 - (-9)$	$-7 + 9$	2
$15 - 10$	$15 + (-10)$	5
$-15 - 10$	$-15 + (-10)$	-25
$15 - (-10)$	$15 + 10$	25
$-15 - (-10)$	$-15 + 10$	-5

Examples 1–3 illustrate all the possible combinations of subtraction with positive and negative numbers. There are no new rules for subtraction. We apply the definition to change each subtraction problem into an equivalent addition problem. The rule for addition can then be used to obtain the correct answer.

EXAMPLE 4   Combine: $-3 + 6 - 2$

SOLUTION   The first step is to change subtraction to addition of the opposite. After that has been done, we add left to right.

$$-3 + 6 - 2 = -3 + 6 + (-2) \qquad \text{Subtracting 2 is equivalent}$$
$$\text{to adding } -2$$
$$= 3 + (-2) \qquad \text{Add left to right}$$
$$= 1$$

EXAMPLE 5   Subtract 3 from −5.

SOLUTION   Subtracting 3 is equivalent to adding −3.

$$-5 - 3 = -5 + (-3) = -8$$

Subtracting 3 from −5 gives us −8.

EXAMPLE 6   Find the difference of $-\dfrac{3}{5}$ and $\dfrac{2}{5}$.

SOLUTION $-\dfrac{3}{5} - \dfrac{2}{5} = -\dfrac{3}{5} + \left(-\dfrac{2}{5}\right)$

$$= -\frac{5}{5}$$

$$= -1$$

EXAMPLE 7   Many of the planes used by the United States during World War II were not pressurized or sealed from outside air. As a result, the temperature inside these planes was the same as the surrounding air temperature outside. Suppose the temperature inside a B-17 Flying Fortress is 50°F at takeoff and then drops to −30°F when the plane reaches its cruising altitude of 28,000 feet. Find the difference in temperature inside this plane at takeoff and at 28,000 feet.

SOLUTION   The temperature at takeoff is 50°F, whereas the temperature at 28,000 feet is −30°F. To find the difference we subtract, with the numbers in the same order as they are given in the problem:

$$50 - (-30) = 50 + 30 = 80$$

The difference in temperature is 80°F.

Subtraction and Taking Away

Some people may believe that the answer to −5 − 9 should be −4 or 4, not −14. If this is happening to you, you are probably thinking of subtraction in terms of taking one number away from another. Thinking of subtraction in this way works well with positive numbers if you always subtract the smaller number from the larger. In algebra, however, we encounter many situations other than this. The definition of subtraction, that $a - b = a + (-b)$ clearly indicates the correct way to use subtraction. That is, when working subtraction problems, you should think "addition of the opposite," not "taking one number away from another."

USING TECHNOLOGY

Calculator Note

Here is how we work the subtraction problem shown in Example 1 on a calculator.

Scientific Calculator: 7 $\boxed{+/-}$ $\boxed{-}$ 2 $\boxed{+/-}$ $\boxed{=}$

Graphing Calculator: $\boxed{(-)}$ 7 $\boxed{-}$ $\boxed{(-)}$ 2 $\boxed{ENT}$

iStockPhoto/©Glenn Frank

GETTING READY FOR CLASS

After reading through the preceding section, respond in your own words and in complete sentences.

1. Write the subtraction problem $5 - 3$ as an equivalent addition problem.

2. Explain the process you would use to subtract 2 from -7.

3. Write an addition problem that is equivalent to the subtraction problem $-20 - (-30)$.

4. To find the difference of -7 and -4 we subtract what number from -7?

Problem Set 9.3

1. $7 - 5$ **2.** $5 - 7$ **3.** $8 - 6$ **4.** $6 - 8$

5. $-3 - 5$ **6.** $-5 - 3$ **7.** $-4 - 1$ **8.** $-1 - 4$

9. $5 - (-2)$ **10.** $2 - (-5)$ **11.** $3 - (-9)$ **12.** $9 - (-3)$

13. $-4 - (-7)$ **14.** $-7 - (-4)$ **15.** $-10 - (-3)$ **16.** $-3 - (-10)$

17. $15 - 18$ **18.** $20 - 32$ **19.** $100 - 113$ **20.** $121 - 21$

21. $-30 - 20$ **22.** $-50 - 60$ **23.** $-79 - 21$ **24.** $-86 - 31$

25. $156 - (-243)$ **26.** $292 - (-841)$ **27.** $-35 - (-14)$ **28.** $-29 - (-4)$

Complete the following tables.

29.

First Number x	Second Number y	the Difference of x and y x − y
8	6	
8	7	
8	8	
8	9	
8	10	

30.

First Number x	Second Number y	the Difference of x and y x − y
10	12	
10	11	
10	10	
10	9	
10	8	

31.

First Number x	Second Number y	the Difference of x and y x − y
8	−6	
8	−7	
8	−8	
8	−9	
8	−10	

32.

First Number x	Second Number y	the Difference of x and y x − y
−10	−12	
−10	−11	
−10	−10	
−10	−9	
−10	−8	

Subtract.

33. $-9.01 - 2.4$ **34.** $-8.23 - 5.4$ **35.** $-0.89 - 1.01$ **36.** $-0.42 - 2.04$

37. $-\dfrac{1}{6} - \dfrac{5}{6}$ **38.** $-\dfrac{4}{7} - \dfrac{3}{7}$ **39.** $\dfrac{5}{12} - \dfrac{5}{6}$ **40.** $\dfrac{7}{15} - \dfrac{4}{5}$

41. $-\dfrac{13}{70} - \dfrac{23}{42}$ **42.** $-\dfrac{17}{60} - \dfrac{17}{90}$

Simplify as much as possible by first changing all subtractions to addition of the opposite and then adding left to right.

43. $4 - 5 - 6$ **44.** $7 - 3 - 2$ **45.** $-8 + 3 - 4$

46. $-10 - 1 + 16$ **47.** $-8 - 4 - 2$ **48.** $-7 - 3 - 6$

49. $-3.4 - 5.6 - 8.5$ **50.** $-2.1 - 3.1 - 4.1$ **51.** $33 - (-22) - 66$

52. $44 - (-11) + 55$ **53.** $\dfrac{1}{2} - \dfrac{1}{3} - \dfrac{1}{4}$ **54.** $\dfrac{1}{5} - \dfrac{1}{6} - \dfrac{1}{7}$

55. $-900 + 400 - (-100)$ **56.** $-300 + 600 - (-200)$

57. Subtract −6 from 5.

58. Subtract 8 from −2.

59. Find the difference of −5 and −1.

60. Find the difference of −7 and −3.

61. Subtract −4 from the sum of −8 and 12.

62. Subtract −7 from the sum of 7 and −12.

63. What number do you subtract from −3 to get −9?

64. What number do you subtract from 5 to get 8?

Estimating

Work Problems 65–74 mentally, without pencil and paper or a calculator.

65. The answer to the problem 52 − 49 is closest to which of the following numbers?
    a. 100   **b.** 0   **c.** −100

66. The answer to the problem −52 − 49 is closest to which of the following numbers?
    a. 100   **b.** 0   **c.** −100

67. The answer to the problem 52 − (−49) is closest to which of the following numbers?
    a. 100   **b.** 0   **c.** −100

68. The answer to the problem −52 − (−49) is closest to which of the following numbers?
    a. 100   **b.** 0   **c.** −100

69. Is the difference −161 − (−62) closer to −200 or −100?

70. Is the difference −553 − 50 closer to −600 or −500?

71. The difference of 37 and 61 is closest to which of the following numbers?
    a. −100   **b.** −20   **c.** 20   **d.** 100

72. The difference of 37 and −61 is closest to which of the following numbers?
    a. −100   **b.** −20   **c.** 20   **d.** 100

73. The difference of −37 and 61 is closest to which of the following numbers?
    a. −100   **b.** −20   **c.** 20   **d.** 100

74. The difference of −37 and −61 is closest to which of the following numbers?
    a. −100   **b.** −20   **c.** 20   **d.** 100

Applying the Concepts

75. Temperature On Monday the temperature reached a high of 28° above 0. That night it dropped to 16° below 0. What is the difference between the high and the low temperatures for Monday?

76. Gambling A gambler loses $35 playing poker one night and then loses another $25 the next night. Express this situation with numbers. How much did the gambler lose?

77. Gas Prices Use the chart to find the difference of the average price of a gallon of gasoline in California and the average price of a gallon of gasoline in Florida.

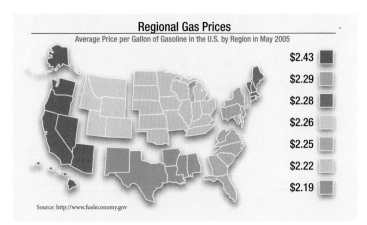

Regional Gas Prices
Average Price per Gallon of Gasoline in the U.S. by Region in May 2005

$2.43
$2.29
$2.28
$2.26
$2.25
$2.22
$2.19

Source: http://www.fueleconomy.gov

78. Checkbook Balance Susan has a balance of $572 in her checking account when she writes a check for $435 to pay the rent. Then she writes another check for $172 for textbooks. Write a subtraction problem that gives the new balance in her checking account. What is the new balance in her checking account?

Repeated below is the table of wind chill temperatures that we used previously. Use it for Problems 79–82.

Wind speed	Air Temperature (°F)							
	30°	25°	20°	15°	10°	5°	0°	−5°
10 mph	16°	10°	3°	−3°	−9°	−15°	−22°	−27°
15 mph	9°	2°	−5°	−11°	−18°	−25°	−31°	−38°
20 mph	4°	−3°	−10°	−17°	−24°	−31°	−39°	−46°
25 mph	1°	−7°	−15°	−22°	−29°	−36°	−44°	−51°
30 mph	−2°	−10°	−18°	−25°	−33°	−41°	−49°	−56°

79. Wind Chill If the temperature outside is 15°F, what is the difference in wind chill temperature between a 15-mile-per-hour wind and a 25-mile-per-hour wind?

80. Wind Chill If the temperature outside is 0°F, what is the difference in wind chill temperature between a 15-mile-per-hour wind and a 25-mile-per-hour wind?

81. Wind Chill Find the difference in temperature between a day in which the air temperature is 20°F and the wind is blowing at 10 miles per hour and a day in which the air temperature is 10°F and the wind is blowing at 20 miles per hour.

82. Wind Chill Find the difference in temperature between a day in which the air temperature is 0°F and the wind is blowing at 10 miles per hour and a day in which the air temperature is −5°F and the wind is blowing at 20 miles per hour.

Use the tables below to work Problems 83–86.

TABLE 1

**RECORD LOW TEMPERATURES
FOR LAKE PLACID, NEW YORK**

MONTH	TEMPERATURE
January	−36°F
February	−30°F
March	−14°F
April	−2°F
May	19°F
June	22°F
July	35°F
August	30°F
September	19°F
October	15°F
November	−11°F
December	−26°F

TABLE 2

**RECORD HIGH TEMPERATURES
FOR LAKE PLACID, NEW YORK**

MONTH	TEMPERATURE
January	54°F
February	59°F
March	69°F
April	82°F
May	90°F
June	93°F
July	97°F
August	93°F
September	90°F
October	87°F
November	67°F
December	60°F

83. **Temperature Difference** Find the difference between the record high temperature and the record low temperature for the month of December.

84. **Temperature Difference** Find the difference between the record high temperature and the record low temperature for the month of March.

85. **Temperature Difference** Find the difference between the record low temperatures of March and December.

86. **Temperature Difference** Find the difference between the record high temperatures of March and December.

Getting Ready for the Next Section

Perform the indicated operations.

87. $3(2)(5)$

88. $5(2)(4)$

89. 6^2

90. 8^2

91. 4^3

92. 3^3

93. $6(3 + 5)$

94. $2(5 + 8)$

95. $3(9 − 2) + 4(7 − 2)$

96. $2(5 − 3) − 7(4 − 2)$

97. $(3 + 7)(6 − 2)$

98. $(6 + 1)(9 − 4)$

99. $\left(\dfrac{2}{3}\right)\left(\dfrac{3}{5}\right)$

100. $\left(\dfrac{3}{4}\right)\left(\dfrac{4}{7}\right)$

101. $\left(\dfrac{7}{8}\right)\left(\dfrac{5}{14}\right)$

102. $\left(\dfrac{5}{6}\right)\left(\dfrac{9}{20}\right)$

103. $(5)(3.4)$

104. $(3)(6.7)$

105. $(0.4)(0.8)$

106. $(0.6)(0.5)$

Multiplication with Negative Numbers

Suppose you buy three shares of a stock on Monday, and by Friday the price per share has dropped $5. How much money have you lost? The answer is $15. Because it is a loss, we can express it as −$15. The multiplication problem below can be used to describe the relationship among the numbers.

SUPERIOR SHIPPING COMPANY
THREE SHARES

3 shares   each loses $5   for a total of −$15

$$3(-5) = -15$$

From this we conclude that it is reasonable to say that the product of a positive number and a negative number is a negative number.

In order to generalize multiplication with negative numbers, recall that we first defined multiplication by whole numbers to be repeated addition. That is:

$$3 \cdot 5 = 5 + 5 + 5$$

Multiplication   Repeated addition

This concept is very helpful when it comes to developing the rule for multiplication problems that involve negative numbers. For the first example we look at what happens when we multiply a negative number by a positive number.

EXAMPLE 1   Multiply: $3(-5)$

SOLUTION   Writing this product as repeated addition, we have

$$3(-5) = (-5) + (-5) + (-5)$$
$$= -10 + (-5)$$
$$= -15$$

The result, −15, is obtained by adding the three negative 5's.

EXAMPLE 2   Multiply: $-3(5)$

SOLUTION   In order to write this multiplication problem in terms of repeated addition, we will have to reverse the order of the two numbers. This is easily done, because multiplication is a commutative operation.

$$-3(5) = 5(-3)$$ Commutative property
$$= (-3) + (-3) + (-3) + (-3) + (-3)$$ Repeated addition
$$= -15$$ Addition

The product of −3 and 5 is −15.

EXAMPLE 3   Multiply: $-3(-5)$

SOLUTION   It is impossible to write this product in terms of repeated addition. We will find the answer to $-3(-5)$ by solving a different problem. Look at the following problem:

$$-3[5 + (-5)] = -3[0] = 0$$

The result is 0, because multiplying by 0 always produces 0. Now we can work the same problem another way and in the process find the answer to $-3(-5)$. Applying the distributive property to the same expression, we have

$$-3[5 + (-5)] = -3(5) + (-3)(-5) \qquad \text{Distributive property}$$
$$= -15 + (?) \qquad\qquad -3(5) = -15$$

The question mark must be $+15$, because we already know that the answer to the problem is 0, and $+15$ is the only number we can add to -15 to get 0. So, our problem is solved:

$$-3(-5) = +15$$

Table 1 gives a summary of what we have done so far in this section.

TABLE 1

Original Numbers Have	For Example	The Answer Is
Same signs	$3(5) = 15$	Positive
Different signs	$-3(5) = -15$	Negative
Different signs	$3(-5) = -15$	Negative
Same signs	$-3(-5) = 15$	Positive

From the examples we have done so far in this section and their summaries in Table 1, we write the following rule for multiplication of positive and negative numbers:

RULE

To multiply any two numbers, we multiply their absolute values.
1. The answer is positive if both the original numbers have the same sign. That is, the product of two numbers with the same sign is positive.
2. The answer is negative if the original two numbers have different signs. The product of two numbers with different signs is negative.

This rule should be memorized. By the time you have finished reading this section and working the problems at the end of the section, you should be fast and accurate at multiplication with positive and negative numbers.

EXAMPLE 4   Find the following products.

a. $2(4)$ **b.** $-2(-4)$ **c.** $2(-4)$ **d.** $-2(4)$

SOLUTION

a. $2(4) = 8$ *Like signs; positive answer*
b. $-2(-4) = 8$ *Like signs; positive answer*
c. $2(-4) = -8$ *Unlike signs; negative answer*
d. $-2(4) = -8$ *Unlike signs; negative answer*

EXAMPLE 5   Simplify $-3(2)(-5)$.

SOLUTION

$$-3(2)(-5) = -6(-5) \quad \text{Multiply } -3 \text{ and } 2 \text{ to get } -6$$
$$= 30$$

EXAMPLE 6   Use the definition of exponents to expand each expression. Then simplify by multiplying.

a. $(-6)^2$ **b.** -6^2 **c.** $(-4)^3$ **d.** -4^3

SOLUTION

a. $(-6)^2 = (-6)(-6)$ *Definition of exponents*
$\quad\quad\quad = 36$ *Multiply*
b. $-6^2 = -6 \cdot 6$ *Definition of exponents*
$\quad\quad\quad = -36$ *Multiply*
c. $(-4)^3 = (-4)(-4)(-4)$ *Definition of exponents*
$\quad\quad\quad = -64$ *Multiply*
d. $-4^3 = -4 \cdot 4 \cdot 4$ *Definition of exponents*
$\quad\quad\quad = -64$ *Multiply*

In Example 6, the base is a negative number in Parts a and c, but not in Parts b and d. We know this is true because of the use of parentheses.

EXAMPLE 7   Simplify: $-4 + 5(-6 + 2)$

SOLUTION   Simplifying inside the parentheses first, we have

$$-4 + 5(-6 + 2) = -4 + 5(-4) \quad \text{Simplify inside parentheses}$$
$$= -4 + (-20) \quad \text{Multiply}$$
$$= -24 \quad \text{Add}$$

EXAMPLE 8   Simplify: $-3(2 - 9) + 4(-7 - 2)$

SOLUTION   We begin by subtracting inside the parentheses:

$$-3(2 - 9) + 4(-7 - 2) = -3(-7) + 4(-9)$$
$$= 21 + (-36)$$
$$= -15$$

> ### USING TECHNOLOGY
>
> *Calculator Note*
>
> Here is how we work a similar problem on a calculator. (The $\times$ key on the first line may, or may not, be necessary. Try your calculator without it and see.)
>
> **Scientific Calculator:** $\boxed{(}\ 3\ \boxed{+/-}\ \boxed{-}\ 7\ \boxed{)}\ \boxed{\times}\ \boxed{(}\ 2\ \boxed{-}\ 6\ \boxed{)}\ \boxed{=}$
>
> **Graphing Calculator:** $\boxed{(}\ \boxed{(-)}\ 3\ \boxed{-}\ 7\ \boxed{)}\ \boxed{(}\ 2\ \boxed{-}\ 6\ \boxed{)}\ \boxed{\text{ENT}}$

EXAMPLE 9   Simplify each expression

a. $\left(\dfrac{2}{3}\right)\left(-\dfrac{3}{5}\right)$ **b.** $\left(-\dfrac{7}{8}\right)\left(-\dfrac{5}{14}\right)$ **c.** $(-5)(3.4)$ **d.** $(-0.4)(-0.8)$

SOLUTION

a. $\left(\dfrac{2}{3}\right)\left(-\dfrac{3}{5}\right) = -\dfrac{6}{15} = -\dfrac{2}{5}$ The rule for multiplication also holds for fractions

b. $\left(-\dfrac{7}{8}\right)\left(-\dfrac{5}{14}\right) = \dfrac{35}{112} = \dfrac{5}{16}$

c. $(-5)(3.4) = -17.0$ The rule for multiplication also holds for decimals

d. $(-0.4)(-0.8) = 0.32$

GETTING READY FOR CLASS

After reading through the preceding section, respond in your own words and in complete sentences.

1. Write the multiplication problem $3(-5)$ as an addition problem.
2. Write the multiplication problem $2(4)$ as an addition problem.
3. If two numbers have the same sign, then their product will have what sign?
4. If two numbers have different signs, then their product will have what sign?

Problem Set 9.4

Find each of the following products. (Multiply.)

1. $7(-8)$ **2.** $-3(5)$ **3.** $-6(10)$ **4.** $4(-8)$

5. $-7(-8)$ **6.** $-4(-7)$ **7.** $-9(-9)$ **8.** $-6(-3)$

9. $-2.1(4.3)$ **10.** $-6.8(5.7)$ **11.** $-\dfrac{4}{5}\left(-\dfrac{15}{28}\right)$ **12.** $-\dfrac{8}{9}\left(-\dfrac{27}{32}\right)$

13. $-12\left(\dfrac{2}{3}\right)$ **14.** $-18\left(\dfrac{5}{6}\right)$ **15.** $3(-2)(4)$ **16.** $5(-1)(3)$

17. $-4(3)(-2)$ **18.** $-4(5)(-6)$ **19.** $-1(-2)(-3)$ **20.** $-2(-3)(-4)$

Use the definition of exponents to expand each of the following expressions. Then multiply according to the rule for multiplication.

21. a. $(-4)^2$ **22. a.** $(-5)^2$ **23. a.** $(-5)^3$ **24. a.** $(-4)^3$

    b. -4^2 **b.** -5^2 **b.** -5^3 **b.** -4^3

25. a. $(-2)^4$ **26. a.** $(-1)^4$

    b. -2^4 **b.** -1^4

Complete the following tables. Remember, if $x = -5$, then $x^2 = (-5)^2 = 25$.

27.

Number x	Square x^2
-3	
-2	
-1	
0	
1	
2	
3	

28.

Number x	Cube x^3
-3	
-2	
-1	
0	
1	
2	
3	

29.

First Number x	Second Number y	Their Product xy
6	2	
6	1	
6	0	
6	-1	
6	-2	

30.

First Number x	Second Number y	Their Product xy
7	4	
7	2	
7	0	
7	-2	
7	-4	

31.

First Number a	Second Number b	Their Product ab
-5	3	
-5	2	
-5	1	
-5	0	
-5	-1	
-5	-2	
-5	-3	

32.

First Number a	Second Number b	Their Product ab
-9	6	
-9	4	
-9	2	
-9	0	
-9	-2	
-9	-4	
-9	-6	

Use the rule for order of operations along with the rules for addition, subtraction, and multiplication to simplify each of the following expressions.

33. $4(-3 + 2)$

34. $7(-6 + 3)$

35. $-10(-2 - 3)$

36. $-5(-6 - 2)$

37. $-3 + 2(5 - 3)$

38. $-7 + 3(6 - 2)$

39. $-7 + 2[-5 - 9]$

40. $-8 + 3[-4 - 1]$

41. $2(-5) + 3(-4)$

42. $6(-1) + 2(-7)$

43. $3(-2)4 + 3(-2)$

44. $2(-1)(-3) + 4(-6)$

45. $(8 - 3)(2 - 7)$

46. $(9 - 3)(2 - 6)$

47. $(2 - 5)(3 - 6)$

48. $(3 - 7)(2 - 8)$

49. $3(5 - 8) + 4(6 - 7)$

50. $-2(8 - 10) + 3(4 - 9)$

51. $-3(4 - 7) - 2(-3 - 2)$

52. $-5(-2 - 8) - 4(6 - 10)$

53. $3(-2)(6 - 7)$

54. $4(-3)(2 - 5)$

55. Find the product of -3, -2, and -1.

56. Find the product of -7, -1, and 0.

57. What number do you multiply by -3 to get 12?

58. What number do you multiply by -7 to get -21?

59. Subtract -3 from the product of -5 and 4.

60. Subtract 5 from the product of -8 and 1.

Applying the Concepts

61. Day Trading Larry is buying and selling stock from his home computer. He owns 100 shares of Oracle Corporation and 50 shares of Gadzoox Networks Inc. On February 28, 2000, those stocks had the gain and loss shown in the table below. What was Larry's net gain or loss for the day on those two stocks?

Stock	Number Of Shares	Gain/Loss
Oracle	100	−2
Gadzoox	50	+8

62. Stock Gain/Loss Amy owns stock that she keeps in her retirement account. She owns 200 shares of Apple Computer and 100 shares of Gap Inc. For the month of February 2000, those stocks had the gain and loss shown in the table below. What was Amy's net gain or loss for the month of February on those two stocks?

Stock	Number Of Shares	Gain/Loss
Apple	200	+14
Gap	100	−5

63. Temperature Change A hot-air balloon is rising to its cruising altitude. Suppose the air temperature around the balloon drops 4 degrees each time the balloon rises 1,000 feet. What is the net change in air temperature around the balloon as it rises from 2,000 feet to 6,000 feet?

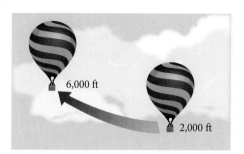

6,000 ft

2,000 ft

64. Temperature Change A small airplane is rising to its cruising altitude. Suppose the air temperature around the plane drops 4 degrees each time the plane increases its altitude by 1,000 feet. What is the net change in air temperature around the plane as it rises from 5,000 feet to 12,000 feet?

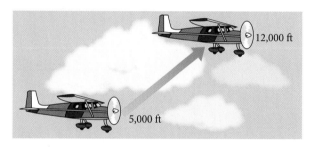

12,000 ft

5,000 ft

Baseball Major league baseball has various player awards at the end of each season. Relief pitchers compete for the Rolaids Relief Award each year. Points are awarded as follows:

Each win (W) earns 2 points   Each loss (L) earns −2 points
Each save (S) earns 3 points   Each blown save (BS) earns −2 points
Each tough save* (TS) earns 4 points

The Rolaids points for someone with 4 wins, 2 losses, 8 saves, 1 tough save, and 3 blown saves would be

$$4(2) + 2(-2) + 8(3) + 1(4) + 3(-2) = 26$$

Use this information to complete the tables for Problems 65 and 66.

65.

NATIONAL LEAGUE						
Name, Team	W	L	S	TS	BS	Pts
Eric Gagne, Los Angeles	2	3	53	2	0	
John Smoltz, Atlanta	0	2	42	3	4	
Billy Wagner, Houston	1	4	40	4	3	
Tim Worrell, San Francisco	4	4	36	2	7	
Joe Borowski, Chicago	2	2	32	2	4	

*A tough save occurs when the pitcher enters the game with the potential tying run on base.

66.

AMERICAN LEAGUE						
Name, Team	W	L	S	TS	BS	Pts
Keith Foulke, Oakland	9	1	39	4	5	
Mariano Rivera, New York	5	2	35	5	6	
Eddie Guardado, Minnesota	3	5	41	0	4	
Troy Percival, Anaheim	0	5	33	0	4	
Jorge Julio, Baltimore	0	7	35	1	8	

Golf One way to give scores in golf is in relation to par, the number of strokes considered necessary to complete a hole or course at the expert level. Scoring this way, if you complete a hole in one stroke less than par, your score is -1, which is called a *birdie*. If you shoot 2 under par, your score is -2, which is called an *eagle*. Shooting 1 over par is a score of $+1$, which is a *bogie*. A *double bogie* is 2 over par, and results in a score of $+2$.

67. **Sergio Garcia's Scorecard** The table below shows the scores Sergio Garcia had on the first round of a PGA tournament. Fill in the last column by multiplying each value by the number of times it occurs. Then add the numbers in the last column to find the total. If par for the course was 72, what was Sergio Garcia's score?

	Value	Number	Product
Eagle	-2	0	
Birdie	-1	7	
Par	0	7	
Bogie	$+1$	3	
Double Bogie	$+2$	1	
		Total:	

68. **Karrie Webb's Scorecard** The table below shows the scores Karrie Webb had on the final round of an LPGA Standard Register Ping Tournament. Fill in the last column by multiplying each value by the number of times it occurs. Then add the numbers in the last column to find the total. If par for the course was 72, what was Karrie Webb's score?

	Value	Number	Product
Eagle	-2	1	
Birdie	-1	5	
Par	0	8	
Bogie	$+1$	3	
Double Bogie	$+2$	1	
		Total:	

Estimating

Work Problems 69–76 mentally, without pencil and paper or a calculator.

69. The product $-32(-522)$ is closest to which of the following numbers?
    a. 15,000   **b.** -500 **c.** $-1,500$ **d.** $-15,000$

70. The product $32(-522)$ is closest to which of the following numbers?
    a. 15,000   **b.** -500 **c.** $-1,500$ **d.** $-15,000$

71. The product $-47(470)$ is closest to which of the following numbers?
    a. 25,000   **b.** 420   **c.** $-2,500$ **d.** $-25,000$

72. The product $-47(-470)$ is closest to which of the following numbers?
    a. 25,000   **b.** 420   **c.** $-2,500$ **d.** $-25,000$

73. The product $-222(-987)$ is closest to which of the following numbers?
    a. 200,000   **b.** 800   **c.** -800 **d.** $-1,200$

74. The sum $-222 + (-987)$ is closest to which of the following numbers?
    a. 200,000   **b.** 800   **c.** -800 **d.** $-1,200$

75. The difference $-222 - (-987)$ is closest to which of the following numbers?
    a. 200,000   **b.** 800   **c.** -800 **d.** $-1,200$

76. The difference $-222 - 987$ is closest to which of the following numbers?
    a. 200,000   **b.** 800   **c.** -800 **d.** $-1,200$

Getting Ready for the Next Section

Perform the indicated operations.

77. $35 \div 5$

78. $32 \div 4$

79. $\dfrac{20}{4}$

80. $\dfrac{30}{5}$

81. $12 - 17$

82. $7 - 11$

83. $\dfrac{6(3)}{2}$

84. $\dfrac{8(5)}{4}$

85. $80 \div 10 \div 2$

86. $80 \div 2 \div 10$

87. $\dfrac{15 + 5(4)}{17 - 12}$

88. $\dfrac{20 + 6(2)}{11 - 7}$

89. $4(10^2) + 20 \div 4$

90. $3(4^2) + 10 \div 5$

Division with Negative Numbers

Suppose four friends invest equal amounts of money in a moving truck to start a small business. After 2 years the truck has dropped $10,000 in value. If we represent this change with the number $-\$10,000$, then the loss to each of the four partners can be found with division:

$$(-\$10,000) \div 4 = -\$2,500$$

From this example it seems reasonable to assume that a negative number divided by a positive number will give a negative answer.

To cover all the possible situations we can encounter with division of negative numbers, we use the relationship between multiplication and division. If we let n be the answer to the problem $12 \div (-2)$, then we know that

$$12 \div (-2) = n \quad \text{and} \quad -2(n) = 12$$

From our work with multiplication, we know that n must be -6 in the multiplication problem above, because -6 is the only number we can multiply -2 by to get 12. Because of the relationship between the two problems above, it must be true that 12 divided by -2 is -6.

The following pairs of problems show more quotients of positive and negative numbers. In each case the multiplication problem on the right justifies the answer to the division problem on the left.

$$
\begin{array}{lll}
6 \div 3 = 2 & \text{because} & 3(2) = 6 \\
6 \div (-3) = -2 & \text{because} & -3(-2) = 6 \\
-6 \div 3 = -2 & \text{because} & 3(-2) = -6 \\
-6 \div (-3) = 2 & \text{because} & -3(2) = -6
\end{array}
$$

These results can be used to write the rule for division with negative numbers.

RULE

To divide two numbers, we divide their absolute values.
1. The answer is positive if both the original numbers have the same sign. That is, the quotient of two numbers with the same signs is positive.
2. The answer is negative if the original two numbers have different signs. That is, the quotient of two numbers with different signs is negative.

EXAMPLE 1 Divide:

a. $12 \div 4$ **b.** $-12 \div 4$ **c.** $12 \div (-4)$ **d.** $-12 \div (-4)$

SOLUTION

a. $12 \div 4 = 3$ *Like signs; positive answer*

b. $-12 \div 4 = -3$ *Unlike signs; negative answer*

c. $12 \div (-4) = -3$ *Unlike signs; negative answer*

d. $-12 \div (-4) = 3$ *Like signs; positive answer*

EXAMPLE 2   Simplify:

a. $\dfrac{20}{5}$ **b.** $\dfrac{-20}{5}$ **c.** $\dfrac{20}{-5}$ **d.** $\dfrac{-20}{-5}$

SOLUTION

a. $\dfrac{20}{5} = 4$ Like signs; positive answer

b. $\dfrac{-20}{5} = -4$ Unlike signs; negative answer

c. $\dfrac{20}{-5} = -4$ Unlike signs; negative answer

d. $\dfrac{-20}{-5} = 4$ Like signs; positive answer

From the examples we have done so far, we can make the following generalization about quotients that contain negative signs:

> If a and b are numbers and b is not equal to 0, then
> $$\frac{-a}{b} = \frac{a}{-b} = -\frac{a}{b} \quad \text{and} \quad \frac{-a}{-b} = \frac{a}{b}$$

The last examples in this section involve more than one operation. We use the rules developed previously in this chapter and the rule for order of operations to simplify each.

EXAMPLE 3   Simplify: $\dfrac{-15 + 5(-4)}{12 - 17}$

SOLUTION   Simplifying above and below the fraction bar, we have

$$\frac{-15 + 5(-4)}{12 - 17} = \frac{-15 + (-20)}{-5} = \frac{-35}{-5} = 7$$

EXAMPLE 4   Simplify: $-4(10^2) + 20 \div (-4)$

SOLUTION   Applying the rule for order of operations, we have

$$
\begin{aligned}
-4(10^2) + 20 \div (-4) &= -4(100) + 20 \div (-4) && \text{Exponents first} \\
&= -400 + (-5) && \text{Multiply and divide} \\
&= -405 && \text{Add}
\end{aligned}
$$

GETTING READY FOR CLASS

After reading through the preceding section, respond in your own words and in complete sentences.

1. Write a multiplication problem that is equivalent to the division problem $-12 \div 4 = -3$.
2. Write a multiplication problem that is equivalent to the division problem $-12 \div (-4) = 3$.
3. If two numbers have the same sign, then their quotient will have what sign?
4. Dividing a negative number by 0 always results in what kind of expression?

Problem Set 9.5

Find each of the following quotients. (Divide.)

1. $-15 \div 5$

2. $15 \div (-3)$

3. $20 \div (-4)$

4. $-20 \div 4$

5. $-30 \div (-10)$

6. $-50 \div (-25)$

7. $\dfrac{-14}{-7}$

8. $\dfrac{-18}{-6}$

9. $\dfrac{12}{-3}$

10. $\dfrac{12}{-4}$

11. $-22 \div 11$

12. $-35 \div 7$

13. $\dfrac{0}{-3}$

14. $\dfrac{0}{-5}$

15. $125 \div (-25)$

16. $-144 \div (-9)$

Complete the following tables.

17.

First Number	Second Number	The Quotient of a and b
a	b	$\dfrac{a}{b}$
100	-5	
100	-10	
100	-25	
100	-50	

18.

First Number	Second Number	The Quotient of a and b
a	b	$\dfrac{a}{b}$
24	-4	
24	-3	
24	-2	
24	-1	

19.

First Number	Second Number	The Quotient of a and b
a	b	$\dfrac{a}{b}$
-100	-5	
-100	5	
100	-5	
100	5	

20.

First Number	Second Number	The Quotient of a and b
a	b	$\dfrac{a}{b}$
-24	-2	
-24	-4	
-24	-6	
-24	-8	

Use any of the rules developed in this chapter and the rule for order of operations to simplify each of the following expressions as much as possible.

21. $\dfrac{4(-7)}{-28}$

22. $\dfrac{6(-3)}{-18}$

23. $\dfrac{-3(-10)}{-5}$

24. $\dfrac{-4(-12)}{-6}$

25. $\dfrac{2(-3)}{6-3}$

26. $\dfrac{2(-3)}{3-6}$

27. $\dfrac{4-8}{8-4}$

28. $\dfrac{9-5}{5-9}$

29. $\dfrac{2(-3)+10}{-4}$

30. $\dfrac{7(-2)-6}{-10}$

31. $\dfrac{2+3(-6)}{4-12}$

32. $\dfrac{3+9(-1)}{5-7}$

33. $\dfrac{6(-7)+3(-2)}{20-4}$

34. $\dfrac{9(-8)+5(-1)}{12-1}$

35. $\dfrac{3(-7)(-4)}{6(-2)}$

36. $\dfrac{-2(4)(-8)}{(-2)(-2)}$

37. $(-5)^2 + 20 \div 4$

38. $6^2 + 36 \div 9$

39. $100 \div (-5)^2$

40. $400 \div (-4)^2$

41. $-100 \div 10 \div 2$

42. $-500 \div 50 \div 10$

43. $-100 \div (10 \div 2)$

44. $-500 \div (50 \div 10)$

45. $(-100 \div 10) \div 2$

46. $(-500 \div 50) \div 10$

457

47. Find the quotient of -25 and 5.

48. Find the quotient of -38 and -19.

49. What number do you divide by -5 to get -7?

50. What number do you divide by 6 to get -7?

51. Subtract -3 from the quotient of 27 and 9.

52. Subtract -7 from the quotient of -72 and -9.

Estimating

Work Problems 53–60 mentally, without pencil and paper or a calculator.

53. Is $397 \div (-401)$ closer to 1 or -1?

54. Is $-751 \div (-749)$ closer to 1 or -1?

55. The quotient $-121 \div 27$ is closest to which of the following numbers?
    a. -150 **b.** -100 **c.** -4 **d.** 6

56. The quotient $1,000 \div (-337)$ is closest to which of the following numbers?
    a. 663   **b.** -3 **c.** -30 **d.** -663

57. Which number is closest to the sum $-151 + (-49)$?
    a. -200 **b.** -100 **c.** 3   **d.** 7,500

58. Which number is closest to the difference $-151 - (-49)$?
    a. -200 **b.** -100 **c.** 3   **d.** 7,500

59. Which number is closest to the product $-151(-49)$?
    a. -200 **b.** -100 **c.** 3   **d.** 7,500

60. Which number is closest to the quotient $-151 \div (-49)$?
    a. -200 **b.** -100 **c.** 3   **d.** 7,500

Applying the Concepts

61. Mean Find the mean of the numbers $-5, 0, 5$.

62. Median Find the median of the numbers $-5, 0, 5$.

63. Averages For the numbers $-5, -4, 0, 2, 2, 2, 3$, find
    a. The mean. **b.** The median. **c.** The mode.

64. Averages For the numbers $-5, -2, -2, 0, 2, 7$, find
    a. The mean. **b.** The median. **c.** The mode.

65. Temperature Line Graph The table below gives the low temperature for each day of one week in White Bear Lake, Minnesota. Draw a line graph of the information in the table.

LOW TEMPERATURES IN WHITE BEAR LAKE, MINNESOTA	
Day	Temperature
Monday	10 °F
Tuesday	8 °F
Wednesday	−5 °F
Thursday	−3 °F
Friday	−8 °F
Saturday	5 °F
Sunday	7 °F

66. Temperature Line Graph The table below gives the low temperature for each day of one week in Fairbanks, Alaska. Draw a line graph of the information in the table.

LOW TEMPERATURES IN FAIRBANKS, ALASKA	
Day	Temperature
Monday	−26 °F
Tuesday	−5 °F
Wednesday	9 °F
Thursday	12 °F
Friday	3 °F
Saturday	−15 °F
Sunday	−20 °F

67. Average Temperatures Use the information in the table in Problem 65 to find
   a. the mean low temperature for the week
   b. the median low temperature for the week

68. Average Temperatures Use the information in the table in Problem 66 to find
   a. the mean low temperature for the week
   b. the median low temperature for the week

Getting Ready for the Next Section

The problems below review some of the properties of addition and multiplication we covered in Chapter 1.

Rewrite each expression using the commutative property of addition or multiplication.

69. $3 + x$ **70.** $4y$

Rewrite each expression using the associative property of addition or multiplication.

71. $5 + (7 + a)$ **72.** $(x + 4) + 6$ **73.** $3(4y)$ **74.** $(3y)8$

Apply the distributive property to each expression.

75. $5(3 + 7)$ **76.** $8(4 + 2)$

Simplify.

77. 6^2 **78.** 12^2 **79.** $100(75)$ **80.** $100(53)$

81. $2(100) + 2(75)$ **82.** $2(100) + 2(53)$ **83. a.** $4 + 3$ **84. a.** $5 + 2$

b. $-5 + 7$ **b.** $-6 + 7$

c. $8 - 1$ **c.** $9 - 1$

d. $-4 - 2$ **d.** $-5 - 3$

e. $3 - 7$ **e.** $2 - 5$

Simplifying Algebraic Expressions 9.6

The woodcut shown here depicts Queen Dido of Carthage around 900 B.C., having an ox hide cut into small strips that will be tied together, making a long rope. The rope will be used to enclose her territory. The question, which has become known as the Queen Dido problem, is: what shape will enclose the largest territory?

To translate the problem into something we are more familiar with, suppose we have 24 yards of fencing that we are to use to build a rectangular dog run. If we want the dog run to have the largest area possible then we want the rectangle, with perimeter 24 yards, that encloses the largest area. The diagram below shows six dog runs, each of which has a perimeter of 24 yards. Notice how the length decreases as the width increases.

Dog Runs with Perimeter = 24 yards

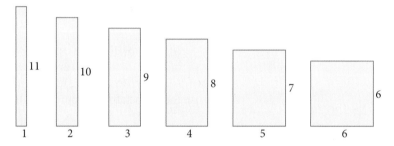

Since area is length times width, we can build a table and a line graph that show how the area changes as we change the width of the dog run.

AREA ENCLOSED BY RECTANGLE OF PERIMETER 24 YARDS	
Width (Yards)	Area (Square Yards)
1	11
2	20
3	27
4	32
5	35
6	36

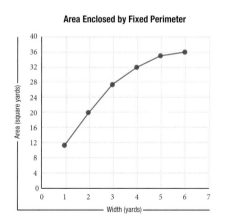

Area Enclosed by Fixed Perimeter

In this section we want to simplify expressions containing variables—that is, algebraic expressions.

To begin let's review how we use the associative properties for addition and multiplication to simplify expressions.

Consider the expression $4(5x)$. We can apply the associative property of multiplication to this expression to change the grouping so that the 4 and the 5 are grouped together, instead of the 5 and the x. Here's how it looks:

$$4(5x) = (4 \cdot 5)x \quad \text{Associative property}$$
$$= 20x \qquad \text{Multiplication: } 4 \cdot 5 = 20$$

We have simplified the expression to $20x$, which in most cases in algebra will be easier to work with than the original expression.

Here are some more examples.

EXAMPLE 1 Simplify $-2(5x)$

SOLUTION

$$-2(5x) = (-2 \cdot 5)x \quad \text{Associative property}$$
$$= -10x \qquad \text{The product of } -2 \text{ and } 5 \text{ is } -10$$

EXAMPLE 2 Simplify $3(-4y)$

SOLUTION

$$3(-4y) = [3(-4)]y \quad \text{Associative property}$$
$$= -12y \qquad 3 \text{ times } -4 \text{ is } -12$$

We can use the associative property of addition to simplify expressions also.

EXAMPLE 3 Simplify $(2x + 5) + 10$

SOLUTION

$$(2x + 5) + 10 = 2x + (5 + 10) \quad \text{Associative property}$$
$$= 2x + 15 \qquad \text{Addition}$$

In Chapter 1 we introduced the distributive property. In symbols it looks like this:

$$a(b + c) = ab + ac$$

Because subtraction is defined as addition of the opposite, the distributive property holds for subtraction as well as addition. That is,

$$a(b - c) = ab - ac$$

We say that multiplication distributes over addition and subtraction. Here are some examples that review how the distributive property is applied to expressions that contain variables.

EXAMPLE 4 Simplify $2(a - 3)$

SOLUTION

$$2(a - 3) = 2(a) - 2(3) \quad \text{Distributive property}$$
$$= 2a - 6 \qquad \text{Multiplication}$$

In Examples 1–3 we simplified expressions such as $4(5x)$ by using the associative property. Here are some examples that use a combination of the associative property and the distributive property.

EXAMPLE 5   Simplify $4(5x + 3)$

SOLUTION

$$
\begin{aligned}
4(5x + 3) &= 4(5x) + 4(3) && \text{Distributive property} \\
&= (4 \cdot 5)x + 4(3) && \text{Associative property} \\
&= 20x + 12 && \text{Multiplication}
\end{aligned}
$$

EXAMPLE 6   Simplify $5(2x + 3y)$

SOLUTION

$$
\begin{aligned}
5(2x + 3y) &= 5(2x) + 5(3y) && \text{Distributive property} \\
&= 10x + 15y && \text{Associative property and} \\
& && \text{multiplication}
\end{aligned}
$$

We can also use the distributive property to simplify expressions like $4x + 3x$. Because multiplication is a commutative operation, we can also rewrite the distributive property like this:

$$ b \cdot a + c \cdot a = (b + c)a $$

Applying the distributive property in this form to the expression $4x + 3x$, we have

$$
\begin{aligned}
4x + 3x &= (4 + 3)x && \text{Distributive property} \\
&= 7x && \text{Addition}
\end{aligned}
$$

Similar Terms

Expressions like $4x$ and $3x$ are called *similar terms* because the variable parts are the same. Some other examples of similar terms are $5y$ and $-6y$ and the terms $7a$, $-13a$, $\frac{3}{4}a$. To simplify an algebraic expression (an expression that involves both numbers and variables), we combine similar terms by applying the distributive property. Table 1 shows several pairs of similar terms and how they can be combined using the distributive property.

TABLE 1				
Original Expression		**Apply Distributive Property**		**Simplified Expression**
$4x + 3x$	=	$(4 + 3)x$	=	$7x$
$7a + a$	=	$(7 + 1)a$	=	$8a$
$-5x + 7x$	=	$(-5 + 7)x$	=	$2x$
$8y - y$	=	$(8 - 1)y$	=	$7y$
$-4a - 2a$	=	$(-4 - 2)a$	=	$-6a$
$3x - 7x$	=	$(3 - 7)x$	=	$-4x$

As you can see from the table, the distributive property can be applied to any combination of positive and negative terms so long as they are similar terms.

Algebraic Expressions Representing Area and Perimeter

Below are a square with a side of length s and a rectangle with a length of l and a width of w. The table that follows the figures gives the formulas for the area and perimeter of each.

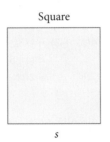

Square

Rectangle

s

l

w

	Square	Rectangle
Area A	s^2	lw
Perimeter P	$4s$	$2l + 2w$

EXAMPLE 7   Find the area and perimeter of a square with a side 6 inches long.

SOLUTION   Substituting 6 for s in the formulas for area and perimeter of a square, we have

$$\text{Area} = A = s^2 = 6^2 = 36 \text{ square inches}$$

$$\text{Perimeter} = P = 4s = 4(6) = 24 \text{ inches}$$

EXAMPLE 8   A soccer field is 100 yards long and 75 yards wide. Find the area and perimeter.

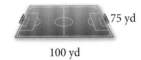

75 yd

100 yd

SOLUTION   Substituting 100 for l and 75 for w in the formulas for area and perimeter of a rectangle, we have

$$\text{Area} = A = lw = 100(75) = 7{,}500 \text{ square yards}$$

$$\text{Perimeter} = P = 2l + 2w = 2(100) + 2(75) = 200 + 150 = 350 \text{ yards}$$

GETTING READY FOR CLASS

After reading through the preceding section, respond in your own words and in complete sentences.

1. Without actually multiplying, how do you apply the associative property to the expression $4(5x)$?

2. What are similar terms?

3. Explain why $2a - a$ is a, rather than 1.

4. Can two rectangles with the same perimeter have different areas?

Problem Set 9.6

Apply the associative property to each expression, and then simplify the result.

1. $5(4a)$ **2.** $8(9a)$ **3.** $6(8a)$ **4.** $3(2a)$

5. $-6(3x)$ **6.** $-2(7x)$ **7.** $-3(9x)$ **8.** $-4(6x)$

9. $5(-2y)$ **10.** $3(-8y)$ **11.** $6(-10y)$ **12.** $5(-5y)$

13. $2 + (3 + x)$ **14.** $9 + (6 + x)$ **15.** $5 + (8 + x)$ **16.** $3 + (9 + x)$

17. $4 + (6 + y)$ **18.** $2 + (8 + y)$ **19.** $7 + (1 + y)$ **20.** $4 + (1 + y)$

21. $(5x + 2) + 4$ **22.** $(8x + 3) + 10$ **23.** $(6y + 4) + 3$ **24.** $(3y + 7) + 8$

25. $(12a + 2) + 19$ **26.** $(6a + 3) + 14$ **27.** $(7x + 8) + 20$ **28.** $(14x + 3) + 15$

Apply the distributive property to each expression, and then simplify.

29. $7(x + 5)$ **30.** $8(x + 3)$ **31.** $6(a - 7)$ **32.** $4(a - 9)$

33. $2(x - y)$ **34.** $5(x - a)$ **35.** $4(5 + x)$ **36.** $8(3 + x)$

37. $3(2x + 5)$ **38.** $8(5x + 4)$ **39.** $6(3a + 1)$ **40.** $4(8a + 3)$

41. $2(6x - 3y)$ **42.** $7(5x - y)$ **43.** $5(7 - 4y)$ **44.** $8(6 - 3y)$

Use the distributive property to combine similar terms. (See Table 1.)

45. $3x + 5x$ **46.** $7x + 8x$ **47.** $3a + a$ **48.** $8a + a$

49. $-2x + 6x$ **50.** $-3x + 9x$ **51.** $6y - y$ **52.** $3y - y$

53. $-8a - 2a$ **54.** $-7a - 5a$ **55.** $4x - 9x$ **56.** $5x - 11x$

Applying the Concepts

Area and Perimeter Find the area and perimeter of each square if the length of each side is as given below.

57. $s = 6$ feet   **58.** $s = 14$ yards   **59.** $s = 9$ inches   **60.** $s = 15$ meters

Area and Perimeter Find the area and perimeter for a rectangle if the length and width are as given below.

61. $l = 20$ inches, $w = 10$ inches   **62.** $l = 40$ yards, $w = 20$ yards

63. $l = 25$ feet, $w = 12$ feet   **64.** $l = 210$ meters, $w = 120$ meters

Temperature Scales In the metric system, the scale we use to measure temperature is the Celsius scale. On this scale water boils at 100 degrees and freezes at 0 degrees. When we write 100 degrees measured on the Celsius scale, we use the notation 100°C, which is read "100 degrees Celsius." If we know the temperature in degrees Fahrenheit, we can convert to degrees Celsius by using the formula

$$C = \frac{5(F - 32)}{9}$$

where F is the temperature in degrees Fahrenheit. Use this formula to find the temperature in degrees Celsius for each of the following Fahrenheit temperatures.

65. 68°F   **66.** 59°F   **67.** 41°F   **68.** 23°F   **69.** 14°F   **70.** 32°F

Chapter 9 Summary

Absolute Value [9.1]

EXAMPLES

1. $|3| = 3$ and $|-3| = 3$

The absolute value of a number is its distance from 0 on the number line. It is the numerical part of a number. The absolute value of a number is never negative.

Opposites [9.1]

2. $-(5) = -5$ and $-(-5) = 5$

Two numbers are called opposites if they are the same distance from 0 on the number line but in opposite directions from 0. The opposite of a positive number is a negative number, and the opposite of a negative number is a positive number.

Addition of Positive and Negative Numbers [9.2]

3.
$$3 + 5 = 8$$
$$-3 + (-5) = -8$$

1. To add two numbers with *the same sign:* Simply add absolute values and use the common sign. If both numbers are positive, the answer is positive. If both numbers are negative, the answer is negative.

2. To add two numbers with *different signs:* Subtract the smaller absolute value from the larger absolute value. The answer has the same sign as the number with the larger absolute value.

Subtraction [9.3]

4.
$$3 - 5 = 3 + (-5) = -2$$
$$-3 - 5 = -3 + (-5) = -8$$
$$3 - (-5) = 3 + 5 = 8$$
$$-3 - (-5) = -3 + 5 = 2$$

Subtracting a number is equivalent to adding its opposite. If a and b represent numbers, then subtraction is defined in terms of addition as follows:

$$a - b = a + (-b)$$

Subtraction   Addition of the opposite

Multiplication with Positive and Negative Numbers [9.4]

5.
$$3(5) = 15$$
$$3(-5) = -15$$
$$-3(5) = -15$$
$$-3(-5) = 15$$

To multiply two numbers, multiply their absolute values.

1. The answer is *positive* if both numbers have the same sign.

2. The answer is *negative* if the numbers have different signs.

Division [9.5]

6. $\dfrac{12}{4} = 3$

$\dfrac{-12}{4} = -3$

$\dfrac{12}{-4} = -3$

$\dfrac{-12}{-4} = 3$

The rule for assigning the correct sign to the answer in a division problem is the same as the rule for multiplication. That is, like signs give a positive answer, and unlike signs give a negative answer.

Simplifying Expressions [9.6]

7. Simplify.
a. $-2(5x) = (-2 \cdot 5)x = -10x$
b. $4(2a - 8) = 4(2a) - 4(8)$
$= 8a - 32$

We simplify algebraic expressions by applying the commutative, associative, and distributive properties.

Combining Similar Terms [9.6]

8. Combine similar terms.
a. $5x + 7x = (5 + 7)x = 12x$
b. $2y - 8y = (2 - 8)y = -6y$

We combine similar terms by applying the distributive property.

Chapter 9 Test Form A

These problems are all taken from examples in your text. Work each problem and check your answers. If you have made a mistake, work the problem again. If you cannot get the correct answer after two tries, look up the correct solution in your text.

1. Give the opposite of each number: 5, 7, 1, −5, −8.

2. Simplify each expression:

    a. $|5|$ **b.** $|-3|$ **c.** $|-7|$

3. Simplify: $10 + (-5) + (-3) + 4$

4. Simplify: $[-3 + (-10)] + [8 + (-2)]$

5. Add: $3.42 + (-6.89)$

6. Subtract: $-7 - 2$

7. Subtract: $12 - (-6)$

8. Combine: $-3 + 6 - 2$

9. Subtract 3 from −5.

10. Multiply:

    a. $2(4)$ **b.** $-2(-4)$ **c.** $2(-4)$ **d.** $-2(4)$

11. Expand and simplify:

    a. $(-6)^2$ **b.** -6^2 **c.** $(-4)^3$ **d.** -4^3

12. Simplify: $-4 + 5(-6 + 2)$

13. Divide:

    a. $12 \div 4$ **b.** $-12 \div 4$ **c.** $12 \div (-4)$ **d.** $-12 \div (-4)$

14. Simplify: $\dfrac{-15 + 5(-4)}{12 - 17}$

15. Simplify: $(2x + 5) + 10$

16. Apply the distributive property: $5(2x + 3y)$

Chapter Test, Form B

For an alternate, more comprehensive, chapter test, go to MathTV.com and select the test and summary for this chapter of the textbook. Click the worksheet labeled Chapter 9 Test, Form B to download it.

Solving Equations

iStockphoto.com © Kameleon007

As we mentioned previously, in the U.S. system, temperature is measured on the Fahrenheit scale, and in the metric system, temperature is measured on the Celsius scale. The table below gives some corresponding temperatures on both temperature scales. The line graph is constructed from the information in the table.

Temperature In Degrees Celsius	Temperature In Degrees Fahrenheit
0°C	32°F
25°C	77°F
50°C	122°F
75°C	167°F
100°C	212°F

The information in the table is *numeric* in nature, whereas the information in the figure is *geometric*. In this chapter we will add a third category for presenting information and writing relationships. Information in this new category is *algebraic* in nature because it is presented using equations and formulas. The following equation is a formula that tells us how to convert from a temperature on the Celsius scale to the corresponding temperature on the Fahrenheit scale.

$$F = \frac{9}{5}C + 32$$

All three items—the table, the line graph, and the equation—show the basic relationship between the two temperature scales. The differences among them are merely differences in form: The table is numeric, the line graph is geometric, and the equation is algebraic.

Success Skills

Dear Student,

Now that you are close to finishing this course, I want to pass on a couple of things that have helped me a great deal with my career. I'll introduce each one with a quote:

Do something for the person you will be 5 years from now.

I have always made sure that I arranged my life so that I was doing something for the person I would be 5 years later. For example, when I was 20 years old, I was in college. I imagined that the person I would be as a 25-year-old, would want to have a college degree, so I made sure I stayed in school. That's all there is to this. It is not a hard, rigid philosophy. It is a soft, behind the scenes, foundation. It does not include ideas such as "Five years from now I'm going to graduate at the top of my class from the best college in the country." Instead, you think, "five years from now I will have a college degree, or I will still be in school working towards it."

This philosophy led to a community college teaching job, writing textbooks, doing videos with the textbooks, then to MathTV and the book you are reading right now. Along the way there were many other options and directions that I didn't take, but all the choices I made were due to keeping the person I would be in 5 years in mind.

It's easier to ride a horse in the direction it is going.

I started my college career thinking that I would become a dentist. I enrolled in all the courses that were required for dental school. When I completed the courses, I applied to a number of dental schools, but wasn't accepted. I kept going to school, and applied again the next year, again, without success. My life was not going in the direction of dental school, even though I had worked hard to put it in that direction. So I did a little inventory of the classes I had taken and the grades I earned, and realized that I was doing well in mathematics. My life was actually going in that direction so I decided to see where that would take me. It was a good decision.

It is a good idea to work hard toward your goals, but it is also a good idea to take inventory every now and then to be sure you are headed in the direction that is best for you.

I wish you good luck with the rest of your college years, and with whatever you decide you want to do as a career.

Pat McKeague
Fall 2010

The Distributive Property and Algebraic Expressions

We recall that the distributive property from Section 1.5 can be used to find the area of a rectangle using two different methods.

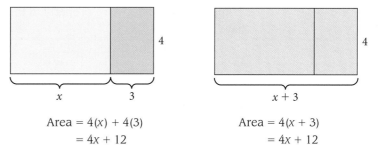

Area = 4(x) + 4(3)   Area = 4(x + 3)
   = 4x + 12   = 4x + 12

Since the areas are equal, the equation $4(x + 3) = 4(x) + 4(3)$ is true.

EXAMPLE 1   Apply the distributive property to the expression

$$5(x + 3)$$

SOLUTION   Distributing the 5 over x and 3, we have

$$5(x + 3) = 5(x) + 5(3) \quad \text{Distributive property}$$
$$= 5x + 15 \quad \text{Multiplication}$$

Remember, $5x$ means "5 times x."

The distributive property can be applied to more complicated expressions involving negative numbers.

EXAMPLE 2   Multiply: $-4(3x + 5)$

SOLUTION   Multiplying both the $3x$ and the 5 by -4, we have

$$-4(3x + 5) = -4(3x) + (-4)5 \quad \text{Distributive property}$$
$$= -12x + (-20) \quad \text{Multiplication}$$
$$= -12x - 20 \quad \text{Definition of subtraction}$$

Notice, first of all, that when we apply the distributive property here, we multiply through by -4. It is important to include the sign with the number when we use the distributive property. Second, when we multiply -4 and $3x$, the result is $-12x$ because

$$-4(3x) = (-4 \cdot 3)x \quad \text{Associative property}$$
$$= -12x \quad \text{Multiplication}$$

We can also use the distributive property to simplify expressions like $4x + 3x$. Because multiplication is a commutative operation, we can rewrite the distributive property like this:

$$b \cdot a + c \cdot a = (b + c)a$$

Applying the distributive property in this form to the expression $4x + 3x$, we have:

$$\begin{aligned} 4x + 3x &= (4 + 3)x \qquad \text{Distributive property}\\ &= 7x \qquad\qquad \text{Addition} \end{aligned}$$

Similar Terms

Note

We are using the word *term* in a different sense here than we did with fractions. (The terms of a fraction are the numerator and the denominator.)

Recall that expressions like $4x$ and $3x$ are called *similar terms* because the variable parts are the same. Some other examples of similar terms are $5y$ and $-6y$, and the terms $7a$, $-13a$, $\frac{3}{4}a$. To simplify an algebraic expression (an expression that involves both numbers and variables), we combine similar terms by applying the distributive property. Table 1 reviews how we combine similar terms using the distributive property.

TABLE 1

Original Expression		Apply Distribution Property		Simplified Expression
$4x + 3x$	=	$(4 + 3)x$	=	$7x$
$7a + a$	=	$(7 + 1)a$	=	$8a$
$-5x + 7x$	=	$(-5 + 7)x$	=	$2x$
$8y - y$	=	$(8 - 1)y$	=	$7y$
$-4a - 2a$	=	$(-4 - 2)a$	=	$-6a$
$3x - 7x$	=	$(3 - 7)x$	=	$-4x$

As you can see from the table, the distributive property can be applied to any combination of positive and negative terms so long as they are similar terms.

EXAMPLE 3   Simplify: $5x - 2 + 3x + 7$

SOLUTION   We begin by changing subtraction to addition of the opposite and applying the commutative property to rearrange the order of the terms. We want similar terms to be written next to each other.

$$\begin{aligned} 5x - 2 + 3x + 7 &= 5x + 3x + (-2) + 7 \qquad \text{Commutative property}\\ &= (5 + 3)x + (-2) + 7 \qquad \text{Distributive property}\\ &= 8x + 5 \qquad\qquad\qquad\quad \text{Addition} \end{aligned}$$

Notice that we take the negative sign in front of the 2 with the 2 when we rearrange terms. How do we justify doing this?

EXAMPLE 4   Simplify: $2(3y + 4) + 5$

SOLUTION   We begin by distributing the 2 across the sum of $3y$ and 4. Then we combine similar terms.

$$
\begin{aligned}
2(3y + 4) + 5 &= 6y + 8 + 5 \qquad \textit{Distributive property} \\
&= 6y + 13 \qquad\quad\ \textit{Add 8 and 5}
\end{aligned}
$$

EXAMPLE 5   Simplify: $2(3x + 1) + 4(2x - 5)$

SOLUTION   Again, we apply the distributive property first; then we combine similar terms. Here is the solution showing only the essential steps:

$$
\begin{aligned}
2(3x + 1) + 4(2x - 5) &= 6x + 2 + 8x - 20 \qquad \textit{Distributive property} \\
&= 14x - 18 \qquad\qquad\ \textit{Combine similar terms}
\end{aligned}
$$

The Value of an Algebraic Expression

An expression such as $3x + 5$ will take on different values depending on what x is. If we were to let x equal 2, the expression $3x + 5$ would become 11. On the other hand, if x is 10, the same expression has a value of 35:

When	$x = 2$		When	$x = 10$
the expression	$3x + 5$		the expression	$3x + 5$
becomes	$3(2) + 5$		becomes	$3(10) + 5$
	$= 6 + 5$			$= 30 + 5$
	$= 11$			$= 35$

Table 2 lists some other algebraic expressions, along with specific values for the variables and the corresponding value of the expression after the variable has been replaced with the given number.

TABLE 2

Original Expression	Value of the Variable	Value of the Expression
$5x + 2$	$x = 4$	$5(4) + 2 = 20 + 2$ $= 22$
$3x - 9$	$x = 2$	$3(2) - 9 = 6 - 9$ $= -3$
$-2a + 7$	$a = 3$	$-2(3) + 7 = -6 + 7$ $= 1$
$6x + 3$	$x = -2$	$6(-2) + 3 = -12 + 3$ $= -9$
$-4y + 9$	$y = -1$	$-4(-1) + 9 = 4 + 9$ $= 13$

FACTS FROM GEOMETRY   *Angles*

An angle is formed by two rays with the same endpoint. The common endpoint is called the *vertex* of the angle, and the rays are called the *sides* of the angle.

In Figure 1, angle θ (theta) is formed by the two rays *OA* and *OB*. The vertex of θ is *O*. Angle θ is also denoted as angle *AOB*, where the letter associated with the vertex is always the middle letter in the three letters used to denote the angle.

Degree Measure The angle formed by rotating a ray through one complete revolution about its endpoint (Figure 2) has a measure of 360 degrees, which we write as 360°.

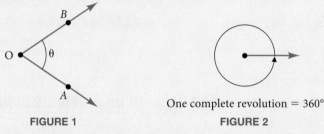

FIGURE 1   **FIGURE 2**

One complete revolution = 360°

One degree of angle measure, written 1°, is $\frac{1}{360}$ of a complete rotation of a ray about its endpoint; there are 360° in one full rotation. (The number 360 was decided upon by early civilizations because it was believed that Earth was at the center of the universe and the sun would rotate once around Earth every 360 days.) Similarly, 180° is half of a complete rotation, and 90° is a quarter of a full rotation. Angles that measure 90° are called *right angles,* and angles that measure 180° are called *straight angles.* If an angle measures between 0° and 90° it is called an *acute angle,* and an angle that measures between 90° and 180° is an *obtuse angle.* Figure 3 illustrates further.

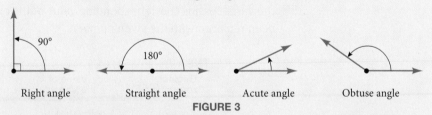

Right angle   Straight angle   Acute angle   Obtuse angle

FIGURE 3

Complementary Angles and Supplementary Angles If two angles add up to 90°, we call them *complementary angles,* and each is called the *complement* of the other. If two angles have a sum of 180°, we call them *supplementary angles,* and each is called the *supplement* of the other. Figure 4 illustrates the relationship between angles that are complementary and angles that are supplementary.

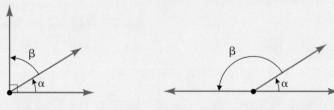

Complementary angles: α + β = 90°   Supplementary angles: α + β = 180°

FIGURE 4

EXAMPLE 6　Find x in each of the following diagrams.

a.

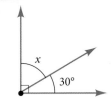

Complementary angles

b.

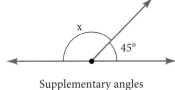

Supplementary angles

SOLUTION　We use subtraction to find each angle.

a. Because the two angles are complementary, we can find x by subtracting 30° from 90°:

$$x = 90° - 30° = 60°$$

We say 30° and 60° are complementary angles. The complement of 30° is 60°.

b. The two angles in the diagram are supplementary. To find x, we subtract 45° from 180°:

$$x = 180° - 45° = 135°$$

We say 45° and 135° are supplementary angles. The supplement of 45° is 135°.

GETTING READY FOR CLASS

After reading through the preceding section, respond in your own words and in complete sentences.

1. What is the distributive property?
2. What property allows $5(x + 3)$ to be rewritten as $5x + 5(3)$?
3. What property allows $3x + 4x$ to be rewritten as $7x$?
4. True or false? The expression $3x$ means 3 multiplied by x.

Problem Set 10.1

For review, use the distributive property to combine each of the following pairs of similar terms.

1. $2x + 8x$

2. $3x + 7x$

3. $6a - 2a$

4. $9a - 3a$

5. $-4y + 5y$

6. $-3y + 10y$

7. $-6x - 2x$

8. $-9x - 4x$

9. $4a - a$

10. $9a - a$

11. $x - 6x$

12. $x - 9x$

Simplify the following expressions by combining similar terms. In some cases the order of the terms must be rearranged first by using the commutative property.

13. $4x + 2x + 3 + 8$

14. $7x + 5x + 2 + 9$

15. $7x - 5x + 6 - 4$

16. $10x - 7x + 9 - 6$

17. $-2a + a + 7 + 5$

18. $-8a + 3a + 12 + 1$

19. $6y - 2y - 5 + 1$

20. $4y - 3y - 7 + 2$

21. $4x + 2x - 8x + 4$

22. $6x + 5x - 12x + 6$

23. $9x - x - 5 - 1$

24. $2x - x - 3 - 8$

25. $2a + 4 + 3a + 5$

26. $9a + 1 + 2a + 6$

27. $3x + 2 - 4x + 1$

28. $7x + 5 - 2x + 6$

29. $12y + 3 + 5y$

30. $8y + 1 + 6y$

31. $4a - 3 - 5a + 2a$

32. $6a - 4 - 2a + 6a$

Simplify.

33. $2(3x + 4) + 8$

34. $2(5x + 1) + 10$

35. $5(2x - 3) + 4$

36. $6(4x - 2) + 7$

37. $8(2y + 4) + 3y$

38. $2(5y + 1) + 2y$

39. $6(4y - 3) + 6y$

40. $5(2y - 6) + 4y$

41. $2(x + 3) + 4(x + 2)$

42. $3(x + 1) + 2(x + 5)$

43. $3(2a + 4) + 7(3a - 1)$

44. $7(2a + 2) + 4(5a - 1)$

Find the value of each of the following expressions when $x = 5$.

45. $2x + 4$

46. $3x + 2$

47. $7x - 8$

48. $8x - 9$

49. $-4x + 1$

50. $-3x + 7$

51. $-8 + 3x$

52. $-7 + 2x$

Find the value of each of the following expressions when $a = -2$.

53. $2a + 5$

54. $3a + 4$

55. $-7a + 4$

56. $-9a + 3$

57. $-a + 10$

58. $-a + 8$

59. $-4 + 3a$

60. $-6 + 5a$

Find the value of each of the following expressions when $x = 3$. You may substitute 3 for x in each expression the way it is written, or you may simplify each expression first and then substitute 3 for x.

61. $3x + 5x + 4$

62. $6x + 8x + 7$

63. $9x + x + 3 + 7$

64. $5x + 3x + 2 + 4$

65. $4x + 3 + 2x + 5$

66. $7x + 6 + 2x + 9$

67. $3x - 8 + 2x - 3$

68. $7x - 2 + 4x - 1$

Use the distributive property to write two equivalent expressions for the area of each figure.

69.

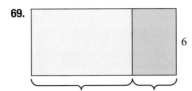

6

x   4

70.

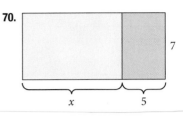

7

x   5

Write an expression for the perimeter of each figure.

71.

x + 1

72.

3x + 2

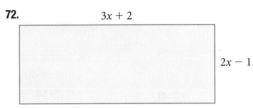

2x − 1

73.

3x + 1

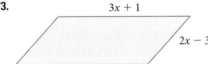

2x − 3

74.

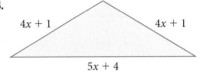

4x + 1   4x + 1

5x + 4

Applying the Concepts

75. Geometry Find the complement and supplement of 25°. Is 25° an acute angle or an obtuse angle?

76. Geometry Find the supplement of 125°. Is 125° an acute angle or an obtuse angle?

77. Temperature and Altitude On a certain day, the temperature on the ground is 72 degrees Fahrenheit, and the temperature at an altitude of A feet above the ground is found from the expression $72 - \frac{A}{300}$. Find the temperature at the following altitudes.

a. 12,000 feet   **b.** 15,000 feet   **c.** 27,000 feet

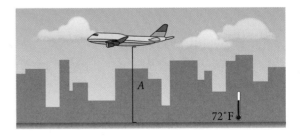

78. Perimeter of a Rectangle As you know, the expression $2l + 2w$ gives the perimeter of a rectangle with length l and width w. The garden below has a width of $3\frac{1}{2}$ feet and a length of 8 feet. What is the length of the fence that surrounds the garden?

3.5 ft

8 ft

79. Cost of Bottled Water A water bottling company charges $7 per month for their water dispenser and $2 for each gallon of water delivered. If you have g gallons of water delivered in a month, then the expression $7 + 2g$ gives the amount of your bill for that month. Find the monthly bill for each of the following deliveries.
  a. 10 gallons   **b.** 20 gallons

80. Cellular Phone Rates A cellular phone company charges $35 per month plus 25 cents for each minute, or fraction of a minute, that you use one of their cellular phones. The expression $\frac{3500 + 25t}{100}$ gives the amount of money, in dollars, you will pay for using one of their phones for t minutes a month. Find the monthly bill for using one of their phones:
  a. 20 minutes in a month   **b.** 40 minutes in a month

Getting Ready for the Next Section

Add.

81. $4 + (-4)$ **82.** $2 + (-2)$ **83.** $-2 + (-4)$ **84.** $-2 + (-5)$

85. $-5 + 2$ **86.** $-3 + 12$ **87.** $\frac{5}{8} + \frac{3}{4}$ **88.** $\frac{5}{6} + \frac{2}{3}$

89. $-\frac{3}{4} + \frac{3}{4}$ **90.** $-\frac{2}{3} + \frac{2}{3}$

Simplify.

91. $x + 0$ **93.** $y + 4 - 6$

92. $y + 0$ **94.** $y + 6 - 2$

Previously we defined complementary angles as two angles whose sum is 90°. If A and B are complementary angles, then

$$A + B = 90°$$

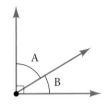

Complementary angles

If we know that $A = 30°$, then we can substitute 30° for A in the formula above to obtain the equation

$$30° + B = 90°$$

In this section we will learn how to solve equations like this one that involve addition and subtraction with one variable.

In general, solving an equation involves finding all replacements for the variable that make the equation a true statement.

DEFINITION

A *solution* for an equation is a number that when used in place of the variable makes the equation a true statement.

For example, the equation $x + 3 = 7$ has as its solution the number 4, because replacing x with 4 in the equation gives a true statement:

$$
\begin{array}{ll}
\text{When} & x = 4 \\
\text{the equation} & x + 3 = 7 \\
\text{becomes} & 4 + 3 = 7 \\
\text{or} & 7 = 7 \qquad \text{A true statement}
\end{array}
$$

EXAMPLE 1 Is $a = -2$ the solution to the equation $7a + 4 = 3a - 2$?

SOLUTION
$$
\begin{array}{ll}
\text{When} & a = -2 \\
\text{the equation} & 7a + 4 = 3a - 2 \\
\text{becomes} & 7(-2) + 4 = 3(-2) - 2 \\
& -14 + 4 = -6 - 2 \\
& -10 = -8 \qquad \text{A false statement}
\end{array}
$$

Because the result is a false statement, we must conclude that $a = -2$ is *not* a solution to the equation $7a + 4 = 3a - 2$.

We want to develop a process for solving equations with one variable. The most important property needed for solving the equations in this section is called the *addition property of equality.* The formal definition looks like this:

Addition Property of Equality

Let A, B, and C represent algebraic expressions.

$$\text{If} \qquad A = B$$
$$\text{then} \qquad A + C = B + C$$

In words: Adding the same quantity to both sides of an equation never changes the solution to the equation.

This property is extremely useful in solving equations. Our goal in solving equations is to isolate the variable on one side of the equation. We want to end up with an expression of the form

$$x = \text{a number}$$

To do so we use the addition property of equality.

EXAMPLE 2 Solve for x: $x + 4 = -2$

SOLUTION We want to isolate x on one side of the equation. If we add -4 to both sides, the left side will be $x + 4 + (-4)$, which is $x + 0$ or just x.

$$
\begin{aligned}
x + 4 &= -2 \\
x + 4 + (-4) &= -2 + (-4) \qquad &\text{Add} -4 \text{ to both sides} \\
x + 0 &= -6 \qquad &\text{Addition} \\
x &= -6 \qquad &x + 0 = x
\end{aligned}
$$

The solution is -6. We can check it if we want to by replacing x with -6 in the original equation:

$$
\begin{aligned}
\text{When} \qquad & x = -6 \\
\text{the equation} \qquad & x + 4 = -2 \\
\text{becomes} \qquad & -6 + 4 = -2 \\
& -2 = -2 \qquad \text{A true statement}
\end{aligned}
$$

EXAMPLE 3 Solve for x: $3x - 2 - 2x = 4 - 9$

SOLUTION Simplifying each side as much as possible, we have

$$
\begin{aligned}
3x - 2 - 2x &= 4 - 9 \\
x - 2 &= -5 \qquad &3x - 2x = x \\
x - 2 + 2 &= -5 + 2 \qquad &\text{Add 2 to both sides} \\
x + 0 &= -3 \qquad &\text{Addition} \\
x &= -3 \qquad &x + 0 = x
\end{aligned}
$$

Note

With some of the equations in this section, you will be able to see the solution just by looking at the equation. But it is important that you show all the steps used to solve the equations anyway. The equations you come across in the future will not be as easy to solve, so you should learn the steps involved very well.

EXAMPLE 4   Solve: $a - \dfrac{3}{4} = \dfrac{5}{8}$

SOLUTION To isolate a we add $\frac{3}{4}$ to each side:

$$a - \frac{3}{4} = \frac{5}{8}$$

$$a - \frac{3}{4} + \frac{3}{4} = \frac{5}{8} + \frac{3}{4}$$

$$a = \frac{11}{8}$$

When solving equations we will leave answers like $\frac{11}{8}$ as improper fractions, rather than change them to mixed numbers.

EXAMPLE 5   Solve: $4(2a - 3) - 7a = 2 - 5$.

SOLUTION We must begin by applying the distributive property to separate terms on the left side of the equation. Following that, we combine similar terms and then apply the addition property of equality.

$4(2a - 3) - 7a = 2 - 5$	*Original equation*
$8a - 12 - 7a = 2 - 5$	*Distributive property*
$a - 12 = -3$	*Simplify each side*
$a - 12 + 12 = -3 + 12$	*Add 12 to each side*
$a = 9$	*Addition*

A Note on Subtraction

Although the addition property of equality is stated for addition only, we can subtract the same number from both sides of an equation as well. Because subtraction is defined as addition of the opposite, subtracting the same quantity from both sides of an equation will not change the solution. If we were to solve the equation in Example 2 using subtraction instead of addition, the steps would look like this:

$x + 4 = -2$	*Original equation*
$x + 4 - 4 = -2 - 4$	*Subtract 4 from each side*
$x = -6$	*Subtraction*

   In my experience teaching algebra, I find that students make fewer mistakes if they think in terms of addition rather than subtraction. So, you are probably better off if you continue to use the addition property just the way we have used it in the examples in this section. But, if you are curious as to whether you can subtract the same number from both sides of an equation, the answer is yes.

GETTING READY FOR CLASS

After reading through the preceding section, respond in your own words and in complete sentences.

1. What is a solution to an equation?

2. True or false? According to the addition property of equality, adding the same value to both sides of an equation will never change the solution to the equation.

3. Show that $x = 5$ is a solution to the equation $3x + 2 = 17$ without solving the equation.

4. True or false? The equations below have the same solution.

Equation 1: $7x + 5 = 19$

Equation 2: $7x + 5 + 3 = 19 + 3$

Problem Set 10.2

Check to see if the number to the right of each of the following equations is the solution to the equation.

1. $2x + 1 = 5; 2$

2. $4x + 3 = 7; 1$

3. $3x + 4 = 19; 5$

4. $3x + 8 = 14; 2$

5. $2x - 4 = 2; 4$

6. $5x - 6 = 9; 3$

7. $2x + 1 = 3x + 3; -2$

8. $4x + 5 = 2x - 1; -6$

9. $x - 4 = 2x + 1; -4$

10. $x - 8 = 3x + 2; -5$

Solve each equation.

11. $x + 2 = 8$

12. $x + 3 = 5$

13. $x - 4 = 7$

14. $x - 6 = 2$

15. $a + 9 = -6$

16. $a + 3 = -1$

17. $x - 5 = -4$

18. $x - 8 = -3$

19. $y - 3 = -6$

20. $y - 5 = -1$

21. $a + \dfrac{1}{3} = -\dfrac{2}{3}$

22. $a + \dfrac{1}{4} = -\dfrac{3}{4}$

23. $x - \dfrac{3}{5} = \dfrac{4}{5}$

24. $x - \dfrac{7}{8} = \dfrac{3}{8}$

25. $y + 7.3 = -2.7$

26. $y + 8.2 = -2.8$

Simplify each side of the following equations before applying the addition property.

27. $x + 4 - 7 = 3 - 10$

28. $x + 6 - 2 = 5 - 12$

29. $x - 6 + 4 = -3 - 2$

30. $x - 8 + 2 = -7 - 1$

31. $3 - 5 = a - 4$

32. $2 - 6 = a - 1$

33. $3a + 7 - 2a = 1$

34. $5a + 6 - 4a = 4$

35. $6a - 2 - 5a = -9 + 1$

36. $7a - 6 - 6a = -3 + 1$

37. $8 - 5 = 3x - 2x + 4$

38. $10 - 6 = 8x - 7x + 6$

The following equations contain parentheses. Apply the distributive property to remove the parentheses, then simplify each side before using the addition property of equality.

39. $2(x + 3) - x = 4$

40. $5(x + 1) - 4x = 2$

41. $-3(x - 4) + 4x = 3 - 7$

42. $-2(x - 5) + 3x = 4 - 9$

43. $5(2a + 1) - 9a = 8 - 6$

44. $4(2a - 1) - 7a = 9 - 5$

45. $-(x + 3) + 2x - 1 = 6$

46. $-(x - 7) + 2x - 8 = 4$

Find the value of x for each of the figures, given the perimeter.

47. $P = 36$

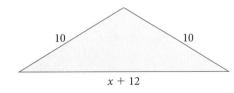

48. $P = 30$

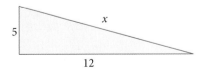

49. $P = 16$

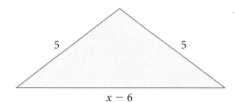

$x - 6$

50. $P = 60$

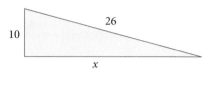

26

10

x

Applying the Concepts

51. Geometry Two angles are complementary angles. If one of the angles is 23°, then solving the equation $x + 23° = 90°$ will give you the other angle. Solve the equation.

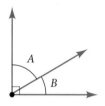

Complementary angles

52. Geometry Two angles are supplementary angles. If one of the angles is 23°, then solving the equation $x + 23° = 180°$ will give you the other angle. Solve the equation.

53. Theater Tickets The El Portal Center for the Arts in North Hollywood, California, holds a maximum of 400 people. The two balconies hold 86 and 89 people each; the rest of the seats are at the stage level. Solving the equation $x + 86 + 89 = 400$ will give you the number of seats on the stage level.
   a. Solve the equation for x.
   b. If tickets on the stage level are $30 each, and tickets in either balcony are $25 each, what is the maximum amount of money the theater can bring in for a show?

54. Geometry The sum of the angles in the triangle on the swing set is 180°. Use this fact to write an equation containing x. Then solve the equation.

Translating Translate each of the following into an equation, and then solve the equation.

55. The sum of x and 12 is 30.

56. The difference of x and 12 is 30.

57. The difference of 8 and 5 is equal to the sum of x and 7.

58. The sum of 8 and 5 is equal to the difference of x and 7.

Getting Ready for the Next Section

Find the reciprocal of each number.

59. 4   **60.** 3   **61.** $\dfrac{1}{2}$ **62.** $\dfrac{1}{3}$ **63.** $\dfrac{2}{3}$ **64.** $\dfrac{3}{5}$

Multiply.

65. $2 \cdot \dfrac{1}{2}$ **66.** $\dfrac{1}{4} \cdot 4$ **67.** $-\dfrac{1}{3}(-3)$ **68.** $-\dfrac{1}{4}(-4)$

69. $\dfrac{3}{2}\left(\dfrac{2}{3}\right)$ **70.** $\dfrac{5}{3}\left(\dfrac{3}{5}\right)$ **71.** $\left(-\dfrac{5}{4}\right)\left(-\dfrac{4}{5}\right)$ **72.** $\left(-\dfrac{4}{3}\right)\left(-\dfrac{3}{4}\right)$

Simplify.

73. $1 \cdot x$ **74.** $1 \cdot a$ **75.** $4x - 11 + 3x$ **76.** $2x - 11 + 3x$

The Multiplication Property of Equality

In this section we will continue to solve equations in one variable. We will again use the addition property of equality, but we will also use another property—the *multiplication property of equality*—to solve the equations in this section. We state the multiplication property of equality and then see how it is used by looking at some examples.

The most popular Internet video download of all time was a *Star Wars* movie trailer. The video was compressed so it would be small enough for people to download over the Internet. In movie theaters, a film plays at 24 frames per second. Over the Internet that number is sometimes cut in half, to 12 frames per second, to make the file size smaller.

We can use the equation $240 = \frac{x}{12}$ to find the number of total frames, x, in a 240-second movie clip that plays at 12 frames per second.

Multiplication Property of Equality

Let A, B, and C represent algebraic expressions, with C not equal to 0.

$$\text{If} \quad A = B$$
$$\text{then} \quad AC = BC$$

In words: Multiplying both sides of an equation by the same nonzero quantity never changes the solution to the equation.

Now, because division is defined as multiplication by the reciprocal, we are also free to divide both sides of an equation by the same nonzero quantity and always be sure we have not changed the solution to the equation.

EXAMPLE 1   Solve for x: $\frac{1}{2}x = 3$

SOLUTION   Our goal here is the same as it was in Section 10.2. We want to isolate x (that is, $1x$) on one side of the equation. We have $\frac{1}{2}x$ on the left side. If we multiply both sides by 2, we will have $1x$ on the left side. Here is how it looks:

$$\frac{1}{2}x = 3$$

$$2\left(\frac{1}{2}x\right) = 2(3) \qquad \text{Multiply both sides by 2}$$

$$x = 6 \qquad \text{Multiplication}$$

To see why $2\left(\frac{1}{2}x\right)$ is equivalent to x, we use the associative property:

$$2\left(\frac{1}{2}x\right) = \left(2 \cdot \frac{1}{2}\right)x \qquad \text{Associative property}$$

$$= 1 \cdot x \qquad\qquad 2 \cdot \frac{1}{2} = 1$$

$$= x \qquad\qquad 1 \cdot x = x$$

Although we will not show this step when solving problems, it is implied.

EXAMPLE 2   Solve for a: $\frac{1}{3}a + 2 = 7$

SOLUTION   We begin by adding -2 to both sides to get $\frac{1}{3}a$ by itself. We then multiply by 3 to solve for a.

$$\frac{1}{3}a + 2 = 7$$

$$\frac{1}{3}a + 2 + (-2) = 7 + (-2) \qquad \text{Add } -2 \text{ to both sides}$$

$$\frac{1}{3}a = 5 \qquad\qquad \text{Addition}$$

$$3 \cdot \frac{1}{3}a = 3 \cdot 5 \qquad\qquad \text{Multiply both sides by 3}$$

$$a = 15 \qquad\qquad \text{Multiplication}$$

We can check our solution to see that it is correct:

$$\text{When} \qquad\qquad a = 15$$

$$\text{the equation} \qquad \frac{1}{3}a + 2 = 7$$

$$\text{becomes} \qquad \frac{1}{3}(15) + 2 = 7$$

$$5 + 2 = 7$$

$$7 = 7 \qquad \text{A true statement}$$

Note The reciprocal of a negative number is also a negative number. Remember, reciprocals are two numbers that have a product of 1. Since 1 is a positive number, any two numbers we multiply to get 1 must both have the same sign. Here are some negative numbers and their reciprocals:

The reciprocal of -2 is $-\frac{1}{2}$.

The reciprocal of -7 is $-\frac{1}{7}$.

The reciprocal of $-\frac{1}{3}$ is -3.

The reciprocal of $-\frac{3}{4}$ is $-\frac{4}{3}$.

The reciprocal of $-\frac{9}{5}$ is $-\frac{5}{9}$.

EXAMPLE 3   Solve: $-\dfrac{4}{5}x = \dfrac{8}{15}$

SOLUTION   The reciprocal of $-\dfrac{4}{5}$ is $-\dfrac{5}{4}$.

$$-\frac{4}{5}x = \frac{8}{15}$$

$$-\frac{5}{4}\left(-\frac{4}{5}x\right) = -\frac{5}{4}\left(\frac{8}{15}\right) \qquad \text{Multiply both sides by } -\frac{5}{4}$$

$$x = -\frac{2}{3} \qquad \text{Multilplication}$$

Many times it is convenient to divide both sides by a nonzero number to solve an equation, as the next example shows.

EXAMPLE 4   Solve for x: $4x = -20$

SOLUTION   If we divide both sides by 4, the left side will be just x, which is what we want. It is okay to divide both sides by 4 because division by 4 is equivalent to multiplication by $\frac{1}{4}$, and the multiplication property of equality states that we can multiply both sides by any number so long as it isn't 0.

$$4x = -20$$

$$\frac{4x}{4} = \frac{-20}{4} \qquad \text{Divide both sides by 4}$$

$$x = -5 \qquad \text{Division}$$

> **Note**
>
> If we multiply each side by $\frac{1}{4}$, the solution looks like this:
>
> $$\frac{1}{4}(4x) = \frac{1}{4}(-20)$$
>
> $$\frac{1}{4} \cdot 4x = -5$$
>
> $$1x = -5$$
>
> $$x = -5$$

Because $4x$ means "4 times x," the factors in the numerator of $\dfrac{4x}{4}$ are 4 and x.

Because the factor 4 is common to the numerator and the denominator, we divide it out to get just x.

EXAMPLE 5   Solve for x: $-3x + 7 = -5$

SOLUTION   We begin by adding -7 to both sides to reduce the left side to $-3x$.

$$-3x + 7 = -5$$

$$-3x + 7 + (-7) = -5 + (-7) \qquad \text{Add } -7 \text{ to both sides}$$

$$-3x = -12 \qquad \text{Addition}$$

$$\frac{-3x}{-3} = \frac{-12}{-3} \qquad \text{Divide both sides by } -3$$

$$x = 4 \qquad \text{Division}$$

With more complicated equations we simplify each side separately before applying the addition or multiplication properties of equality. The next example illustrates.

EXAMPLE 6   Solve for x: $5x - 8x + 3 = 4 - 10$

SOLUTION   We combine similar terms to simplify each side and then solve as usual.

$$5x - 8x + 3 = 4 - 10$$
$$-3x + 3 = -6 \qquad \text{Simplify each side}$$
$$-3x + 3 + (-3) = -6 + (-3) \qquad \text{Add } -3 \text{ to both sides}$$
$$-3x = -9 \qquad \text{Addition}$$
$$\frac{-3x}{-3} = \frac{-9}{-3} \qquad \text{Divide both sides by } -3$$
$$x = 3 \qquad \text{Division}$$

COMMON MISTAKE

Before we end this section we should mention a very common mistake made by students when they first begin to solve equations. It involves trying to subtract away the number in front of the variable—like this:

$$7x = 21$$

$$7x - 7 = 21 - 7 \qquad \text{Add } -7 \text{ to both sides}$$
$$x = 14 \longleftarrow \text{Mistake}$$

The mistake is not in trying to subtract 7 from both sides of the equation. The mistake occurs when we say $7x - 7 = x$. It just isn't true. We can add and subtract only similar terms. The numbers $7x$ and 7 are not similar, because one contains x and the other doesn't. The correct way to do the problem is like this:

$$7x = 21$$
$$\frac{7x}{7} = \frac{21}{7} \qquad \text{Divide both sides by 7}$$
$$x = 3 \qquad \text{Division}$$

GETTING READY FOR CLASS

After reading through the preceding section, respond in your own words and in complete sentences.

1. True or false? Multiplying both sides of an equation by the same non-zero quantity will never change the solution to the equation.

2. If we were to multiply the right side of an equation by 2, then the left side should be multiplied by _____.

3. Dividing both sides of the equation $4x = -20$ by 4 is the same as multiplying both sides by what number?

4. To solve the equation $-5x + 6 = -14$, would you begin by dividing both sides by -5 or by adding -6 to each side?

Problem Set 10.3

Use the multiplication property of equality to solve each of the following equations. In each case show all the steps.

1. $\frac{1}{4}x = 2$ **2.** $\frac{1}{3}x = 7$ **3.** $\frac{1}{2}x = -3$ **4.** $\frac{1}{5}x = -6$

5. $-\frac{1}{3}x = 2$ **6.** $-\frac{1}{3}x = 5$ **7.** $-\frac{1}{6}x = -1$ **8.** $-\frac{1}{2}x = -4$

9. $\frac{3}{4}y = 12$ **10.** $\frac{2}{3}y = 18$ **11.** $3a = 48$ **12.** $2a = 28$

13. $-\frac{3}{5}x = \frac{9}{10}$ **14.** $-\frac{4}{5}x = -\frac{8}{15}$ **15.** $5x = -35$ **16.** $7x = -35$

17. $-8y = 64$ **18.** $-9y = 27$ **19.** $-7x = -42$ **20.** $-6x = -42$

Using the addition property of equality first, solve each of the following equations.

21. $3x - 1 = 5$ **22.** $2x + 4 = 6$ **23.** $-4a + 3 = -9$

24. $-5a + 10 = 50$ **25.** $6x - 5 = 19$ **26.** $7x - 5 = 30$

27. $\frac{1}{3}a + 3 = -5$ **28.** $\frac{1}{2}a + 2 = -7$ **29.** $-\frac{1}{4}a + 5 = 2$

30. $-\frac{1}{5}a + 3 = 7$ **31.** $2x - 4 = -20$ **32.** $3x - 5 = -26$

33. $\frac{2}{3}x - 4 = 6$ **34.** $\frac{3}{4}x - 2 = 7$ **35.** $-11a + 4 = -29$

36. $-12a + 1 = -47$ **37.** $-3y - 2 = 1$ **38.** $-2y - 8 = 2$

39. $-2x - 5 = -7$ **40.** $-3x - 6 = -36$

Simplify each side of the following equations first, then solve.

41. $2x + 3x - 5 = 7 + 3$ **42.** $4x + 5x - 8 = 6 + 4$

43. $4x - 7 + 2x = 9 - 10$ **44.** $5x - 6 + 3x = -6 - 8$

45. $3a + 2a + a = 7 - 13$ **46.** $8a - 6a + a = 8 - 14$

47. $5x + 4x + 3x = 4 - 8$ **48.** $4x + 8x - 2x = 15 - 10$

49. $5 - 18 = 3y - 2y + 1$ **50.** $7 - 16 = 4y - 3y + 2$

Find the value of x for each of the figures, given the perimeter.

51. $P = 72$ **52.** $P = 96$

2x

3x

53. $P = 80$

3x

2x

54. $P = 64$

5x

3x

Applying the Concepts

55. Basketball Kendra plays basketball for her high school. In one game she scored 21 points total, with a combination of free throws, field goals, and three-pointers. Each free throw is worth 1 point, each field goal is 2 points, and each three-pointer is worth 3 points. If she made 1 free throw and 4 field goals, then solving the equation

$$1 + 2(4) + 3x = 21$$

will give us the number of three-pointers she made. Solve the equation to find the number of three-point shots Kendra made.

56. Break-Even Point The El Portal Center for the Arts is showing a movie to raise money for a local charity. The cost to put on the event is $1,840, which includes rent on the Center and movie, insurance, and wages for the paid attendants. If tickets cost $8 each, then solving the equation $8x = 1,840$ gives the number of tickets they must sell in order to cover their costs. This number is called the break-even point. Solve the equation for x to find the break-even point.

Translations Translate each sentence below into an equation, then solve the equation.

57. The sum of $2x$ and 5 is 19.

58. The sum of 8 and $3x$ is 2.

59. The difference of $5x$ and 6 is -9.

60. The difference of 9 and $6x$ is 21.

Getting Ready for the Next Section

Perform the indicated operation.

61. $2.52 + 3.78$

62. $0.98 + 0.42$

63. $-3.78 + 3.78$

64. $-2.4 + 2.4$

65. $2(6.3)$

66. $5(10.7)$

67. $1.40 \div 8$

68. $0.95 \div 5$

Apply the distributive property to each of the following expressions.

69. $2(3a - 8)$

70. $4(2a - 5)$

71. $-3(5x - 1)$

72. $-2(7x - 3)$

Simplify each of the following expressions as much as possible.

73. $3(y - 5) + 6$

74. $5(y + 3) + 7$

75. $6(2x - 1) + 4x$

76. $8(3x - 2) + 4x$

Linear Equations in One Variable 10.4

The Rhind Papyrus is an ancient Egyptian document, created around 1650 BC, that contains some mathematical riddles. One problem on the Rhind Papyrus asked the reader to find a quantity such that when it is added to one-fourth of itself the sum is 15. The equation that describes this situation is

$$x + \frac{1}{4}x = 15$$

As you can see, this equation contains a fraction. One of the topics we will discuss in this section is how to solve equations that contain fractions.

In this chapter we have been solving what are called *linear equations in one variable.* They are equations that contain only one variable, and that variable is always raised to the first power and never appears in a denominator. Here are some examples of linear equations in one variable:

$$3x + 2 = 17, \quad 7a + 4 = 3a - 2, \quad 2(3y - 5) = 6$$

Because of the work we have done in the first three sections of this chapter, we are now able to solve any linear equation in one variable. The steps outlined below can be used as a guide to solving these equations.

Note

Once you have some practice at solving equations, these steps will seem almost automatic. Until that time, it is a good idea to pay close attention to these steps.

Steps to Solve a Linear Equation in One Variable

Step 1 Simplify each side of the equation as much as possible. This step is done using the commutative, associative, and distributive properties.

Step 2 Use the addition property of equality to get all *variable terms* on one side of the equation and all *constant terms* on the other, then combine like terms. A *variable term* is any term that contains the variable. A *constant term* is any term that contains only a number.

Step 3 Use the multiplication property of equality to get the variable by itself on one side of the equation.

Step 4 Check your solution in the original equation if you think it is necessary.

EXAMPLE 1 Solve: $3(x + 2) = -9$

SOLUTION We begin by applying the distributive property to the left side:

Step 1
$$3(x + 2) = -9$$
$$3x + 6 = -9 \qquad \text{Distributive property}$$

Step 2
$$3x + 6 + (-6) = -9 + (-6) \qquad \text{Add } -6 \text{ to both sides}$$
$$3x = -15 \qquad \text{Addition}$$

Step 3
$$\frac{3x}{3} = \frac{-15}{3} \qquad \text{Divide both sides by 3}$$
$$x = -5 \qquad \text{Division}$$

This general method of solving linear equations involves using the two properties developed in Sections 10.2 and 10.3. We can add any number to both sides of an equation or multiply (or divide) both sides by the same nonzero number and always be sure we have not changed the solution to the equation. The equations may change in form, but the solution to the equation stays the same. Looking back to Example 1, we can see that each equation looks a little different from the preceding one. What is interesting, and useful, is that each of the equations says the same thing about x. They all say that x is -5. The last equation, of course, is the easiest to read. That is why our goal is to end up with x isolated on one side of the equation.

EXAMPLE 2   Solve: $4a + 5 = 2a - 7$

SOLUTION   Neither side can be simplified any further. What we have to do is get the variable terms ($4a$ and $2a$) on the same side of the equation. We can eliminate the variable term from the right side by adding $-2a$ to both sides:

$$4a + 5 = 2a - 7$$

Step 2
$$4a + (-2a) + 5 = 2a + (-2a) - 7 \qquad \text{Add } -2a \text{ to both sides}$$
$$2a + 5 = -7 \qquad \text{Addition}$$
$$2a + 5 + (-5) = -7 + (-5) \qquad \text{Add } -5 \text{ to both sides}$$
$$2a = -12 \qquad \text{Addition}$$

Step 3
$$\frac{2a}{2} = \frac{-12}{2} \qquad \text{Divide by 2}$$
$$a = -6 \qquad \text{Division}$$

EXAMPLE 3   Solve: $2(x - 4) + 5 = -11$

SOLUTION   We begin by applying the distributive property to multiply 2 and $x - 4$:

$$2(x - 4) + 5 = -11$$

Step 1
$$2x - 8 + 5 = -11 \qquad \text{Distributive property}$$
$$2x - 3 = -11 \qquad \text{Addition}$$

Step 2
$$2x - 3 + 3 = -11 + 3 \qquad \text{Add 3 to both sides}$$
$$2x = -8 \qquad \text{Addition}$$

Step 3
$$\frac{2x}{2} = \frac{-8}{2} \qquad \text{Divide by 2}$$
$$x = -4 \qquad \text{Division}$$

EXAMPLE 4   Solve: $5(2x - 4) + 3 = 4x - 5$

SOLUTION   We apply the distributive property to multiply 5 and $2x - 4$. We then combine similar terms and solve as usual:

$$\textbf{Step 1} \begin{cases} 5(2x - 4) + 3 = 4x - 5 & \text{\small\it Distributive property} \\ 10x - 20 + 3 = 4x - 5 & \text{\small\it Simplify the left side} \\ 10x - 17 = 4x - 5 \end{cases}$$

$$\textbf{Step 2} \begin{cases} 10x + (-4x) - 17 = 4x + (-4x) - 5 & \text{\small\it Add} -4x \text{\small\it\ to both sides} \\ 6x - 17 = -5 & \text{\small\it Addition} \\ 6x - 17 + 17 = -5 + 17 & \text{\small\it Add 17 to both sides} \\ 6x = 12 & \text{\small\it Addition} \end{cases}$$

$$\textbf{Step 3} \begin{cases} \dfrac{6x}{6} = \dfrac{12}{6} & \text{\small\it Divide by 6} \\ x = 2 & \text{\small\it Division} \end{cases}$$

Equations Involving Fractions

We will now solve some equations that involve fractions. Because integers are usually easier to work with than fractions, we will begin each problem by clearing the equation we are trying to solve of all fractions. To do this we will use the multiplication property of equality to multiply each side of the equation by the LCD for all fractions appearing in the equation. Here is an example.

EXAMPLE 5   Solve the equation $\dfrac{x}{2} + \dfrac{x}{6} = 8$.

SOLUTION   The LCD for the fractions $\frac{x}{2}$ and $\frac{x}{6}$ is 6. It has the property that both 2 and 6 divide it evenly. Therefore, if we multiply both sides of the equation by 6, we will be left with an equation that does not involve fractions.

$$6\left(\frac{x}{2} + \frac{x}{6}\right) = 6(8) \qquad \text{\small\it Multiply each side by 6}$$

$$6\left(\frac{x}{2}\right) + 6\left(\frac{x}{6}\right) = 6(8) \qquad \text{\small\it Apply the distributive property}$$

$$3x + x = 48 \qquad \text{\small\it Multiplication}$$

$$4x = 48 \qquad \text{\small\it Combine similar terms}$$

$$x = 12 \qquad \text{\small\it Divide each side by 4}$$

We could check our solution by substituting 12 for x in the original equation. If we do so, the result is a true statement. The solution is 12.

As you can see from Example 5, the most important step in solving an equation that involves fractions is the first step. In that first step we multiply both sides of the equation by the LCD for all the fractions in the equation. After we have done so, the equation is clear of fractions because the LCD has the property that all the denominators divide it evenly.

EXAMPLE 6   Solve the equation $2x + \dfrac{1}{2} = \dfrac{3}{4}$.

SOLUTION   This time the LCD is 4. We begin by multiplying both sides of the equation by 4 to clear the equation of fractions.

$$4\left(2x + \frac{1}{2}\right) = 4\left(\frac{3}{4}\right) \qquad \text{Multiply each side by the LCD, 4}$$

$$4(2x) + 4\left(\frac{1}{2}\right) = 4\left(\frac{3}{4}\right) \qquad \text{Apply the distributive property}$$

$$8x + 2 = 3 \qquad \text{Multiplication}$$

$$8x = 1 \qquad \text{Add } -2 \text{ to each side}$$

$$x = \frac{1}{8} \qquad \text{Divide each side by 8}$$

EXAMPLE 7   Solve for x: $\dfrac{3}{x} + 2 = \dfrac{1}{2}$. (Assume x is not 0.)

SOLUTION   This time the LCD is $2x$. Following the steps we used in Examples 5 and 6, we have

$$2x\left(\frac{3}{x} + 2\right) = 2x\left(\frac{1}{2}\right) \qquad \text{Multiply through by the LCD, } 2x$$

$$2x\left(\frac{3}{x}\right) + 2x(2) = 2x\left(\frac{1}{2}\right) \qquad \text{Distributive property}$$

$$6 + 4x = x \qquad \text{Multiplication}$$

$$6 = -3x \qquad \text{Add } -4x \text{ to each side}$$

$$-2 = x \qquad \text{Divide each side by } -3$$

GETTING READY FOR CLASS

After reading through the preceding section, respond in your own words and in complete sentences.

1. Apply the distributive property to the expression $3(x + 4)$.

2. Write the equation that results when $-4a$ is added to both sides of the equation below.

$$6a + 9 = 4a - 3$$

3. Solve the equation $2x + \dfrac{1}{2} = \dfrac{3}{4}$ by first adding $-\dfrac{1}{2}$ to each side. Compare your answer with the solution to the equation shown in Example 6.

4. Find the LCD for the fractions in the equation $\dfrac{x}{2} + \dfrac{x}{6} = 8$.

Problem Set 10.4

Solve each equation using the methods shown in this section.

1. $5(x + 1) = 20$

2. $4(x + 2) = 24$

3. $6(x - 3) = -6$

4. $7(x - 2) = -7$

5. $2x + 4 = 3x + 7$

6. $5x + 3 = 2x + (-3)$

7. $7y - 3 = 4y - 15$

8. $3y + 5 = 9y + 8$

9. $12x + 3 = -2x + 17$

10. $15x + 1 = -4x + 20$

11. $6x - 8 = -x - 8$

12. $7x - 5 = -x - 5$

13. $7(a - 1) + 4 = 11$

14. $3(a - 2) + 1 = 4$

15. $8(x + 5) - 6 = 18$

16. $7(x + 8) - 4 = 10$

17. $2(3x - 6) + 1 = 7$

18. $5(2x - 4) + 8 = 38$

19. $10(y + 1) + 4 = 3y + 7$

20. $12(y + 2) + 5 = 2y - 1$

21. $4(x - 6) + 1 = 2x - 9$

22. $7(x - 4) + 3 = 5x - 9$

23. $2(3x + 1) = 4(x - 1)$

24. $7(x - 8) = 2(x - 13)$

25. $3a + 4 = 2(a - 5) + 15$

26. $10a + 3 = 4(a - 1) + 1$

27. $9x - 6 = -3(x + 2) - 24$

28. $8x - 10 = -4(x + 3) + 2$

29. $3x - 5 = 11 + 2(x - 6)$

30. $5x - 7 = -7 + 2(x + 3)$

Solve each equation by first finding the LCD for the fractions in the equation and then multiplying both sides of the equation by it. (Assume x is not 0 in Problems 39–46.)

31. $\dfrac{x}{3} + \dfrac{x}{6} = 5$

32. $\dfrac{x}{2} - \dfrac{x}{4} = 3$

33. $\dfrac{x}{5} - x = 4$

34. $\dfrac{x}{3} + x = 8$

35. $3x + \dfrac{1}{2} = \dfrac{1}{4}$

36. $3x - \dfrac{1}{3} = \dfrac{1}{6}$

37. $\dfrac{x}{3} + \dfrac{1}{2} = -\dfrac{1}{2}$

38. $\dfrac{x}{2} + \dfrac{4}{3} = -\dfrac{2}{3}$

39. $\dfrac{4}{x} = \dfrac{1}{5}$

40. $\dfrac{2}{3} = \dfrac{6}{x}$

41. $\dfrac{3}{x} + 1 = \dfrac{2}{x}$

42. $\dfrac{4}{x} + 3 = \dfrac{1}{x}$

43. $\dfrac{3}{x} - \dfrac{2}{x} = \dfrac{1}{5}$

44. $\dfrac{7}{x} + \dfrac{1}{x} = 2$

45. $\dfrac{1}{x} - \dfrac{1}{2} = -\dfrac{1}{4}$

46. $\dfrac{3}{x} - \dfrac{4}{5} = -\dfrac{1}{5}$

Find the value of x for each of the figures, given the perimeter.

47. $P = 36$

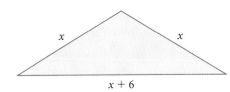

48. $P = 30$

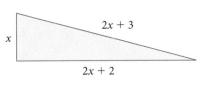

49. $P = 16$

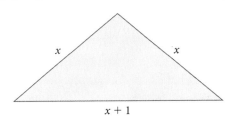

50. $P = 60$

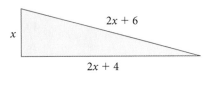

Applying the Concepts

The problems below contain a variety of the equations we have solved in this chapter.

51. Geometry The figure below shows part of a room. From a point on the floor, the angle of elevation to the top of the window is 45°, while the angle of elevation to the ceiling above the window is 58°. Solving either of the equations $58 - x = 45$ or $45 + x = 58$ will give us the number of degrees in the angle labeled $x°$. Solve both equations.

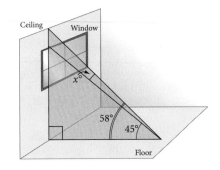

52. Stock Market In the year 2000, one share of Yahoo.com rose from $177\frac{1}{8}$ at the beginning of the week to a high of $183\frac{1}{4}$ at the end of the week. Solving the equation $177\frac{1}{8} + x = 183\frac{1}{4}$ for x will yield the increase in the stock price.
  a. Solve the equation for x.
  b. If you owned 240 shares of this stock, how much would your investment have increased during the week?

53. Rhind Papyrus As we mentioned in the introduction to this section, the Rhind Papyrus was created around 1650 BC and contains the riddle "What quantity when added to one-fourth of itself becomes 15?" This riddle can be solved by finding x in the equation below. Solve this equation.

$$x + \frac{1}{4}x = 15$$

54. Test Average Justin earns scores of 75, 83, 77, and 81 on four tests in his math class. The final exam has twice the weight of a test. Justin wants to end the semester with an average of 80 for the course. Solving the equation below will give the lowest score Justin can earn on the final exam and still end the course with an 80 average. Solve the equation.

$$\frac{75 + 83 + 77 + 81 + 2x}{6} = 80$$

Getting Ready for the Next Section

Write the mathematical expressions that are equivalent to each of the following English phrases.

55. The sum of a number and 2

56. The sum of a number and 5

57. Twice a number

58. Three times a number

59. Twice the sum of a number and 6

60. Three times the sum of a number and 8

61. The difference of x and 4

62. The difference of 4 and x

63. The sum of twice a number and 5

64. The sum of three times a number and 4

Applications

As you begin reading through the examples in this section, you may find yourself asking why some of these problems seem so contrived. The title of the section is "Applications," but many of the problems here don't seem to have much to do with real life. You are right about that. Example 5 is what we refer to as an "age problem." Realistically, it is not the kind of problem you would expect to find if you choose a career in which you use algebra. However, solving age problems is good practice for someone with little experience with application problems, because the solution process has a form that can be applied to all similar age problems.

To begin this section we list the steps used in solving application problems. We call this strategy the *Blueprint for Problem Solving*. It is an outline that will overlay the solution process we use on all application problems.

> ## Blueprint for Problem Solving
>
> **Step 1** **Read** the problem, and then mentally **list** the items that are known and the items that are unknown.
>
> **Step 2** **Assign a variable** to one of the unknown items. (In most cases this will amount to letting x equal the item that is asked for in the problem.) Then **translate** the other **information** in the problem to expressions involving the variable.
>
> **Step 3** **Reread** the problem, and then **write an equation,** using the items and variables listed in Steps 1 and 2, that describes the situation.
>
> **Step 4** **Solve the equation** found in Step 3.
>
> **Step 5** **Write** your **answer** using a complete sentence.
>
> **Step 6** **Reread** the problem, and **check** your solution with the original words in the problem.

There are a number of substeps within each of the steps in our blueprint. For instance, with Steps 1 and 2 it is always a good idea to draw a diagram or picture if it helps you to visualize the relationship between the items in the problem.

Number Problems

EXAMPLE 1   The sum of a number and 2 is 8. Find the number.

SOLUTION   Using our blueprint for problem solving as an outline, we solve the problem as follows:

Step 1   *Read* the problem, and then mentally *list* the items that are known and the items that are unknown.

> *Known items:*   The numbers 2 and 8
> *Unknown item:*   The number in question

Step 2   *Assign a variable* to one of the unknown items. Then *translate* the other *information* in the problem to expressions involving the variable.

> Let x = the number asked for in the problem
> Then "The sum of a number and 2" translates to $x + 2$.

Step 3   *Reread* the problem, and then *write an equation,* using the items and variables listed in Steps 1 and 2, that describes the situation.

> With all word problems, the word "is" translates to = .
>
> The sum of x and 2 is 8.
> $x + 2 = 8$

Step 4   *Solve the equation* found in Step 3.

$$x + 2 = 8$$
$$x + 2 + (-2) = 8 + (-2) \qquad \text{Add } -2 \text{ to each side}$$
$$x = 6$$

Step 5   *Write* your *answer* using a complete sentence.

> The number is 6.

Step 6   *Reread* the problem, and *check* your solution with the original words in the problem.

> The sum of 6 and 2 is 8.   A true statement

   To help with other problems of the type shown in Example 1, here are some common English words and phrases and their mathematical translations.

English	Algebra	
The sum of a and b	$a + b$	
The difference of a and b	$a - b$	
The product of a and b	$a \cdot b$	
The quotient of a and b	$\dfrac{a}{b}$	
Of	$\cdot$	(multiply)
Is	$=$	(equals)
A number	x	
4 more than x	$x + 4$	
4 times x	$4x$	
4 less than x	$x - 4$	

You may find some examples and problems in this section and the problem set that follows that you can solve without using algebra or our blueprint. It is very important that you solve those problems using the methods we are showing here. The purpose behind these problems is to give you experience using the blueprint as a guide to solving problems written in words. Your answers are much less important than the work that you show in obtaining your answer.

EXAMPLE 2   If 5 is added to the sum of twice a number and three times the number, the result is 25. Find the number.

SOLUTION

Step 1 *Read and list.*

Known items: The numbers 5 and 25, twice a number, and three times a number

Unknown item: The number in question

Step 2 *Assign a variable and translate the information.*

Let x = the number asked for in the problem.
Then "The sum of twice a number and three times the number" translates to $2x + 3x$.

Step 3 *Reread and write an equation.*

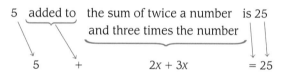

Step 4 *Solve the equation.*

$$5 + 2x + 3x = 25$$

$5x + 5 = 25$	Simplify the left side
$5x + 5 + (-5) = 25 + (-5)$	Add -5 to both sides
$5x = 20$	Addition
$\dfrac{5x}{5} = \dfrac{20}{5}$	Divide by 5
$x = 4$	

Step 5 *Write your answer.*

The number is 4.

Step 6 *Reread and check.*

Twice 4 is 8, and three times 4 is 12. Their sum is $8 + 12 = 20$. Five added to this is 25. Therefore, 5 added to the sum of twice 4 and three times 4 is 25.

Geometry Problems

EXAMPLE 3   The length of a rectangle is three times the width. The perimeter is 72 centimeters. Find the width and the length.

SOLUTION

Step 1 *Read and list.*

> Known items: The length is three times the width.
>   The perimeter is 72 centimeters.
> Unknown items: The length and the width

Step 2 *Assign a variable, and translate the information.* We let x = the width. Because the length is three times the width, the length must be $3x$. A picture will help.

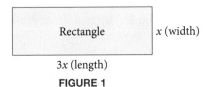

Rectangle x (width)

$3x$ (length)

FIGURE 1

Step 3 *Reread and write an equation.* Because the perimeter is the sum of the sides, it must be $x + x + 3x + 3x$ (the sum of the four sides). But the perimeter is also given as 72 centimeters. Hence,

$$x + x + 3x + 3x = 72$$

Step 4 *Solve the equation.*

$$x + x + 3x + 3x = 72$$
$$8x = 72$$
$$x = 9$$

Step 5 *Write your answer.* The width, x, is 9 centimeters. The length, $3x$, must be 27 centimeters.

Step 6 *Reread and check.* From the diagram below, we see that these solutions check:

Perimeter is 72   Length = 3 × Width

$9 + 9 + 27 + 27 = 72$ $27 = 3 \cdot 9$

27

9   9

27

FIGURE 2

Next we review some facts about triangles that we introduced in a previous chapter.

FACTS FROM GEOMETRY

Labeling Triangles and the Sum of the Angles in a Triangle
One way to label the important parts of a triangle is to label the vertices with capital letters and the sides with small letters, as shown in Figure 3.

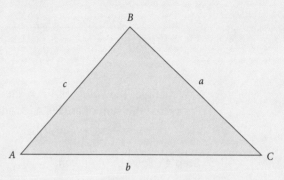

FIGURE 3

In Figure 3, notice that side a is opposite vertex A, side b is opposite vertex B, and side c is opposite vertex C. Also, because each vertex is the vertex of one of the angles of the triangle, we refer to the three interior angles as A, B, and C.

In any triangle, the sum of the interior angles is 180°. For the triangle shown in Figure 3, the relationship is written

$$A + B + C = 180°$$

EXAMPLE 4   The angles in a triangle are such that one angle is twice the smallest angle, while the third angle is three times as large as the smallest angle. Find the measure of all three angles.

SOLUTION

Step 1 *Read and list.*

Known items:	The sum of all three angles is 180°, one angle is twice the smallest angle, and the largest angle is three times the smallest angle.
Unknown items:	The measure of each angle

Step 2 *Assign a variable and translate information.* Let x be the smallest angle, then $2x$ will be the measure of another angle, and $3x$ will be the measure of the largest angle.

Step 3 *Reread and write an equation.* When working with geometric objects, drawing a generic diagram will sometimes help us visualize what it is that we are asked to find. In Figure 4, we draw a triangle with angles A, B, and C.

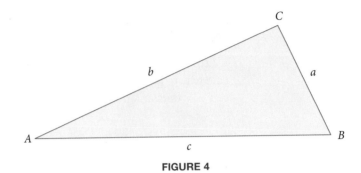

FIGURE 4

We can let the value of $A = x$, the value of $B = 2x$, and the value of $C = 3x$. We know that the sum of angles A, B, and C will be 180°, so our equation becomes

$$x + 2x + 3x = 180°$$

Step 4 *Solve the equation.*

$$x + 2x + 3x = 180°$$
$$6x = 180°$$
$$x = 30°$$

Step 5 *Write the answer.*

The smallest angle A measures 30°
Angle B measures $2x$, or $2(30°) = 60°$
Angle C measures $3x$, or $3(30°) = 90°$

Step 6 *Reread and check.* The angles must add to 180°:

$$A + B + C = 180°$$
$$30° + 60° + 90° = 180°$$
$$180° = 180° \quad \textit{Our answers check}$$

Age Problem

EXAMPLE 5 Jo Ann is 22 years older than her daughter Stacey. In six years the sum of their ages will be 42. How old are they now?

SOLUTION
Step 1 *Read and list:*

Known items: Jo Ann is 22 years older than Stacey. Six years from now their ages will add to 42.

Unknown items: Their ages now

Step 2 *Assign a variable and translate the information.* Let x = Stacey's age now. Because Jo Ann is 22 years older than Stacey, her age is $x + 22$.

Step 3 *Reread and write an equation.* As an aid in writing the equation we use the following table:

	Now	In Six years
Stacey	x	$x + 6$
Jo Ann	$x + 22$	$x + 28$

Their ages in six years will be their ages now plus 6

Because the sum of their ages six years from now is 42, we write the equation as

$$(x + 6) + (x + 28) = 42$$

*Stacey's   Jo Ann's
age in   age in
6 years   6 years*

Step 4 *Solve the equation.*

$$x + 6 + x + 28 = 42$$
$$2x + 34 = 42$$
$$2x = 8$$
$$x = 4$$

Step 5 *Write your answer.* Stacey is now 4 years old, and Jo Ann is $4 + 22 = 26$ years old.

Step 6 *Reread and check.* To check, we see that in six years, Stacey will be 10, and Jo Ann will be 32. The sum of 10 and 32 is 42, which checks. ▓

Car Rental Problem

▓ **EXAMPLE 6**   A car rental company charges $11 per day and 16 cents per mile for their cars. If a car were rented for 1 day and the charge was $25.40, how many miles was the car driven?

SOLUTION
Step 1 *Read and list.*

Known items:	Charges are $11 per day and 16 cents per mile. Car is rented for 1 day. Total charge is $25.40.
Unknown items:	How many miles the car was driven

Step 2 *Assign a variable and translate information.* If we let x = the number of miles driven, then the charge for the number of miles driven will be $0.16x$, the cost per mile times the number of miles.

Step 3 *Reread and write an equation.* To find the total cost to rent the car, we add 11 to $0.16x$. Here is the equation that describes the situation:

$$\underbrace{\$11 \text{ per day}}_{} + \underbrace{16 \text{ cents per mile}}_{} = \underbrace{\text{Total cost}}_{}$$

$$11 \quad + \quad 0.16x \quad = \quad 25.40$$

Step 4   *Solve the equation.* To solve the equation, we add -11 to each side and then divide each side by 0.16.

$$11 + (-11) + 0.16x = 25.40 + (-11) \qquad \text{Add } -11 \text{ to each side}$$
$$0.16x = 14.40$$
$$\frac{0.16x}{0.16} = \frac{14.40}{0.16} \qquad \text{Divide each side by 0.16}$$
$$x = 90 \qquad 14.40 \div 0.16 = 90$$

Step 5   *Write the answer.* The car was driven 90 miles.

Step 6   *Reread and check.* The charge for 1 day is $11. The 90 miles adds $90(\$0.16) = \$14.40$ to the 1-day charge. The total is $\$11 + \$14.40 = \$25.40$, which checks with the total charge given in the problem.

Coin Problem

EXAMPLE 7   Diane has $1.60 in dimes and nickels. If she has 7 more dimes than nickels, how many of each coin does she have?

SOLUTION

Step 1   *Read and list.*

Known items:	We have dimes and nickels. There are 7 more dimes than nickels, and the total value of the coins is $1.60.
Unknown items:	How many of each type of coin Diane has

Step 2   *Assign a variable and translate information.* If we let $x =$ the number of nickels, then the number of dimes must be $x + 7$, because Diane has 7 more dimes than nickels. Because each nickel is worth 5 cents, the amount of money she has in nickels is $0.05x$. Similarly, because each dime is worth 10 cents, the amount of money she has in dimes is $0.10(x + 7)$. Here is a table that summarizes what we have so far:

	Nickels	Dimes
Number of	x	$x + 7$
Value of	$0.05x$	$0.10(x + 7)$

Step 3   *Reread and write an equation.* Because the total value of all the coins is $1.60, the equation that describes this situation is

$$\underbrace{\text{Amount of money in nickels}}_{0.05x} + \underbrace{\text{Amount of money in dimes}}_{0.10(x+7)} = \underbrace{\text{Total amount of money}}_{1.60}$$

Step 4   *Solve the equation.* This time, let's show only the essential steps in the solution.

$$0.05x + 0.10x + 0.70 = 1.60 \qquad \text{Distributive property}$$
$$0.15x + 0.70 = 1.60 \qquad \text{Add } 0.05x \text{ and } 0.10x \text{ to get } 0.15x$$
$$0.15x = 0.90 \qquad \text{Add } -0.70 \text{ to each side}$$
$$x = 6 \qquad \text{Divide each side by } 0.15$$

Step 5   *Write the answer.* Because $x = 6$, Diane has 6 nickels. To find the number of dimes, we add 7 to the number of nickels (she has 7 more dimes than nickels). The number of dimes is $6 + 7 = 13$.

Step 6   *Reread and check.* Here is a check of our results.

6 nickels are worth 6($0.05) = $0.30
13 dimes are worth 13($0.10) = $1.30
The total value is $1.60

GETTING READY FOR CLASS

After reading through the preceding section, respond in your own words and in complete sentences.

1. What is the first step in solving a word problem?
2. Write a mathematical expression equivalent to the phrase "the sum of x and ten."
3. Write a mathematical expression equivalent to the phrase "twice the sum of a number and ten."
4. Suppose the length of a rectangle is three times the width. If we let x represent the width of the rectangle, what expression do we use to represent the length?

Problem Set 10.5

Write each of the following English phrases in symbols using the variable x.

1. The sum of x and 3

2. The difference of x and 2

3. The sum of twice x and 1

4. The sum of three times x and 4

5. Five x decreased by 6

6. Twice the sum of x and 5

7. Three times the sum of x and 1

8. Four times the sum of twice x and 1

9. Five times the sum of three x and 4

10. Three x added to the sum of twice x and 1

Use the six steps in the "Blueprint for Problem Solving" to solve the following word problems. You may recognize the solution to some of them by just reading the problem. In all cases, be sure to assign a variable and write the equation used to describe the problem. Write your answer using a complete sentence.

Number Problems

11. The sum of a number and 3 is 5. Find the number.

12. If 2 is subtracted from a number, the result is 4. Find the number.

13. The sum of twice a number and 1 is −3. Find the number.

14. If three times a number is increased by 4, the result is −8. Find the number.

15. When 6 is subtracted from five times a number, the result is 9. Find the number.

16. Twice the sum of a number and 5 is 4. Find the number.

17. Three times the sum of a number and 1 is 18. Find the number.

18. Four times the sum of twice a number and 6 is −8. Find the number.

19. Five times the sum of three times a number and 4 is −10. Find the number.

20. If the sum of three times a number and two times the same number is increased by 1, the result is 16. Find the number.

Geometry Problems

21. The length of a rectangle is twice its width. The perimeter is 30 meters. Find the length and the width.

22. The width of a rectangle is 3 feet less than its length. If the perimeter is 22 feet, what is the width?

23. The perimeter of a square is 32 centimeters. What is the length of one side?

24. Two sides of a triangle are equal in length, and the third side is 10 inches. If the perimeter is 26 inches, how long are the two equal sides?

25. Two angles in a triangle are equal, and their sum is equal to the third angle in the triangle. What are the measures of each of the three interior angles?

26. One angle in a triangle measures twice the smallest angle, while the largest angle is six times the smallest angle. Find the measures of all three angles.

27. The smallest angle in a triangle is $\frac{1}{3}$ as large as the largest angle. The third angle is twice the smallest angle. Find the three angles.

28. One angle in a triangle is half the largest angle, but three times the smallest. Find all three angles.

Age Problems

29. Pat is 20 years older than his son Patrick. In 2 years, the sum of their ages will be 90. How old are they now?

	Now	In 2 Years
Patrick	x	
Pat		

30. Diane is 23 years older than her daughter Amy. In 5 years, the sum of their ages will be 91. How old are they now?

	Now	In 5 Years
Amy	x	
Diane		

31. Dale is 4 years older than Sue. Five years ago the sum of their ages was 64. How old are they now?

32. Pat is 2 years younger than his wife, Wynn. Ten years ago the sum of their ages was 48. How old are they now?

Renting a Car

33. A car rental company charges $10 a day and 16 cents per mile for their cars. If a car were rented for 1 day for a total charge of $23.92, how many miles was it driven?

34. A car rental company charges $12 a day and 18 cents per mile to rent their cars. If the total charge for a 1-day rental were $33.78, how many miles was the car driven?

35. A rental company charges $9 per day and 15 cents a mile for their cars. If a car were rented for 2 days for a total charge of $40.05, how many miles was it driven?

36. A car rental company charges $11 a day and 18 cents per mile to rent their cars. If the total charge for a 2-day rental were $61.60, how many miles was it driven?

Coin Problems

37. Mary has $2.20 in dimes and nickels. If she has 10 more dimes than nickels, how many of each coin does she have?

38. Bob has $1.65 in dimes and nickels. If he has 9 more nickels than dimes, how many of each coin does he have?

39. Suppose you have $9.60 in dimes and quarters. How many of each coin do you have if you have twice as many quarters as dimes?

40. A collection of dimes and quarters has a total value of $2.75. If there are 3 times as many dimes as quarters, how many of each coin is in the collection?

Miscellaneous Problems

41. Magic Square The sum of the numbers in each row, each column, and each diagonal of the square shown here is 15. Use this fact, along with the information in the first column of the square, to write an equation containing the variable x, then solve the equation to find x. Next, write and solve equations that will give you y and z.

x	1	y
3	5	7
4	z	2

42. Magic Square The sum of the numbers in each row, each column, and each diagonal of the square shown here is 3. Use this fact, along with the information in the second row of the square, to write an equation containing the variable a, then solve the equation to find a. Next, write and solve an equation that will allow you to find the value of b. Next, write and solve equations that will give you c and d.

4	d	b
a	1	3
0	c	-2

43. Wages JoAnn works in the publicity office at the state university. She is paid $14 an hour for the first 35 hours she works each week and $21 an hour for every hour after that. If she makes $574 one week, how many hours did she work?

44. Ticket Sales Stacey is selling tickets to the school play. The tickets are $6 for adults and $4 for children. She sells twice as many adult tickets as children's tickets and brings in a total of $112. How many of each kind of ticket did she sell?

Dance Lessons Ike and Nancy Lara give western dance lessons at the Elks Lodge on Sunday nights. The lessons cost $3 for members of the lodge and $5 for nonmembers. Half of the money collected for the lessons is paid to Ike and Nancy. The Elks Lodge keeps the other half. One Sunday night Ike counts 36 people in the dance lesson. Use this information to work Problems 45 through 48.

45. What is the least amount of money Ike and Nancy could make?

46. What is the largest amount of money Ike and Nancy could make?

47. At the end of the evening, the Elks Lodge gives Ike and Nancy a check for $80 to cover half of the receipts. Can this amount be correct?

48. Besides the number of people in the dance lesson, what additional information does Ike need to know in order to always be sure he is being paid the correct amount?

Getting Ready for the Next Section

Simplify.

49. $\dfrac{5}{9}(95 - 32)$

50. $\dfrac{5}{9}(77 - 32)$

51. Find the value of $90 - x$ when $x = 25$.

52. Find the value of $180 - x$ when $x = 25$.

53. Find the value of $2x + 6$ when $x = -2$

54. Find the value of $2x + 6$ when $x = 0$.

Solve.

55. $40 = 2l + 12$

56. $80 = 2l + 12$

57. $6 + 3y = 4$

58. $8 + 3y = 4$

Evaluating Formulas

In mathematics a formula is an equation that contains more than one variable. The equation $P = 2w + 2l$ is an example of a formula. This formula tells us the relationship between the perimeter P of a rectangle, its length l, and its width w.

There are many formulas with which you may be familiar already. Perhaps you have used the formula $d = r \cdot t$ to find out how far you would go if you traveled at 50 miles an hour for 3 hours. If you take a chemistry class while you are in college, you will certainly use the formula that gives the relationship between the two temperature scales, Fahrenheit and Celsius, that we mentioned at the beginning of this chapter.

$$F = \frac{9}{5}C + 32$$

Although there are many kinds of problems we can work using formulas, we will limit ourselves to those that require only substitutions. The examples that follow illustrate this type of problem.

EXAMPLE 1 The perimeter P of a rectangular livestock pen is 40 feet. If the width w is 6 feet, find the length.

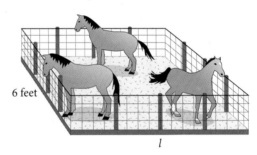

6 feet

l

SOLUTION First we substitute 40 for P and 6 for w in the formula $P = 2l + 2w$. Then we solve for l:

When $P = 40$ and $w = 6$
the formula $P = 2l + 2w$
becomes $40 = 2l + 2(6)$
or $40 = 2l + 12$ Multiply 2 and 6
             $28 = 2l$ Add -12 to each side
             $14 = l$ Multiply each side by $\frac{1}{2}$

To summarize our results, if a rectangular pen has a perimeter of 40 feet and a width of 6 feet, then the length must be 14 feet.

EXAMPLE 2   Use the formula $C = \frac{5}{9}(F - 32)$ to find C when F is 95 degrees.

SOLUTION   Substituting 95 for F in the formula gives us the following:

$$\text{When} \qquad F = 95$$

$$\text{the formula} \quad C = \frac{5}{9}(F - 32)$$

$$\text{becomes} \qquad C = \frac{5}{9}(95 - 32)$$

$$= \frac{5}{9}(63)$$

$$= \frac{5}{9} \cdot \frac{63}{1}$$

$$= \frac{315}{9}$$

$$= 35$$

A temperature of 95 degrees Fahrenheit is the same as a temperature of 35 degrees Celsius.

EXAMPLE 3   Use the formula $y = 2x + 6$ to find y when x is -2.

SOLUTION   Proceeding as we have in the previous examples, we have:

$$\text{When} \qquad x = -2$$

$$\text{the formula} \quad y = 2x + 6$$

$$\text{becomes} \qquad y = 2(-2) + 6$$

$$= -4 + 6$$

$$= 2$$

In some cases evaluating a formula also involves solving an equation, as the next example illustrates.

EXAMPLE 4   Find y when x is 3 in the formula $2x + 3y = 4$.

SOLUTION   First we substitute 3 for x; then we solve the resulting equation for y.

$$\text{When} \qquad\qquad x = 3$$

$$\text{the equation} \qquad 2x + 3y = 4$$

$$\text{becomes} \qquad 2(3) + 3y = 4$$

$$6 + 3y = 4$$

$$3y = -2 \qquad \text{Add } -6 \text{ to each side}$$

$$y = -\frac{2}{3} \qquad \text{Divide each side by 3}$$

Rate Equation

Now we will look at some problems that use what is called the *rate equation*. You use this equation on an intuitive level when you are estimating how long it will take you to drive long distances. For example, if you drive at 50 miles per hour for 2 hours, you will travel 100 miles. Here is the rate equation:

$$\text{Distance} = \text{rate} \cdot \text{time, or } d = r \cdot t$$

The rate equation has two equivalent forms, one of which is obtained by solving for r, while the other is obtained by solving for t. Here they are:

$$r = \frac{d}{t} \text{ and } t = \frac{d}{r}$$

The rate in this equation is also referred to as *average speed.*

EXAMPLE 5   At 1 p.m. Jordan leaves her house and drives at an average speed of 50 miles per hour to her sister's house. She arrives at 4 P.M.

a. How many hours was the drive to her sister's house?
b. How many miles from her sister does Jordan live?

SOLUTION **a.** If she left at 1:00 P.M. and arrived at 4:00 P.M. we simply subtract 1 from 4 for an answer of 3 hours.

b. We are asked to find a distance in miles given a rate of 50 miles per hour and a time of 3 hours. We will use the rate equation, $d = r \cdot t$, to solve this. We have:

$$d = 50 \text{ miles per hour} \cdot 3 \text{ hours}$$
$$d = 50(3)$$
$$d = 150 \text{ miles}$$

Notice that we were asked to find a distance in miles, so our answer has a unit of miles. When we are asked to find a time, our answer will include a unit of time, like days, hours, minutes, or seconds.

When we are asked to find a rate, our answer will include units of rate, like miles per hour, feet per second, problems per minute, and so on.

FACTS FROM GEOMETRY

Earlier we defined complementary angles as angles that add to 90°. That is, if x and y are complementary angles, then

$$x + y = 90°$$

If we solve this formula for y, we obtain a formula equivalent to our original formula:

$$y = 90° - x$$

Because y is the complement of x, we can generalize by saying that the complement of angle x is the angle $90° - x$. By a similar reasoning process, we can say that the supplement of angle x is the angle $180° - x$. To summarize, if x is an angle, then

the complement of x is $90° - x$, and
the supplement of x is $180° - x$

If you go on to take a trigonometry class, you will see these formulas again.

EXAMPLE 6   Find the complement and the supplement of 25°.

SOLUTION   We can use the formulas above with $x = 25°$.

The complement of 25° is $90° - 25° = 65°$.
The supplement of 25° is $180° - 25° = 155°$.

GETTING READY FOR CLASS

After reading through the preceding section, respond in your own words and in complete sentences.

1. What is a formula?
2. How do you solve a formula for one of its variables?
3. What are complementary angles?
4. What is the formula that converts temperature on the Celsius scale to temperature on the Fahrenheit scale?

Problem Set 10.6

The formula for the area A of a rectangle with length l and width w is $A = l \cdot w$. Find A if:

1. $l = 32$ feet and $w = 22$ feet

2. $l = 22$ feet and $w = 12$ feet

3. $l = \dfrac{3}{2}$ inch and $w = \dfrac{3}{4}$ inch

4. $l = \dfrac{3}{5}$ inch and $w = \dfrac{3}{10}$ inch

The formula $G = H \cdot R$ tells us how much gross pay G a person receives for working H hours at an hourly rate of pay R. In Problems 5-8, find G.

5. $H = 40$ hours and $R = \$6$

6. $H = 36$ hours and $R = \$8$

7. $H = 30$ hours and $R = \$9\dfrac{1}{2}$

8. $H = 20$ hours and $R = \$6\dfrac{3}{4}$

Because there are 3 feet in every yard, the formula $F = 3 \cdot Y$ will convert Y yards into F feet. In Problems 9-12, find F.

9. $Y = 4$ yards

10. $Y = 8$ yards

11. $Y = 2\dfrac{2}{3}$ yards

12. $Y = 6\dfrac{1}{3}$ yards

If you invest P dollars (P is for *principal*) at simple interest rate R for T years, the amount of interest you will earn is given by the formula $I = P \cdot R \cdot T$. In Problems 13 and 14, find I.

13. $P = \$1{,}000$, $R = \dfrac{7}{100}$, and $T = 2$ years

14. $P = \$2{,}000$, $R = \dfrac{6}{100}$, and $T = 2\dfrac{1}{2}$ years

In Problems 15-18, use the formula $P = 2w + 2l$ to find P.

15. $w = 10$ inches and $l = 19$ inches

16. $w = 12$ inches and $l = 22$ inches

17. $w = \dfrac{3}{4}$ foot and $l = \dfrac{7}{8}$ foot

18. $w = \dfrac{1}{2}$ foot and $l = \dfrac{3}{2}$ feet

We have mentioned the two temperature scales, Fahrenheit and Celsius, a number of times in this book. Table 1 is intended to give you a more intuitive idea of the relationship between the two temperatures scales.

TABLE 1

COMPARING TWO TEMPERATURE SCALES

Situation	Temperature Fahrenheit	Temperature Celsius
Water freezes	32°F	0°C
Room temperature	68°F	20°C
Normal body temperature	$98\dfrac{3}{5}$°F	37°C
Water boils	212°F	100°C
Bake cookies	365°F	185°C

Table 2 gives the formulas, in both symbols and words, that are used to convert between the two scales.

TABLE 2

FORMULAS FOR CONVERTING BETWEEN TEMPERATURE SCALES

To Convert From	Formula in Symbols	Formula in Words
Fahrenheit to Celsius	$C = \frac{5}{9}(F - 32)$	Subtract 32, then multiply by $\frac{5}{9}$.
Celsius to Fahrenheit	$F = \frac{9}{5}C + 32$	Multiply by $\frac{9}{5}$, then add 32.

19. Let $F = 212$ in the formula $C = \frac{5}{9}(F - 32)$, and solve for C. Does the value of C agree with the information in Table 1?

20. Let $C = 100$ in the formula $F = \frac{9}{5}C + 32$, and solve for F. Does the value of F agree with the information in Table 1?

21. Let $F = 68$ in the formula $C = \frac{5}{9}(F - 32)$, and solve for C. Does the value of C agree with the information in Table 1?

22. Let $C = 37$ in the formula $F = \frac{9}{5}C + 32$, and solve for F. Does the value of F agree with the information in Table 1?

23. Find C when F is 32°.

24. Find C when F is −4°.

25. Find F when C is −15°.

26. Find F when C is 35°.

Maximum Heart Rate In exercise physiology, a person's maximum heart rate, in beats per minute, is found by subtracting his age in years from 220. So, if A represents your age in years, then your maximum heart rate is

$$M = 220 - A$$

Use this formula to complete the following tables.

27.

Age (Years)	Maximum Heart Rate (Beats per Minute)
18	
19	
20	
21	
22	
23	

28.

Age (Years)	Maximum Heart Rate (Beats per Minute)
15	
20	
25	
30	
35	
40	

Training Heart Rate A person's training heart rate, in beats per minute, is the person's resting heart rate plus 60% of the difference between maximum heart rate and his resting heart rate. If resting heart rate is R and maximum heart rate is M, then the formula that gives training heart rate is

$$T = R + \frac{3}{5}(M - R)$$

Use this formula along with the results of Problems 27 and 28 to fill in the following two tables.

29. For a 20-year-old person

Resting Heart Rate (Beats Per Minute)	Training Heart Rate (Beats Per Minute)
60	
62	
64	
68	
70	
72	

30. For a 40-year-old person

Resting Heart Rate (Beats Per Minute)	Training Heart Rate (Beats Per Minute)
60	
62	
64	
68	
70	
72	

Use the rate equation $d = r \cdot t$ to solve Problems 31 and 32.

31. At 2:30 P.M. Shelly leaves her house and drives at an average speed of 55 miles per hour to her sister's house. She arrives at 6:30 P.M.
  a. How many hours was the drive to her sister's house?
  b. How many miles from her sister does Shelly live?

32. At 1:30 P.M. Cary leaves his house and drives at an average speed of 65 miles per hour to his brother's house. He arrives at 5:30 P.M.
  a. How many hours was the drive to his brother's house?
  b. How many miles from his brother's house does Cary live?

Use the rate equation $r = \dfrac{d}{t}$ to solve Problems 33 and 34.

33. At 2:30 P.M. Brittney leaves her house and drives 260 miles to her sister's house. She arrives at 6:30 P.M.
  a. How many hours was the drive to her sister's house?
  b. What was Brittney's average speed?

34. At 8:30 A.M. Ethan leaves his house and drives 220 miles to his brother's house. He arrives at 12:30 P.M.
  a. How many hours was the drive to his brother's house?
  b. What was Ethan's average speed?

As you know, the volume V enclosed by a rectangular solid with length l, width w, and height h is $V = l \cdot w \cdot h$. In Problems 35-38, find V if:

35. $l = 6$ inches, $w = 12$ inches, and $h = 5$ inches

36. $l = 16$ inches, $w = 22$ inches, and $h = 15$ inches

37. $l = 6$ yards, $w = \dfrac{1}{2}$ yard, and $h = \dfrac{1}{3}$ yard

38. $l = 30$ yards, $w = \dfrac{5}{2}$ yards, and $h = \dfrac{5}{3}$ yards

Suppose $y = 3x - 2$. In Problems 39–44, find y if:

39. $x = 3$ **40.** $x = -5$ **41.** $x = -\dfrac{1}{3}$ **42.** $x = \dfrac{2}{3}$

43. $x = 0$ **44.** $x = 5$

Suppose $x + y = 5$. In Problems 45–50, find x if:

45. $y = 2$ **46.** $y = -2$ **47.** $y = 0$ **48.** $y = 5$ **49.** $y = -3$ **50.** $y = 3$

Suppose $x + y = 3$. In Problems 51–56, find y if:

51. $x = 2$ **52.** $x = -2$ **53.** $x = 0$ **54.** $x = 3$

55. $x = \dfrac{1}{2}$ **56.** $x = -\dfrac{1}{2}$

Suppose $4x + 3y = 12$. In Problems 57–62, find y if:

57. $x = 3$ **58.** $x = -5$ **59.** $x = -\dfrac{1}{4}$ **60.** $x = \dfrac{3}{2}$

61. $x = 0$ **62.** $x = -3$

Suppose $4x + 3y = 12$. In Problems 63-68, find x if:

63. $y = 4$ **64.** $y = -4$ **65.** $y = -\dfrac{1}{3}$ **66.** $y = \dfrac{5}{3}$

67. $y = 0$ **68.** $y = -3$

Find the complement and supplement of each angle.

69. $45°$ **70.** $75°$ **71.** $31°$ **72.** $59°$

Applying the Concepts

73. **Digital Video** The biggest video download of all time was a *Star Wars* movie trailer. The video was compressed so it would be small enough for people to download over the Internet. A formula for estimating the size, in kilobytes, of a compressed video is

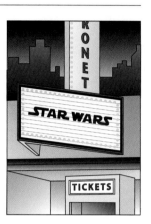

$$S = \frac{height \cdot width \cdot fps \cdot time}{35,000}$$

where *height* and *width* are in pixels, *fps* is the number of frames per second the video is to play (television plays at 30 fps), and *time* is given in seconds.

a. Estimate the size in kilobytes of the *Star Wars* trailer that has a height of 480 pixels, has a width of 216 pixels, plays at 30 fps, and runs for 150 seconds.

b. Estimate the size in kilobytes of the *Star Wars* trailer that has a height of 320 pixels, has a width of 144 pixels, plays at 15 fps, and runs for 150 seconds.

74. **Fermat's Last Theorem** The postage stamp shows Fermat's last theorem, which states that if *n* is an integer greater than 2, then there are no positive integers *x*, *y*, and *z* that will make the formula $x^n + y^n = z^n$ true.

Use the formula $x^n + y^n = z^n$ to

a. Find *x* if $n = 1$, $y = 7$, and $z = 15$.

b. Find *y* if $n = 1$, $x = 23$, and $z = 37$.

Estimating Vehicle Weight If you can measure the area that the tires on your car contact the ground, and you know the air pressure in the tires, then you can estimate the weight of your car, in pounds, with the following formula:

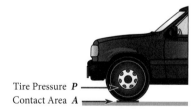

Tire Pressure **P**
Contact Area **A**

$$W = APN$$

where *W* is the vehicle's weight in pounds, *A* is the average tire contact area with a hard surface in square inches, *P* is the air pressure in the tires in pounds per square inch (psi, or lb/in²), and *N* is the number of tires.

75. What is the approximate weight of a car if the average tire contact area is a rectangle 6 inches by 5 inches and if the air pressure in the tires is 30 psi?

76. What is the approximate weight of a car if the average tire contact area is a rectangle 5 inches by 4 inches, and the tire pressure is 30 psi?

Chapter 10 Summary

Combining Similar Terms [10.1]

1. $7x + 2x = (7 + 2)x$
$\qquad = 9x$

Two terms are similar terms if they have the same variable part. The expressions $7x$ and $2x$ are similar because the variable part in each is the same. Similar terms are combined by using the distributive property.

Finding the Value of an Algebraic Expression [10.1]

2. When $x = 5$, the expression $2x + 7$ becomes
$2(5) + 7 = 10 + 7 = 17$

An algebraic expression is a mathematical expression that contains numbers and variables. Expressions that contain a variable will take on different values depending on the value of the variable.

The Solution to an Equation [10.2]

A solution to an equation is a number that, when used in place of the variable, makes the equation a true statement.

The Addition Property of Equality [10.2]

3. We solve $x - 4 = 9$ by adding 4 to each side.
$x - 4 = 9$
$x - 4 + 4 = 9 + 4$
$x + 0 = 13$
$x = 13$

Let A, B, and C represent algebraic expressions.

$$\text{If} \qquad A = B$$
$$\text{then} \qquad A + C = B + C$$

In words: Adding the same quantity to both sides of an equation will not change the solution.

The Multiplication Property of Equality [10.3]

4. Solve $\frac{1}{3}x = 5$.

$\frac{1}{3}x = 5$

$3 \cdot \frac{1}{3}x = 3 \cdot 5$

$x = 15$

Let A, B, and C represent algebraic expressions with C not equal to 0.

$$\text{If} \qquad A = B$$
$$\text{then} \qquad AC = BC$$

In words: Multiplying both sides of an equation by the same nonzero number will not change the solution to the equation. This property holds for division as well.

Steps Used to Solve a Linear Equation in One Variable [10.4]

5. We solve $x - 4 = 9$ by adding 4 to each side.
$$x - 4 = 9$$
$$x - 4 + 4 = 9 + 4$$
$$x + 0 = 13$$
$$x = 13$$

Step 1 Simplify each side of the equation.

Step 2 Use the addition property of equality to get all variable terms on one side and all constant terms on the other side.

Step 3 Use the multiplication property of equality to get just one x isolated on either side of the equation.

Step 4 Check the solution in the original equation if necessary.

If the original equation contains fractions, you can begin by multiplying each side by the LCD for all fractions in the equation.

6. When $w = 8$ and $l = 13$ the formula $P = 2w + 2l$ becomes
$$P = 2 \cdot 8 + 2 \cdot 13$$
$$= 16 + 26$$
$$= 42$$

Evaluating Formulas [10.6]

In mathematics, a formula is an equation that contains more than one variable. For example, the formula for the perimeter of a rectangle is $P = 2l + 2w$. We evaluate a formula by substituting values for all but one of the variables and then solving the resulting equation for that variable.

Chapter 10 Test Form A

1. Simplify: $5x - 2 + 3x + 7$

2. Multiply: $-4(3x + 5)$

3. Apply the distributive property: $5(x + 3)$

4. Is $a = -2$ a solution to the expression $7a + 4 = 3a - 2$?

5. Solve for x: $3x - 2 - 2x = 4 - 9$

6. Solve: $-3x + 7 = -5$

7. Solve: $4x = -20$

8. Solve: $\frac{1}{3}a + 2 = 7$

9. Solve: $2x + \frac{1}{2} = \frac{3}{4}$

10. Solve: $2(x - 4) + 5 = -11$

11. Solve: $3(x + 2) = -9$

12. The sum of a number and 2 is 8. Find the number.

13. If 5 is added to the sum of twice a number and three times that number, the result is 25. Find the number.

14. Jo Ann is 22 years older than her daughter Stacey. In six years the sum of their ages will be 42. How old are they now?

15. The perimeter P, of a rectangular livestock pen is 40 feet. If the width is 6 feet, find the length.

16. Use $C = \frac{5}{9}(F - 32)$ to find C when F is 95 degrees.

Chapter Test, Form B

For an alternate, more comprehensive, chapter test, go to MathTV.com and select the test and summary for this chapter of the textbook. Click the worksheet labeled Chapter 10 Test, Form B to download it.

Chapter 1

PROBLEM SET 1.1

1. 8 ones, 7 tens **3.** 5 ones, 4 tens **5.** 8 ones, 4 tens, 3 hundreds **7.** 8 ones, 0 tens, 6 hundreds

9. 8 ones, 7 tens, 3 hundreds, 2 thousands **11.** 9 ones, 6 tens, 5 hundreds, 3 thousands, 7 ten thousands, 2 hundred thousands

13. Ten thousands **15.** Hundred millions **17.** Ones **19.** Hundred thousands **21.** $600 + 50 + 8$ **23.** $60 + 8$

25. $4,000 + 500 + 80 + 7$ **27.** $30,000 + 2,000 + 600 + 70 + 4$ **29.** $3,000,000 + 400,000 + 60,000 + 2,000 + 500 + 70 + 7$

31. $400 + 7$ **33.** $30,000 + 60 + 8$ **35.** $3,000,000 + 4,000 + 8$ **37.** Twenty-nine **39.** Forty **41.** Five hundred seventy-three

43. Seven hundred seven **45.** Seven hundred seventy **47.** Twenty-three thousand, five hundred forty

49. Three thousand, four **51.** Three thousand, forty **53.** One hundred four million, sixty-five thousand, seven hundred eighty

55. Five billion, three million, forty thousand, eight **57.** Two million, five hundred forty-six thousand, seven hundred thirty-one

59. 325 **61.** 5,432 **63.** 86,762 **65.** 2,000,200 **67.** 2,002,200 **69.** Thousands **71.** $100,000 + 6,000 + 700 + 20 + 1$

73. 275,000,000 **75.** One hundred twenty-seven million

PROBLEM SET 1.2

1. 15 **3.** 14 **5.** 24 **7.** 15 **9.** 20 **11.** 68 **13.** 98 **15.** 7,297 **17.** 6,487 **19.** 96 **21.** 7,449 **23.** 65 **25.** 102 **27.** 875

29. 829 **31.** 10,391 **33.** 16,204 **35.** 155,554 **37.** 111,110 **39.** 17,391 **41.** 14,892 **43.** 180 **45.** 2,220 **47.** 18,285

49.

First Number a	Second Number b	Their Sum $a + b$
61	38	99
63	36	99
65	34	99
67	32	99

51.

First Number a	Second Number b	Their Sum $a + b$
9	16	25
36	64	100
81	144	225
144	256	400

53. $9 + 5$ **55.** $8 + 3$ **57.** $4 + 6$

59. $1 + (2 + 3)$ **61.** $2 + (1 + 6)$ **63.** $(1 + 9) + 1$ **65.** $4 + (n + 1)$ **67.** $n = 4$ **69.** $n = 5$ **71.** $n = 8$ **73.** $n = 8$

75. The sum of 4 and 9 **77.** The sum of 8 and 1 **79.** The sum of 2 and 3 is 5. **81. a.** $5 + 2$ **b.** $8 + 3$ **83. a.** $m + 1$ **b.** $m + n$

85. 12 in. **87.** 16 ft **89.** 26 yd **91.** 18 in. **93.** 34 gallons **95.** \$349

PROBLEM SET 1.3

1. 40 **3.** 50 **5.** 50 **7.** 80 **9.** 460 **11.** 470 **13.** 56,780 **15.** 4,500 **17.** 500 **19.** 800 **21.** 900 **23.** 1,100 **25.** 5,000

27. 39,600 **29.** 5,000 **31.** 10,000 **33.** 1,000 **35.** 658,000 **37.** 510,000 **39.** 3,789,000

41. – 47.

49. \$2,500,000 **51.** 4,265,997 babies

	Original Number	Rounded to the Nearest		
		Ten	Hundred	Thousand
41.	7,821	7,820	7,800	8,000
43.	5,999	6,000	6,000	6,000
45.	10,985	10,990	11,000	11,000
47.	99,999	100,000	100,000	100,000

53. 2,300,000 babies **55.** \$15,200 **57.** \$31,000 **59.** 1,200 **61.** 1,900 **63.** 58,000

65.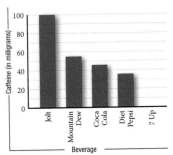

67.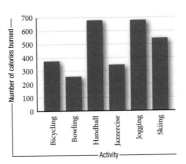

PROBLEM SET 1.4

1. 32   **3.** 22   **5.** 10   **7.** 111   **9.** 312   **11.** 403   **13.** 1,111   **15.** 4,544   **17.** 15   **19.** 33   **21.** 5   **23.** 33   **25.** 95   **27.** 152

29. 274   **31.** 488   **33.** 538   **35.** 163   **37.** 1,610   **39.** 46,083

41.

First Number a	Second Number b	The Difference of a and b $a - b$
25	15	10
24	16	8
23	17	6
22	18	4

43.

First Number a	Second Number b	The Difference of a and b $a - b$
400	256	144
400	144	256
225	144	81
225	81	144

45. The difference of 10 and 2   **47.** The difference of a and 6   **49.** The difference of 8 and 2 is 6.   **51.** $8 - 3$ **53.** $y - 9$

55. $3 - 2 = 1$ **57.** $255   **59.** 33 feet   **61.** 168 students   **63.** $574   **65.** $350

67. a.

Year	Sales ($ millions)
1996	386
1997	518
1998	573
1999	819

b. $301,000,000

PROBLEM SET 1.5

1. 300   **3.** 600   **5.** 3,000   **7.** 5,000   **9.** 21,000   **11.** 81,000   **13.** 100   **15.** 228   **17.** 36   **19.** 1,440   **21.** 950   **23.** 1,725

25. 121   **27.** 1,552   **29.** 4,200   **31.** 66,248   **33.** 279,200   **35.** 12,321   **37.** 106,400   **39.** 198,592   **41.** 612,928   **43.** 333,180

45. 18,053,805   **47.** 263,646,976

49.

First Number a	Second Number b	Their product ab
11	11	121
11	22	242
22	22	484
22	44	968

51.

First Number a	Second Number b	Their product ab
25	10	250
25	100	2,500
25	1,000	25,000
25	10,000	250,000

21. $\frac{3}{5}$ **23.** 9   **25.** 1   **27.** 8   **29.** $\frac{1}{15}$ **31.** $\frac{4}{9}$ **33.** $\frac{9}{16}$ **35.** $\frac{1}{4}$ **37.** $\frac{8}{27}$ **39.** $\frac{1}{2}$ **41.** $\frac{9}{100}$ **43.** 3   **45.** 24   **47.** 4   **49.** 9
51. $\frac{3}{10}$; numerator should be 3, not 4.

53. a.

Number x	Square x^2
1	1
2	4
3	9
4	16
5	25
6	36
7	49
8	64

b. Either *larger* or *greater* will work.   **55.** 14   **57.** 14   **59.** 133 in²   **61.** $\frac{4}{9}$ ft²
63. 3 yd²   **65.** 3,476 students   **67.** 126,500 ft³   **69.** About 2,121 children
71. 2   **73.** 3   **75.** 2   **77.** 5   **79.** 3   **81.** $\frac{4}{3}$ **83.** 3   **85.** $\frac{1}{7}$

PROBLEM SET 2.4

1. $\frac{15}{4}$ **3.** $\frac{4}{3}$ **5.** 9   **7.** 200   **9.** $\frac{3}{8}$ **11.** 1   **13.** $\frac{49}{64}$ **15.** $\frac{3}{4}$ **17.** $\frac{15}{16}$ **19.** $\frac{1}{6}$ **21.** 6   **23.** $\frac{5}{18}$ **25.** $\frac{9}{2}$ **27.** $\frac{2}{9}$ **29.** 9   **31.** $\frac{4}{5}$
33. $\frac{15}{22}$ **35.** 40   **37.** $\frac{7}{10}$ **39.** 13   **41.** 12   **43.** 186   **45.** 646   **47.** $\frac{3}{5}$ **49.** 40   **51.** $3 \div \frac{1}{5} = 3 \cdot \frac{5}{1} = 3 \cdot 5$ **53.** 14 blankets
55. 48 bags   **57.** 6   **59.** $\frac{14}{32} = \frac{7}{16}$ **61.** $\frac{3}{6}$ **63.** $\frac{9}{6}$ **65.** $\frac{4}{12}$ **67.** $\frac{8}{12}$ **69.** $\frac{14}{30}$ **71.** $\frac{18}{30}$ **73.** $\frac{12}{24}$ **75.** $\frac{4}{24}$ **77.** $\frac{15}{36}$

CHAPTER 2 TEST

1. $\frac{15}{20}$ **2.** $\frac{9x}{12x}$ **3.** $2^2 \cdot 3 \cdot 5$ **4.** $\frac{1}{7}$ **5.** $\frac{5}{6}$ **6.** $\frac{3}{40}$ **7.** $\frac{1}{3}$ **8.** $\frac{1}{6}$ **9.** 6   **10.** $\frac{1}{3}$

Chapter 3

PROBLEM SET 3.1

1. $\frac{2}{3}$ **3.** $\frac{1}{4}$ **5.** $\frac{1}{2}$ **7.** $\frac{1}{3}$ **9.** $\frac{3}{2}$ **11.** $\frac{x+6}{2}$ **13.** $\frac{4}{5}$ **15.** $\frac{10}{3}$

17.

First Number a	Second Number b	The Sum of a and b $a+b$
$\frac{1}{2}$	$\frac{1}{3}$	$\frac{5}{6}$
$\frac{1}{3}$	$\frac{1}{4}$	$\frac{7}{12}$
$\frac{1}{4}$	$\frac{1}{5}$	$\frac{9}{20}$
$\frac{1}{5}$	$\frac{1}{6}$	$\frac{11}{30}$

19.

First Number a	Second Number b	The Sum of a and b $a+b$
$\frac{1}{12}$	$\frac{1}{2}$	$\frac{7}{12}$
$\frac{1}{12}$	$\frac{1}{3}$	$\frac{5}{12}$
$\frac{1}{12}$	$\frac{1}{4}$	$\frac{1}{3}$
$\frac{1}{12}$	$\frac{1}{6}$	$\frac{1}{4}$

21. $\frac{7}{9}$ **23.** $\frac{7}{3}$ **25.** $\frac{7}{4}$ **27.** $\frac{7}{6}$ **29.** $\frac{9}{20}$ **31.** $\frac{7}{10}$ **33.** $\frac{19}{24}$ **35.** $\frac{13}{60}$ **37.** $\frac{31}{100}$ **39.** $\frac{67}{144}$ **41.** $\frac{29}{35}$ **43.** $\frac{949}{1,260}$ **45.** $\frac{13}{420}$
47. $\frac{41}{24}$ **49.** $\frac{53}{60}$ **51.** $\frac{5}{4}$ **53.** $\frac{88}{9}$ **55.** $\frac{3}{4}$ **57.** $\frac{1}{4}$ **59.** 19   **61.** 3   **63.** $\frac{1}{4} < \frac{3}{8} < \frac{1}{2} < \frac{3}{4}$ **65.** $\frac{160}{63}$ **67.** $\frac{5}{8}$ **69.** $\frac{9}{2}$ pints
71. $\frac{5}{8}(2,120) = 5(265) = \$1,325$ **73.** $\frac{2}{5}$

75.

Grade	Number of Students	Fraction of Students
A	5	$\frac{1}{8}$
B	8	$\frac{1}{5}$
C	20	$\frac{1}{2}$
below C	7	$\frac{7}{40}$
Total	40	1

77. 10 lots   **79.** $\frac{3}{2}$ in.   **81.** $\frac{9}{5}$ ft   **83.** $\frac{7}{3}$ **85.** 3   **87.** 59   **89.** $\frac{16}{8}$
91. $\frac{8}{8}$ **93.** $\frac{11}{4}$ **95.** $\frac{17}{8}$ **97.** $\frac{9}{8}$ **99.** 2 R 3   **101.** 8 R 16

PROBLEM SET 3.2

1. $\frac{14}{3}$ **3.** $\frac{21}{4}$ **5.** $\frac{13}{8}$ **7.** $\frac{47}{3}$ **9.** $\frac{104}{21}$ **11.** $\frac{427}{33}$ **13.** $1\frac{1}{8}$ **15.** $4\frac{3}{4}$ **17.** $4\frac{5}{6}$ **19.** $3\frac{1}{4}$ **21.** $4\frac{1}{27}$ **23.** $28\frac{8}{15}$

25. a. $3\frac{2}{15}$ **b.** $1\frac{19}{20}$ **27.** \$6 **29.** $\frac{71}{12}$ **31.**

33. $\frac{1,359}{10}$ **35.** $\frac{11}{4}$ **37.** $\frac{37}{8}$

39. $\frac{14}{5}$ **41.** $\frac{9}{40}$ **43.** $\frac{3}{8}$ **45.** $\frac{32}{35}$

47. $\frac{4}{7}$

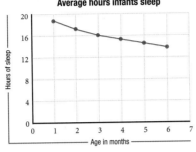

Average hours infants sleep

(Graph: Hours of sleep vs. Age in months)

PROBLEM SET 3.3

1. $5\frac{1}{10}$ **3.** $13\frac{2}{3}$ **5.** $6\frac{93}{100}$ **7.** $5\frac{5}{6}$ **9.** $9\frac{3}{4}$ **11.** $3\frac{1}{5}$ **13.** $12\frac{1}{2}$ **15.** $9\frac{9}{20}$ **17.** $\frac{32}{45}$ **19.** $1\frac{2}{3}$ **21.** 4 **23.** $4\frac{3}{10}$ **25.** $\frac{1}{10}$

27. $3\frac{1}{5}$ **29.** $2\frac{1}{8}$ **31.** $7\frac{1}{2}$ **33.** $\frac{11}{13}$ **35.** $5\frac{1}{2}$ cups **37.** $\frac{1}{3} \times 2\frac{1}{2} = \frac{5}{6}$ cup **39.** $1\frac{1}{3}$ **41.** $1,087\frac{1}{5}$ cents **43.** $163\frac{3}{4}$ mi

45. $4\frac{1}{2}$ yd **47.** $2\frac{1}{4}$ yd² **49.** $\frac{3}{4} < \frac{5}{4} < 1\frac{1}{2} < 2\frac{1}{8}$

51. Can 1 contains $157\frac{1}{2}$ calories, whereas Can 2 contains $87\frac{1}{2}$ calories. Can 1 contains 70 more calories than Can 2.

53. Can 1 contains 1,960 milligrams of sodium, whereas Can 2 contains 1,050 milligrams of sodium. Can 1 contains 910 more milligrams of sodium than Can 2.

55. a. $\frac{10}{15}$ **b.** $\frac{3}{15}$ **c.** $\frac{9}{15}$ **d.** $\frac{5}{15}$ **57. a.** $\frac{5}{20}$ **b.** $\frac{12}{20}$ **c.** $\frac{18}{20}$ **d.** $\frac{2}{20}$ **59.** $\frac{13}{15}$ **61.** $\frac{14}{9} = 1\frac{5}{9}$ **63.** $\frac{3}{5}$

PROBLEM SET 3.4

1. $5\frac{4}{5}$ **3.** $12\frac{2}{5}$ **5.** $3\frac{4}{9}$ **7.** 12 **9.** $1\frac{3}{8}$ **11.** $14\frac{1}{6}$ **13.** $4\frac{1}{12}$ **15.** $2\frac{1}{12}$ **17.** $26\frac{7}{12}$ **19.** 12 **21.** $2\frac{1}{2}$ **23.** $8\frac{6}{7}$ **25.** $3\frac{3}{8}$

27. $10\frac{4}{15}$ **29.** $2\frac{1}{15}$ **31.** 9 **33.** $18\frac{1}{10}$ **35.** 14 **37.** 17 **39.** $24\frac{1}{24}$ **41.** $27\frac{6}{7}$ **43.** $37\frac{3}{20}$ **45.** $6\frac{1}{4}$ **47.** $9\frac{7}{10}$ **49.** $5\frac{1}{2}$

51. $\frac{2}{3}$ **53.** $1\frac{11}{12}$ **55.** $3\frac{11}{12}$ **57.** $5\frac{19}{20}$ **59.** $5\frac{1}{2}$ **61.** $\frac{13}{24}$ **63.** $3\frac{1}{2}$ **65.** $5\frac{29}{40}$ **67.** $12\frac{1}{4}$ in. **69.** $\frac{1}{8}$ mi **71.** $31\frac{1}{6}$ in.

73. NFL: $P = 306\frac{2}{3}$ yd, Canadian: $P = 350$ yd, Arena: $P = 156\frac{2}{3}$ yd **75. a.** $2\frac{1}{2}$ **b.** \$250 **77.** \$300 **79.** $4\frac{63}{64}$ **81.** 2

83. $\frac{11}{8} = 1\frac{3}{8}$ **85.** $3\frac{5}{8}$

PROBLEM SET 3.5

1. 7 **3.** 7 **5.** 2 **7.** 35 **9.** $\frac{7}{8}$ **11.** $8\frac{1}{3}$ **13.** $\frac{11}{36}$ **15.** $3\frac{2}{3}$ **17.** $6\frac{3}{8}$ **19.** $4\frac{5}{12}$ **21.** $\frac{8}{9}$ **23.** $\frac{1}{2}$ **25.** $1\frac{1}{10}$ **27.** 5 **29.** $\frac{3}{5}$

31. $\frac{7}{11}$ **33.** 5 **35.** $\frac{17}{28}$ **37.** $1\frac{7}{16}$ **39.** $\frac{13}{22}$ **41.** $\frac{5}{22}$ **43.** $\frac{15}{16}$ **45.** $1\frac{5}{17}$ **47.** $\frac{3}{29}$ **49.** $1\frac{34}{67}$ **51.** $\frac{346}{441}$ **53.** $5\frac{2}{5}$ **55.** 8

57. $115\frac{2}{3}$ yd

CHAPTER 3 TEST

1. $\frac{11}{12}$ **2.** $\frac{13}{30}$ **3.** $\frac{59}{9}$ **4.** $2\frac{3}{4}$ **5.** $\frac{44}{5} = 8\frac{4}{5}$ **6.** $\frac{4}{7}$ **7.** $\frac{113}{9} = 12\frac{5}{9}$ **8.** $4\frac{5}{7}$ **9.** $\frac{85}{6} = 14\frac{1}{6}$ **10.** $2\frac{4}{5}$

Chapter 4

PROBLEM SET 4.1

1. Three tenths **3.** Fifteen thousandths **5.** Three and four tenths **7.** Fifty-two and seven tenths **9.** $405\frac{36}{100}$ **11.** $9\frac{9}{1,000}$

13. $1\frac{234}{1,000}$ **15.** $\frac{305}{100,000}$ **17.** Tens **19.** Tenths **21.** Hundred thousandths **23.** Ones **25.** Hundreds **27.** 0.55 **29.** 6.9

31. 11.11 **33.** 100.02 **35.** 3,000.003

37. – 45.

	Number	Whole Number	Tenth	Hundredth	Thousandth
			Rounded to the Nearest		
37.	47.5479	48	47.5	47.55	47.548
39.	0.8175	1	.8	.82	.818
41.	0.1562	0	0.2	0.16	0.156
43.	2,789.3241	2,789	2,789.3	2,789.32	2,789.324
45.	99.9999	100	100.0	100.00	100.000

47. Three and eleven hundredths; two and five tenths   **49.** 186,282.40   **51.** Fifteen hundredths   **53.**

55. a. $<$ **b.** $>$ **57.** $0.002 < 0.005 < 0.02 < 0.025 < 0.05 < 0.052$ **59.** 7.451, 7.54   **61.** $\frac{1}{4}$

63. $\frac{1}{8}$ **65.** $\frac{5}{8}$ **67.** $\frac{7}{8}$ **69.** 9.99   **71.** 10.05   **73.** 0.05   **75.** 0.01   **77.** $6\frac{31}{100}$ **79.** $6\frac{23}{50}$

81. $18\frac{123}{1,000}$

PRICE OF 1 GALLON OF REGULAR GASOLINE

Date	Price (Dollars)
4/5/04	2.126
4/12/04	2.157
4/19/04	2.148
4/26/04	2.124

PROBLEM SET 4.2

1. 6.19   **3.** 1.13   **5.** 6.29   **7.** 9.042   **9.** 8.021   **11.** 11.7843   **13.** 24.343   **15.** 24.111   **17.** 258.5414   **19.** 666.66

21. 11.11   **23.** 3.57   **25.** 4.22   **27.** 120.41   **29.** 44.933   **31.** 7.673   **33.** 530.865   **35.** 27.89   **37.** 35.64   **39.** 411.438   **41.** 6

43. 1   **45.** 3.1   **47.** 5.9   **49.** 3.272   **51.** 4.001   **53.** $116.82   **55.** $1,571.10   **57.** 4.5 in.   **59.** $5.43   **61.** 6.42 seconds

63. a. Less **b.** More **c.** 15,000 megawatts   **65.** 2.9 seconds   **67.** 3 in.   **69.** $2.15; two $1 bills, one dime, one nickel   **71.** 3.875

73. $\frac{3}{10}$ **75.** $\frac{271}{10,000}$ **77.** $4\frac{9}{10}$ **79.** 1,872   **81.** $\frac{3}{20,000}$ **83.** $\frac{7}{50}$ **85.** 210,366   **87.** 1,656

PROBLEM SET 4.3

1. 0.28   **3.** 0.028   **5.** 0.0027   **7.** 0.78   **9.** 0.792   **11.** 0.0156   **13.** 24.29821   **15.** 0.03   **17.** 187.85   **19.** 0.002   **21.** 27.96

23. 0.43   **25.** 49,940   **27.** 9,876,540   **29.** 1.89   **31.** 0.0025   **33.** 5.1106   **35.** 7.3485   **37.** 4.4   **39.** 2.074   **41.** 3.58   **43.** 187.4

45. 116.64   **47.** 20.75   **49.** 0.126   **51.** Moves it two places to the right   **53.** $1381.38   **55.** $7.10   **57.** $44.40   **59.** $293.04

61. 8,509 mm²   **63.** 1.18 in²   **65.** 1,879   **67.** 1,516 R 4   **69.** 298   **71.** 34.8   **73.** 49.896   **75.** 825

PROBLEM SET 4.4

1. 19.7   **3.** 6.2   **5.** 5.2   **7.** 11.04   **9.** 4.8   **11.** 9.7   **13.** 2.63   **15.** 4.24   **17.** 2.55   **19.** 1.35   **21.** 6.5   **23.** 9.9   **25.** 0.05

27. 89   **29.** 2.2   **31.** 1.35   **33.** 16.97   **35.** 0.25   **37.** 2.71   **39.** 11.69   **41.** 3.98   **43.** 5.98   **45.** 7.5 mi

47.

Rank Name	Number of Tournaments	Total Earnings	Average per Tournament
1. Annika Sorenstam	14	$1,957,200	$139,800
2. Paula Creamer	20	$1,332,254	$66,613
3. Cristie Kerr	18	$1,297,864	$72,104
4. Lorena Ochoa	18	$1,156,542	$64,252
5. Jeong Jang	21	$950,709	$45,272

49. $5.65/hr   **51.** 22.4 mi   **53.** 5 hr   **55.** 7 min   **57.** 2.73   **59.** 0.13   **61.** 0.77778   **63.** 307.20607   **65.** 0.70945   **67.** $\frac{3}{4}$ **69.** $\frac{2}{3}$

71. $\frac{3}{8}$ **73.** $\frac{19}{50}$ **75.** $\frac{6}{10}$ **77.** $\frac{60}{100}$ **79.** $\frac{12}{15}$ **81.** $\frac{60}{15}$ **83.** $\frac{18}{15}$ **85.** 0.75   **87.** 0.875

PROBLEM SET 4.5

1. 0.125 **3.** 0.625 **5.**

Fraction	$\frac{1}{4}$	$\frac{2}{4}$	$\frac{3}{4}$	$\frac{4}{4}$
Decimal	0.25	0.5	0.75	1

7.

Fraction	$\frac{1}{6}$	$\frac{2}{6}$	$\frac{3}{6}$	$\frac{4}{6}$	$\frac{5}{6}$	$\frac{6}{6}$
Decimal	$0.1\overline{6}$	$0.\overline{3}$	0.5	$0.\overline{6}$	$0.8\overline{3}$	1

9. 0.48 **11.** 0.4375 **13.** 0.92 **15.** 0.27 **17.** 0.09 **19.** 0.28

21.

Decimal	0.125	0.250	0.375	0.500	0.625	0.750	0.875
Fraction	$\frac{1}{8}$	$\frac{1}{4}$	$\frac{3}{8}$	$\frac{1}{2}$	$\frac{5}{8}$	$\frac{3}{4}$	$\frac{7}{8}$

23. $\frac{3}{20}$ **25.** $\frac{2}{25}$ **27.** $\frac{3}{8}$ **29.** $5\frac{3}{5}$ **31.** $5\frac{3}{50}$

33. $1\frac{11}{50}$ **35.** 2.4 **37.** 3.98 **39.** 3.02 **41.** 0.3 **43.** 0.072 **45.** 0.8 **47.** 1 **49.** 0.25 **51.** $8.42 **53.** $38.66 **55.** 9 in.

57. 104.625 calories **59.** $10.38 **61.** Yes **63.** 36 **65.** 25 **67.** 125 **69.** 9 **71.** $\frac{1}{81}$ **73.** $\frac{25}{36}$ **75.** 0.25 **77.** 1.44 **79.** 25

81. 100

PROBLEM SET 4.6

1. 8 **3.** 9 **5.** 6 **7.** 5 **9.** 15 **11.** 48 **13.** 45 **15.** 48 **17.** 15 **19.** 1 **21.** 78 **23.** 9 **25.** $\frac{4}{7}$ **27.** $\frac{3}{4}$ **29.** False **31.** True

33. 10 in. **35.** 13 ft **37.** 6.40 in. **39.** 17.49 m **41.** 30 ft **43.** 25 ft **45.** See Chapter Introduction **47.** 1.1180 **49.** 11.1803

51. 3.46 **53.** 11.18 **55.** 0.58 **57.** 0.58 **59.** 12.124 **61.** 9.327 **63.** 12.124 **65.** 12.124 **67.**

Height h(feet)	Distance d(miles)
10	4
50	9
90	12
130	14
170	16
190	17

CHAPTER 4 TEST

1. .0035 **2. a.** Four tenths **b.** Four hundreths **c.** Four thousandths **3.** 11.142 **4.** 2.8186 **5.** 13.298 **6.** 0.2115 **7.** 8.7

8. 8.25 **9.** $0.\overline{27}$ **10.** $\frac{19}{50}$ **11.** 0.684 **12.** 7 **13.** 4 **14.** $\frac{5}{9}$ **15.** 17.3205

Chapter 5

PROBLEM SET 5.1

1. $\frac{4}{3}$ **3.** $\frac{16}{3}$ **5.** $\frac{2}{5}$ **7.** $\frac{1}{2}$ **9.** $\frac{3}{1}$ **11.** $\frac{7}{6}$ **13.** $\frac{7}{5}$ **15.** $\frac{5}{7}$ **17.** $\frac{8}{5}$ **19.** $\frac{1}{3}$ **21.** $\frac{1}{10}$ **23.** $\frac{3}{25}$ **25. a.** $\frac{1}{2}$ **b.** $\frac{1}{3}$ **c.** $\frac{2}{3}$

27. a. $\frac{13}{8}$ **b.** $\frac{1}{4}$ **c.** $\frac{3}{8}$ **d.** $\frac{13}{3}$ **29. a.** $\frac{68}{63}$ **b.** $\frac{9}{8}$ **c.** $\frac{7}{12}$ **d.** $\frac{24}{17}$ **31. a.** $\frac{3}{4}$ **b.** 12 **c.** $\frac{3}{4}$ **33.** $\frac{2,408}{2,314} \approx 1.04$ **35.** $\frac{4,722}{2,408} \approx 1.96$

37. 40 **39.** 0.2 **41.** 0.695 **43.** 3.98 **45.** 368 **47.** 0.065 **49.** 0.025

PROBLEM SET 5.2

1. 55 mi/hr **3.** 84 km/hr **5.** 0.2 gal/sec **7.** 12 L/min **9.** 19 mi/gal **11.** $4\frac{1}{3}$ mi/L **13.** 16¢ per ounce **15.** 4.95¢ per ounce

17. Dry Baby: 34.7¢/diaper, Happy Baby: 31.6¢/diaper, Happy Baby is better buy **19.** 7.7 tons/year **21.** 10.8¢ per day

23. 9.3 mi/gal **25.** $64 **27.** $16,000 **29.** $n = 6$ **31.** $n = 4$ **33.** $n = 4$ **35.** $n = 65$

PROBLEM SET 5.3

1. 35 **3.** 18 **5.** 14 **7.** n **9.** y **11.** $n = 2$ **13.** $x = 7$ **15.** $y = 7$ **17.** $n = 8$ **19.** $a = 8$ **21.** $x = 2$ **23.** $y = 1$

25. $a = 6$ **27.** $n = 5$ **29.** $x = 3$ **31.** $n = 7$ **33.** $y = 1$ **35.** $y = 9$ **37.** $n = \frac{7}{2} = 3\frac{1}{2}$ **39.** $x = \frac{7}{2} = 3\frac{1}{2}$

41. $a = \frac{12}{5} = 2\frac{2}{5}$ **43.** $y = \frac{4}{7}$ **45.** $y = \frac{10}{13}$ **47.** $x = \frac{5}{2} = 2\frac{1}{2}$ **49.** $n = \frac{3}{2} = 1\frac{1}{2}$ **51.** $\frac{3}{4}$ **53.** 1.2 **55.** 6.5

PROBLEM SET 5.4

1. Means: 3, 5; extremes: 1, 15; products: 15 **3.** Means: 25, 2; extremes: 10, 5; products: 50

5. Means: $\frac{1}{2}$, 4; extremes: $\frac{1}{3}$, 6; products: 2 **7.** Means: 5, 1; extremes: 0.5, 10; products: 5 **9.** 10 **11.** $\frac{12}{5}$ **13.** $\frac{3}{2}$ **15.** $\frac{10}{9}$
17. 7 **19.** 14 **21.** 18 **23.** 6 **25.** 40 **27.** 50 **29.** 108 **31.** 3 **33.** 1 **35.** 0.25 **37.** 108 **39.** 65 **41.** 41 **43.** 108 **45.** 20
47. 297.5 **49.** 450 **51.** 5

PROBLEM SET 5.5

1. 329 mi **3.** 360 points **5.** 15 pt **7.** 427.5 mi **9.** 900 eggs **11.** 435 in. = 36.25 ft **13.** $119.70 **15.** 265 g **17.** 91.3 liters
19. 60,113 people **21.** 2 **23.** 147 **25.** 20 **27.** 147

PROBLEM SET 5.6

1. $h = 9$ **3.** $y = 14$ **5.** $x = 12$ **7.** $a = 25$ **9.** $y = 32$

11.

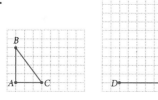

13.

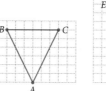

15.

17.

19. 45 in. **21.** 16.25 in. **23.** 960 pixels **25.** 1,440 pixels **27.** 57 ft **29.** 4 ft

CHAPTER 5 TEST

1. $\frac{1}{3}$ **2.** $\frac{3}{2}$ **3.** 62.5 miles per hour **4.** 18 miles/gallon **5.** $a = 6$ **6.** n **7.** $b = 30$ **8.** $n = 0.15$ **9.** $y = 6.5$
10. 450 miles **11.** 297.5 miles **12.** 147 feet **13.** 20

Chapter 6

PROBLEM SET 6.1

1. $\frac{20}{100}$ **3.** $\frac{60}{100}$ **5.** $\frac{24}{100}$ **7.** $\frac{65}{100}$ **9.** 0.23 **11.** 0.92 **13.** 0.09 **15.** 0.034 **17.** 0.0634 **19.** 0.009 **21.** 23% **23.** 92%
25. 45% **27.** 3% **29.** 60% **31.** 80% **33.** 27% **35.** 123% **37.** $\frac{3}{5}$ **39.** $\frac{3}{4}$ **41.** $\frac{1}{25}$ **43.** $\frac{53}{200}$ **45.** $\frac{7,187}{10,000}$ **47.** $\frac{3}{400}$
49. $\frac{1}{16}$ **51.** $\frac{1}{3}$ **53.** 50% **55.** 75% **57.** $33\frac{1}{3}$% **59.** 80% **61.** 87.5% **63.** 14% **65.** 325% **67.** 150% **69.** 48.8%
71. 0.50; 0.75 **73. a.** $\frac{27}{50}, \frac{7}{25}, \frac{3}{20}, \frac{3}{100}$ **b.** 0.54, 0.28, 0.15, 0.03 **c.** About 2 times as likely. **75.** 20%
77. Liberal Arts: 16%, Science & Math: 12%, Engineering: 25%, Business: 15%, Architecture & Environmental Design: 10%,
Agriculture: 22%
79. 78.4% **81.** 11.8% **83.** 72.2% **85.** 8.3%; 0.2% **87.** 18.5 **89.** 10.875 **91.** 0.5 **93.** 62.5 **95.** 0.5

PROBLEM SET 6.2

1. 8 **3.** 24 **5.** 20.52 **7.** 7.37 **9.** 50% **11.** 10% **13.** 25% **15.** 75% **17.** 64 **19.** 50 **21.** 925 **23.** 400 **25.** 5.568
27. 120 **29.** 13.72 **31.** 22.5 **33.** 50% **35.** 942.684 **37.** 97.8 **39.** What number is 25% of 350? **41.** What percent of 24 is 16?
43. 46 is 75% of what number? **45.** 4.8% calories from fat; healthy **47.** 50% calories from fat; not healthy **49.** 0.80 **51.** 0.76
53. 48

PROBLEM SET 6.3

1. 70% **3.** 84% **5.** 45 mL **7.** 18.2 acres for farming; 9.8 acres are not available for farming **9.** 3,000 students **11.** 400 students

13. 1,664 female students **15.** 31.25% **17.** 50% **19.** About 19.2 million **21.** 33 **23.** 8,685 **25.** 136 **27.** 0.05 **29.** 15,300
31. 0.15

PROBLEM SET 6.4

1. $52.50 **3.** $2.70; $47.70 **5.** $150; $156 **7.** 5% **9.** $2,820 **11.** $200 **13.** 14% **15.** $11.93 **17.** 4.5% **19.** $3,995
21. 1,100 **23.** 75 **25.** 0.16 **27.** 4 **29.** 396 **31.** 415.8

PROBLEM SET 6.5

1. $24,610 **3.** $3,510 **5.** $13,200 **7.** 10% **9.** 20% **11.** 21% **13.** $45; $255 **15.** $381.60 **17.** $46,595.88
19. a. 51.9% **b.** 7.8% **21.** 140 **23.** 4 **25.** 152.25 **27.** 3,434.7 **29.** 10,150 **31.** 10,456.78 **33.** 2,140 **35.** 3,210

PROBLEM SET 6.6

1. $2,160 **3.** $665 **5.** $8,560 **7.** $2,160 **9.** $5 **11.** $813.33 **13.** $5,618
15. $8,407.56, Some answers may vary in the hundredths column depending on whether rounding is done in the intermediate steps.
17. $974.59 **19. a.** $13,468.55 **b.** $13,488.50 **c.** $12,820.37 **d.** $12,833.59

CHAPTER 6 TEST

1. a. 0.37 **b.** 0.68 **c.** 1.2 **d.** 0.008 **2. a.** 27% **b.** 489% **c.** 50% **d.** 9% **3.** 37.5% **4.** 9.45 **5.** 40.9% **6.** 117.2
7. 80% **8.** 5% **9.** $8,685 **10.** 16% **11.** $23,100 **12.** $2,140 **13.** $3,434.70 **14.** $913.50

Chapter 7

PROBLEM SET 7.1

1. 60 in. **3.** 120 in. **5.** 6 ft **7.** 162 in. **9.** $2\frac{1}{4}$ **11.** 13,200 ft **13.** $1\frac{1}{3}$ yd **15.** 1,800 cm **17.** 4,800 m **19.** 50 cm
21. 0.248 km **23.** 670 mm **25.** 34.98 m **27.** 6.34 dm **29.** 20 yd **31.** 80 in. **33.** 244 cm **35.** 65 mm **37.** 2,960 chains
39. 120,000 μm **41.** 7,920 ft **43.** 80.7 ft/sec **45.** 19.5 mi/hr **47.** 1,023 mi/hr **49.** $18,216 **51.** $157.50 **53.** 3,965,280 ft
55. 179,352 in. **57.** 2.7 mi **59.** 18,094,560 ft **61.** 144 **63.** 8 **65.** 1,000 **67.** 3,267,000 **69.** 6 **71.** 0.4 **73.** 405
75. 450 **77.** 45 **79.** 2,200 **81.** 607.5

PROBLEM SET 7.2

1. 432 in² **3.** 2 ft² **5.** 1,306,800 ft² **7.** 1,280 acres **9.** 3 mi² **11.** 108 ft² **13.** 1,700 mm² **15.** 28,000 cm² **17.** 0.0012 m²
19. 500 m² **21.** 700 a **23.** 3.42 ha
25. NFL: $A = 5,333\frac{1}{3}$ sq yd = 1.1 acres; Canadian: $A = 7,150$ sq yd = 1.48 acres; Arena: $A = 1,416\frac{2}{3}$ sq yd = 0.29 acres
27. 30 a **29.** 5,500 bricks **31.** 135 ft² **33.** 48 fl oz **35.** 8 qt **37.** 20 pt **39.** 480 fl oz **41.** 8 gal **43.** 6 qt **45.** 9 yd³
47. 5,000 mL **49.** 0.127 L **51.** 4,000,000 mL **53.** 14,920 L **55.** 16 cups **57.** 34,560 in³ **59.** 48 glasses **61.** 20,288,000 acres
63. 3,230.93 mi² **65.** 23.35 gal **67.** 21,492 ft³ **69.** 285,795,000 ft³ **71.** 192 **73.** 6,000 **75.** 300,000 **77.** 12,500 **79.** 12.5

PROBLEM SET 7.3

1. 128 oz **3.** 4,000 lb **5.** 12 lb **7.** 0.9 T **9.** 32,000 oz **11.** 56 oz **13.** 13,000 lb **15.** 2,000 g **17.** 40 mg **19.** 200,000 cg
21. 508 cg **23.** 4.5 g **25.** 47.895 cg **27.** 1.578 g **29.** 0.42 kg **31.** 48 g **33.** 4 g **35.** 9.72 g **37.** 120 g **39.** 3 L **41.** 1.5 L
43. 20.32 **45.** 6.36 **47.** 50 **49.** 56.8 **51.** 122 **53.** 248 **55.** 38.9

PROBLEM SET 7.4

1. 15.24 cm **3.** 13.12 ft **5.** 6.56 yd **7.** 32,200 m **9.** 5.98 yd² **11.** 24.7 acres **13.** 8,195 mL **15.** 2.12 qt **17.** 75.8 L
19. 339.6 g **21.** 33 lb **23.** 365°F **25.** 30°C **27.** 3.94 in. **29.** 7.62 m **31.** 46.23 L **33.** 17.67 oz **35.** Answers will vary.
37. 91.46 m **39.** 20.90 m² **41.** 88.55 km/hr **43.** 2.03 m **45.** 38.3°C **47.** 75 **49.** 82 **51.** 3.25 **53.** 22 **55.** 41 **57.** 48
59. 195 **61.** 3.27 **63.** $90.00

PROBLEM SET 7.5

1. a. 270 min **b.** 4.5 hr **3. a.** 320 min **b.** 5.33 hr **5. a.** 390 sec **b.** 6.5 min **7. a.** 320 sec **b.** 5.33 min
9. a. 40 oz **b.** 2.5 lb **11. a.** 76 oz **b.** 4.75 lb **13. a.** 54 in. **b.** 4.5 ft **15. a.** 69 in. **b.** 5.75 ft **17. a.** 9 qt **b.** 2.25 gal
19. 11 hr **21.** 22 ft 4 in. **23.** 11 lb **25.** 5 hr 40 min **27.** 3 hr 47 min **29.** 52 min

31.

Triathlete	Swim Time (Hr:Min:Sec)	Bike Time (Hr:Min:Sec)	Run Time (Hr:Min:Sec)	Total Time (Hr:Min:Sec)
Peter Reid	0:50:36	4:40:04	2:47:38	8:18:18
Lori Bowden	0:56:51	5:09:00	3:02:10	9:08:01

33. 00:06:15 **35.** $104 **37.** 10 hr
39. $150 **41.** $6

CHAPTER 7 TEST

1. 3,600 in. **2.** .025 m **3.** 3.65 dm **4.** 144 in² **5.** 1,000,000 mm² **6.** 450 gal. **7.** 192 oz. **8.** 300,000 cg **9.** 50 mi
10. 122 in³ **11.** 38.9 **12. a.** 195 min **b.** 3.25 hr **13.** 7 hr 41 min

Chapter 8

PROBLEM SET 8.1

1. 32 in. **3.** 260 yd **5.** 36 in. **7.** $15\frac{3}{4}$ in. **9.** 76 in. **11.** 168 ft **13.** 18.28 in. **15.** 25.12 in. **17.** 178.5 in. **19.** 37.68 in.
21. 24,492 mi **23.** 388 mm **25.** 83.21 mm **27.** 4.5 in. **29.** 9 in. **31.** $w = 1$ and $l = 5$; $w = 2$ and $l = 4$; $w = 3$ and $l = 3$
33. Yes, when $w = l = 5$ ft **35.** 2 in. **37.** 9 **39.** 1.64 **41.** 660.19 **43.** 30.96

PROBLEM SET 8.2

1. 25 cm² **3.** 336 m² **5.** 60 ft² **7.** 2,200 ft² **9.** 945 cm² **11.** 50.24 in² **13.** 22.28 in² **15. a.** 100 in² **b.** 314 in²
17. 133 in² **19.** $\frac{4}{9}$ ft² **21.** 124 tiles **23.** 1.18 in² **25.** 8,509 mm²
27. The area increases from 25 ft² to 49 ft², which is an increase of 24 ft². **29.** 551.27 mm²

31. b.

PERIMETERS OF SQUARES	
Length of each side (in Centimeters)	Perimeter (in Centimeters)
1	4
2	8
3	12
4	16

AREAS OF SQUARES	
Length of each side (in Centimeters)	Area (in Square cm)
1	1
2	4
3	9
4	16

33. 22.6 cm **35.** 24 **37.** 72 **39.** 4.71 **41.** 547

PROBLEM SET 8.3

1. 96 cm² **3.** 378 ft² **5.** 270 in² **7.** 125.6 ft² **9.** 75.36 ft² **11.** 50.24 mi² **13.** 191.04 in² **15.** 197.82 ft² **17.** 1,268.95 mm²
19. 216 m² **21. a.** 1.354 in. **b.** 93.5 in² **c.** Yes **23.** 288 ft² **25.** 60,288,000 mi² **27.** 282.6 cm² **29.** 27 **31.** $\frac{2}{3}$ **33.** 0.294
35. 785

PROBLEM SET 8.4

1. 64 cm³ **3.** 420 ft³ **5.** 162 in³ **7.** 100.48 ft³ **9.** 50.24 ft³ **11.** 33.49 mi³ **13.** 226.08 ft³ **15.** 113.03 in³
17. 1,102.53 mm³ **19.** 85.27 cm³ **21.** 426.51 in³ **23.** 49.63 cm³ **25.** 142.72 ft³ **27.** 46,208 cm³

CHAPTER 8 TEST

1. 26 yards **2. a.** 73 mm **b.** 3.27 in **3.** 10 cm² **4.** 660 mm² **5.** 94 in² **6.** 4.71 in² **7.** 225 in³ **8.** 47.1 cm³

Chapter 9

PROBLEM SET 9.1

1. 4 is less than 7.   **3.** 5 is greater than -2.   **5.** -10 is less than -3.   **7.** 0 is greater than -4.   **9.** $30 > -30$ **11.** $-10 < 0$
13. $-3 > -15$ **15.** $3 < 7$ **17.** $7 > -5$ **19.** $-6 < 0$ **21.** $-12 < -2$ **23.** $-\frac{1}{2} > -\frac{3}{4}$ **25.** $-0.75 < 0.25$
27. $-0.1 < -0.01$ **29.** $-3 < |6|$ **31.** $15 > |-4|$ **33.** $|-2| < |-7|$ **35.** 2   **37.** 100   **39.** 8   **41.** 231   **43.** $\frac{3}{4}$ **45.** 200
47. 8   **49.** 231   **51.** -3 **53.** 2   **55.** -75 **57.** 0   **59.** 0.123   **61.** $-\frac{7}{8}$ **63.** 2   **65.** 8   **67.** -2 **69.** -8 **71.** 0   **73.** Positive
75. -100 **77.** -20 **79.** -360 **81.** $-61°$ F, $-51°$ F   **83.** $-5°$ F, $-15°$ F   **85.** $-7°$ F   **87.** $10°$ F and 25-mph wind

89.

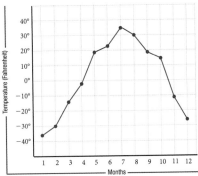

91. 25   **93.** 5   **95.** 6   **97.** 19   **99.** $\frac{1}{4}$ **101.** $\frac{3}{4}$ **103.** 7   **105.** 3.47

PROBLEM SET 9.2

1. 5   **3.** 1   **5.** -2 **7.** -6 **9.** 4   **11.** 4   **13.** -9 **15.** 15   **17.** -3 **19.** -11 **21.** -7 **23.** -3 **25.** -16 **27.** -8
29. -127 **31.** 49   **33.** 34   **35.**

First Number	Second Number	Their Sum
a	b	a + b
5	-3	2
5	-4	1
5	-5	0
5	-6	-1
5	-7	-2

37.

First Number	Second Number	Their Sum
x	y	x + y
-5	-3	-8
-5	-4	-9
-5	-5	-10
-5	-6	-11
-5	-7	-12

39. -4 **41.** 10   **43.** -445 **45.** 107   **47.** -1 **49.** -20 **51.** -17 **53.** -50 **55.** -7 **57.** 3   **59.** 50   **61.** -73 **63.** -11
65. 17   **67.** -3.8 **69.** 14.4   **71.** -9.89 **73.** -1 **75.** $-\frac{2}{7}$ **77.** $-\frac{3}{5}$ **79.** -0.86 **81.** -4.2 **83.** $-\frac{1}{4}$ **85.** $-1\frac{1}{3}$ **87.** -21
89. -5 **91.** -4 **93.** 7   **95.** 10   **97.** a   **99.** b   **101.** d   **103.** c   **105.** \$10   **107.** $\$74 + (-\$141) = -\$67$ **109.** $3 + (-5) = -2$
111. -7 and 13   **113.** -2 **115.** 4   **117.** $-\frac{2}{5}$ **119.** 30   **121.** -60.3 **123.** 2   **125.** 3

PROBLEM SET 9.3

1. 2   **3.** 2   **5.** -8 **7.** -5 **9.** 7   **11.** 12   **13.** 3   **15.** -7 **17.** -3 **19.** -13 **21.** -50 **23.** -100 **25.** 399   **27.** -21

29.

First Number	Second Number	Their Difference
x	y	x − y
8	6	2
8	7	1
8	8	0
8	9	-1
8	10	-2

31.

First Number	Second Number	Their Difference
x	y	x − y
8	-6	14
8	-7	15
8	-8	16
8	-9	17
8	-10	18

33. -11.41 **35.** -1.9 **37.** -1 **39.** $-\frac{5}{12}$ **41.** $-\frac{11}{15}$ **43.** -7 **45.** -9 **47.** -14 **49.** -17.5 **51.** -11 **53.** $-\frac{1}{12}$

55. -400 **57.** 11 **59.** -4 **61.** 8 **63.** 6 **65.** b **67.** a **69.** -100 **71.** b **73.** a **75.** $44°$ **77.** $0.18

79. $-11 - (-22) = 11° F$ **81.** $3 - (-24) = 27° F$ **83.** $60 - (-26) = 86° F$ **85.** $-14 - (-26) = 12° F$ **87.** 30 **89.** 36

91. 64 **93.** 48 **95.** 41 **97.** 40 **99.** $\frac{2}{5}$ **101.** $\frac{5}{16}$ **103.** 17 **105.** 0.32

PROBLEM SET 9.4

1. -56 **3.** -60 **5.** 56 **7.** 81 **9.** -9.03 **11.** $\frac{3}{7}$ **13.** -8 **15.** -24 **17.** 24 **19.** -6 **21. a.** 16 **b.** -16

23. a. -125 **b.** -125 **25. a.** 16 **b.** -16

27.

Number x	Square x^2
-3	9
-2	4
-1	1
0	0
1	1
2	4
3	9

29.

First Number x	Second Number y	Their Product xy
6	2	12
6	1	6
6	0	0
6	-1	-6
6	-2	-12

31.

First Number a	Second Number b	Their Product ab
-5	3	-15
-5	2	-10
-5	1	-5
-5	0	0
-5	-1	5
-5	-2	10
-5	-3	15

33. -4 **35.** 50 **37.** 1 **39.** -35 **41.** -22 **43.** -30 **45.** -25 **47.** 9

49. -13 **51.** 19 **53.** 6 **55.** -6 **57.** -4 **59.** -17 **61.** A gain of $200 **63.** $-16°$

65.

NATIONAL LEAGUE						
Name, Team	W	L	S	TS	BS	Pts
Eric Gagne, Los Angeles	2	3	53	2	0	165
John Smoltz, Atlanta	0	2	42	3	4	126
Billy Wagner, Houston	1	4	40	4	3	124
Tim Worrell, San Francisco	4	4	36	2	7	102
Joe Borowski, Chicago	2	2	32	2	4	96

67. If par for the course is 72, then his score was $72 - 2 = 70$.

	Value	Number	Product
Eagle	-2	0	0
Birdie	-1	7	-7
Par	0	7	0
Bogie	$+1$	3	$+3$
Double Bogie	$+2$	1	$+2$
		Total:	-2

69. a **71.** d **73.** a **75.** b **77.** 7 **79.** 5 **81.** -5 **83.** 9

85. 4 **87.** 7 **89.** 405

PROBLEM SET 9.5

1. -3 **3.** -5 **5.** 3 **7.** 2 **9.** -4 **11.** -2 **13.** 0 **15.** -5

17.

First Number a	Second Number b	The Quotient of a and b $\frac{a}{b}$
100	-5	-20
100	-10	-10
100	-25	-4
100	-50	-2

19.

First Number a	Second Number b	The Quotient of a and b $\frac{a}{b}$
-100	-5	20
-100	5	-20
100	-5	-20
100	5	20

21. 1 **23.** -6 **25.** -2 **27.** -1 **29.** -1 **31.** 2 **33.** -3 **35.** -7 **37.** 30 **39.** 4 **41.** -5 **43.** -20 **45.** -5 **47.** -5

49. 35 **51.** 6 **53.** -1 **55.** c **57.** a **59.** d **61.** 0 **63. a.** 0 **b.** 2 **c.** 2 **65.**

67. a. $2°$ F **b.** $5°$ F **69.** $x + 3$ **71.** $(5 + 7) + a$ **73.** $(3 \cdot 4)y$

75. $5(3) + 5(7)$ **77.** 36 **79.** 7,500 **81.** 350

83. a. 7 **b.** 2 **c.** 7 **d.** -6 **e.** -4

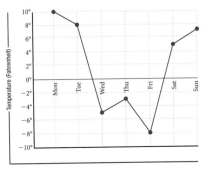

PROBLEM SET 9.6

1. $20a$ **3.** $48a$ **5.** $-18x$ **7.** $-27x$ **9.** $-10y$ **11.** $-60y$ **13.** $5 + x$ **15.** $13 + x$ **17.** $10 + y$ **19.** $8 + y$ **21.** $5x + 6$

23. $6y + 7$ **25.** $12a + 21$ **27.** $7x + 28$ **29.** $7x + 35$ **31.** $6a - 42$ **33.** $2x - 2y$ **35.** $20 + 4x$ **37.** $6x + 15$ **39.** $18a + 6$

41. $12x - 6y$ **43.** $35 - 20y$ **45.** $8x$ **47.** $4a$ **49.** $4x$ **51.** $5y$ **53.** $-10a$ **55.** $-5x$ **57.** $A = 36$ ft²; $P = 24$ ft

59. $A = 81$ in²; $P = 36$ in. **61.** $A = 200$ in²; $P = 60$ in. **63.** $A = 300$ ft²; $P = 74$ ft **65.** $20°$ C **67.** $5°$ C **69.** $-10°$ C

CHAPTER 9 TEST

1. $-5, -7, -1, 5, 8$ **2. a.** 5 **b.** 3 **c.** 7 **3.** 6 **4.** -7 **5.** -3.47 **6.** -9 **7.** 18 **8.** 1 **9.** -8 **10. a.** 8 **b.** 8 **c.** -8

d. -8 **11. a.** 36 **b.** -36 **c.** -64 **d.** -64 **12.** -24 **13. a.** 3 **b.** -3 **c.** -3 **d.** 3 **14.** 7 **15.** $2x + 15$ **16.** $10x + 15y$

Chapter 10

PROBLEM SET 10.1

1. $10x$ **3.** $4a$ **5.** y **7.** $-8x$ **9.** $3a$ **11.** $-5x$ **13.** $6x + 11$ **15.** $2x + 2$ **17.** $-a + 12$ **19.** $4y - 4$ **21.** $-2x + 4$

23. $8x - 6$ **25.** $5a + 9$ **27.** $-x + 3$ **29.** $17y + 3$ **31.** $a - 3$ **33.** $6x + 16$ **35.** $10x - 11$ **37.** $19y + 32$ **39.** $30y - 18$

41. $6x + 14$ **43.** $27a + 5$ **45.** 14 **47.** 27 **49.** -19 **51.** 7 **53.** 1 **55.** 18 **57.** 12 **59.** -10 **61.** 28 **63.** 40 **65.** 26

67. 4 **69.** $6(x + 4) = 6x + 24$ **71.** $4x + 4$ **73.** $10x - 4$ **75.** Complement: $65°$; supplement: $155°$; acute

77. a. $32°$F **b.** $22°$F **c.** $-18°$F **79. a.** $27 **b.** $47 **81.** 0 **83.** -6 **85.** -3 **87.** $\frac{11}{8}$ **89.** 0 **91.** x **93.** $y - 2$

PROBLEM SET 10.2

1. Yes **3.** Yes **5.** No **7.** Yes **9.** No **11.** 6 **13.** 11 **15.** -15 **17.** 1 **19.** -3 **21.** -1 **23.** $\frac{7}{5}$ **25.** -10 **27.** -4

29. -3 **31.** 2 **33.** -6 **35.** -6 **37.** -1 **39.** -2 **41.** -16 **43.** -3 **45.** 10 **47.** $x = 4$ **49.** $x = 12$ **51.** $67°$

53. a. 225 **b.** $11,125 **55.** Equation: $x + 12 = 30$; $x = 18$ **57.** Equation: $8 - 5 = x + 7$; $x = -4$ **59.** $\frac{1}{4}$ **61.** 2 **63.** $\frac{3}{2}$

65. 1 **67.** 1 **69.** 1 **71.** 1 **73.** x **75.** $7x - 11$

PROBLEM SET 10.3

1. 8 **3.** -6 **5.** -6 **7.** 6 **9.** 16 **11.** 16 **13.** $-\frac{3}{2}$ **15.** -7 **17.** -8 **19.** 6 **21.** 2 **23.** 3 **25.** 4 **27.** -24 **29.** 12

31. -8 **33.** 15 **35.** 3 **37.** -1 **39.** 1 **41.** 3 **43.** 1 **45.** -1 **47.** $-\frac{1}{3}$ **49.** -14 **51.** $x = 9$ **53.** $x = 8$

55. 4 three-pointers **57.** $2x + 5 = 19$; $x = 7$ **59.** $5x - 6 = -9$; $x = -\frac{3}{5}$ **61.** 6.3 **63.** 0 **65.** 12.6 **67.** 0.175 **69.** $6a - 16$

71. $-15x + 3$ **73.** $3y - 9$ **75.** $16x - 6$

PROBLEM SET 10.4

1. 3 **3.** 2 **5.** -3 **7.** -4 **9.** 1 **11.** 0 **13.** 2 **15.** -2 **17.** 3 **19.** -1 **21.** 7 **23.** -3 **25.** 1 **27.** -2 **29.** 4

31. 10 **33.** -5 **35.** $-\frac{1}{12}$ **37.** -3 **39.** 20 **41.** -1 **43.** 5 **45.** 4 **47.** $x = 10$ **49.** $x = 5$ **51.** $x = 13$ **53.** 12 **55.** $x + 2$

57. $2x$ **59.** $2(x + 6)$ **61.** $x - 4$ **63.** $2x + 5$

PROBLEM SET 10.5

1. $x + 3$ **3.** $2x + 1$ **5.** $5x - 6$ **7.** $3(x + 1)$ **9.** $5(3x + 4)$ **11.** The number is 2.   **13.** The number is -2.

15. The number is 3.   **17.** The number is 5.   **19.** The number is -2.   **21.** The length is 10 m and the width is 5 m.

23. The length of one side is 8 cm.   **25.** The measures of the angles are 45°, 45°, and 90°.   **27.** The angles are 30°, 60°, and 90°.

29. Patrick is 33 years old, and Pat is 53 years old.   **31.** Sue is 35 years old, and Dale is 39 years old.   **33.** 87 mi   **35.** 147 mi

37. 8 nickels, 18 dimes   **39.** 16 dimes, 32 quarters   **41.** $x = 8, y = 6, z = 9$ **43.** 39 hours   **45.** $54   **47.** Yes   **49.** 35   **51.** 65

53. 2   **55.** 14   **57.** $-\frac{2}{3}$

PROBLEM SET 10.6

1. 704 ft²   **3.** $\frac{9}{8}$ in²   **5.** $240   **7.** $285   **9.** 12 ft   **11.** 8 ft   **13.** $140   **15.** 58 in.   **17.** $3\frac{1}{4} = \frac{13}{4}$ ft   **19.** $C = 100$°C; yes

21. $C = 20$°C; yes   **23.** 0°C   **25.** 5°F   **27.**

Age (Years)	Maximum Heart Rate (Beats per Minute)
18	202
19	201
20	200
21	199
22	198
23	197

29.

Resting Heart Rate (Beats per Minute)	Training Heart Rate (Beats per Minute)
60	144
62	144.8
64	145.6
68	146.4
70	147.2
72	148

31. a. 4 hrs **b.** 220 miles   **33. a.** 4 hours **b.** 65 mph   **35.** 360 in³   **37.** 1 yd³   **39.** $y = 7$ **41.** $y = -3$ **43.** $y = -2$

45. $x = 3$ **47.** $x = 5$ **49.** $x = 8$ **51.** $y = 1$ **53.** $y = 3$ **55.** $y = \frac{5}{2}$ **57.** $y = 0$ **59.** $y = \frac{13}{3}$ **61.** $y = 4$ **63.** $x = 0$

65. $x = \frac{13}{4}$ **67.** $x = 3$ **69.** Complement: 45°; supplement: 135°   **71.** Complement: 59°; supplement: 149°

73. a. 13,330 kilobytes **b.** 2,962 kilobytes   **75.** 3,600 lb

CHAPTER 10 TEST

1. $8x + 5$ **2.** $-12x - 20$ **3.** $5x + 15$ **4.** No   **5.** $x = -3$ **6.** $x = 4$ **7.** -5 **8.** $a = 15$ **9.** $x = \frac{1}{8}$ **10.** $x = -4$

11. $x = -5$ **12.** 6   **13.** 4   **14.** Stacey, 4; Jo Ann 26   **15.** 14 ft   **16.** 35°

Index